Ice diving operations

Walt "Butch" Hendrick
Andrea Zaferes

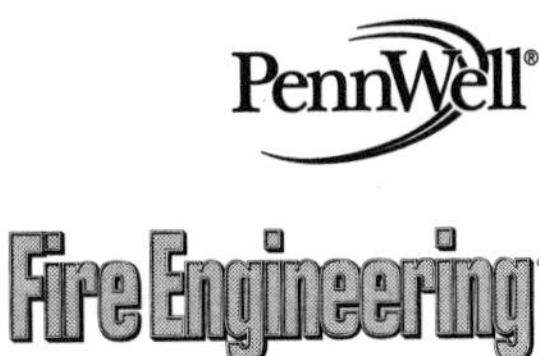

PennWell Corporation
1421 South Sheridan/P.O. Box 1260
Tulsa, Oklahoma 74112-6600 USA

800.752.9764
+1.918.831.9421
sales@pennwell.com
www.pennwell-store.com
www.pennwell.com

Cover design by Clark Bell
Book design by Amy Spehar
Jared Wicklund, Supervising Editor
Indexed by Phyllis Linn

Library of Congress Cataloging-in-Publication Data Pending

Hendrick, Walt, 1947–
Ice diving operations / by Walt Hendrick and Andrea Zafares.
p. cm.
Includes bibliographical references ( ) and index.
ISBN 0-87814-843-4
1. Scuba diving—Safety measures. 2. Scuba diving accidents. 3. Search and rescue operations. I. Zafares, Andrea, 1965– II. Title.
GV838.674.S74 H46 2003
797.2'3—dc21

2002155708

Printed in the United States of America

1 2 3 4 5 07 06 05 04 03

Contents

chapter five

chapter six

chapter seven

chapter eight

chapter thirteen

chapter fourteen

Figures

Tables

Photos

(All photos by authors except as noted)

Preface

Ice Diving Operations was written to help in the ice diving operation-training process. This book is not a substitute for hands-on training from a certified recreational or public safety ice diving instructor.

To get the most out of the information in this book, simply start at the beginning and read each chapter in succession. That way, early chapters can build a foundation for later ones, thereby ensuring your understanding of the techniques, equipment, and reasoning behind them. Because each part of the book builds on what precedes it, skipping chapters or reading only parts of the book may cause confusion. As you read, use the knowledge of this methodology to answer the questions at the end of each chapter.

If you are reading this book as a public safety diver, then you will find it highly useful to have first read *Public Safety Diving*[1] to provide an important foundation of logic for much of the information provided in this book. *Public Safety Diving* also provides the necessary information on public safety diving (PSD) specific procedures and equipment that are applicable for ice and non-ice search operations. The information presented in *Public Safety Diving* is not what you will find in other books, articles, or training programs, so reading it is important even if you have read other related books.

It will also be useful to read *Surface Ice Rescue*[2], since any ice diving situation could include the need for surface ice rescue skills. Ice strength can be extremely variable, so anyone working on ice should be prepared to manage an accidental immersion situation.

If you are reading this book as a recreational ice diver or ice instructor, then you may find much of the information in *Public Safety Diving* valuable. For example, both recreational and PSD ice diving depend on the use of hard tethering with harnesses, quick-release pony bottles, and certified tenders, all of which are important parts of standard PSD. Although these items are presented in both books, some topics are covered in greater detail in *Public Safety Diving.*

Ice Diving Operations provides a minimal amount of techniques and procedures and a set list of equipment necessary to perform safe ice dives. There are many different ways of doing things, but we elected to include only what was optimal and most effective. We felt it was not helpful to include less safe or less effective procedures, techniques, or equipment choices.

The diving industry, particularly the public safety diving industry, is continually embroiled in a battle for or against establishing a set of general standards for operating procedures and guidelines. The anti camp says that different bodies of water require different procedures and guidelines. The pro camp holds that every type of public safety operation, such as hazmat, high angle, firefighting, and EMS, have national and local standards to protect public safety personnel safety and liability, so why should diving be exempt? The pro camp says that diving a frozen blackwater lake in Utah requires the same basic safety standards as diving a frozen blackwater lake in New York. Sure there may be a few differences between managing a hazmat situation with an over-turned tractor trailer on a Florida highway as there would be managing that situation on a California highway, but the basics are still the same. Diving operations are no different.

Do some ice divers still dive with octopuses? Of course they do, but it is not a safe equipment choice, and therefore will not be provided as an option in this book. The logic for why this is so will be explained. Some ice divers still dive without hard tethering to a harness with a locking carabiner; they dive with hand loops or BCD attachment points. As too many ice diving fatalities have demonstrated, diving without proper hard tethering and pony bottles is dangerous; so both are listed as required equipment in this book.

Everything presented in this book has been tested and proven in literally thousands of ice dives, and is further backed by more than twenty years of teaching and actual search and rescue/recovery diving experience. Many of the standards, procedures, guidelines, and equipment choices presented in this book have been developed, tested, and proven in non-ice public safety diving or in surface ice rescue situations with literally tens of thousands of dives around the world. What we have discovered comes from our experience in more than forty years and fifteen countries. Most often, members of the anti camp state that there are too many differences between different locations to implement a universal set of standards and procedures. This severely localized point of view does not match the experience of teaching in hundreds, let alone thousands of different bodies of water with thousands of divers from different backgrounds.

This book provides standards, guidelines, procedures, and techniques that will apply to the greatest majority of sport and public safety ice divers and incur the least amount of risk with the greatest amount of effectiveness. Many of the safety standards are the same for both sport and public safety ice operations. The sections on searching for bodies and evidence and for operating on less than four inches of ice specifically address the needs of public safety, not recreational, ice divers. If a section in this book does not specify whether it is for sport or working divers, then the information applies to both groups of divers.

For public safety divers, the main difficulty is getting access to a point at which tenders can place themselves and where divers can enter the water and reach the victim's point of entry. Most public safety ice divers take a recreational ice diving course and then feel qualified to perform public safety ice diving operations. But recreational ice diving classes are only conducted on ice that is strong enough to support standing divers and tenders. The problem is that if the object or victim fell through the ice, then so will divers and tenders as they attempt to cross the ice. That is the real world for public safety ice teams.

Therefore, the procedures presented for public safety ice divers in this book deal with the real-world problem of thin, unsupportive ice. Recreational ice diving operations have the time to set up elaborate ice shacks to dive from, which is rarely an option for public safety teams with limited budgets and time. Ice shelters cannot be presented as mandatory gear for teams that perform in rescue modes. The issue of entanglement prevention and management is critical for public safety divers who are bottom dwellers in blackwater. Mid-water recreational ice divers do not face the same entanglement situations. However, pony bottles, certified tenders, distance-marked tether lines, 90%-ready divers with backup divers, and hard tethering are, from our experience, mandatory for sport and working divers.

Mixed gas diving is not included in this book because our experience has taught us that it is not useful for more than 95% of this book's readership. Nitrox is used to decrease the risk of decompression sickness when air tables are used or to increase dive times. In general, ice divers should not have dive profiles in which decompression sickness is a serious concern, because they should not be diving that deep or for that long. This is especially true for public safety divers who have the additional risk factors of low or zero visibility, thin ice, night operations, significant entanglement concerns, and perhaps rescue-mode stress. Few recreational or public safety divers have the training, skills, or necessary equipment for a mixed gas operation. We do not believe that public safety divers should dive deeper than sixty feet without extensive training and experience with the right equipment. Recreational ice divers should consider a similar maximum depth. Therefore mixed gases for ice dives below 130 feet will not be addressed in this book.

The medical section is covered in detail for two reasons. First, ice divers cannot assume that their local EMS, or even hospital system, is well educated on cold water drowning and immersion hypothermia patient management. Ice divers need to be educated to provide on-scene patient first-aid. Secondly, the medical section can be used to educate pre-hospital and hospital care workers.

We describe some specific equipment, such as the MARSARS ice board and the LGS buoyancy compensator, and do so because there is no other equipment that we have found that meets the same needs. Other ice transport devices are available, and we have worked with all that we could find. However, nothing works as well in as many varied conditions as the MARSARS ice board. There is no equipment that even comes close. Therefore we felt it was important to describe such equipment in this book.

There are hundreds of procedures that are important to diving in general, but if they are not specific to ice diving, then it is unlikely they will be covered in this book. Although we would have loved to make this a 1,000-page book, such a project would not have been feasible from the publishing and purchasing end. So, for example, the technique of wearing a standard face mask strap under a neoprene hood is important, but such techniques will not be presented here. Such a technique does not apply to divers wearing drysuits with dry hoods or full face mask divers, nor is it specific to ice.

Some foundation information is repeated in this book because the reality is that some readers will read chapters out of order or may have long durations between chapters. Also, as we have learned in training, some information is important enough to repeat.

We hope you enjoy the book, and that you continually seek hands-on training to refresh and update your skills and knowledge. Remember, the only requirement on every dive, no matter what the mission, is to go home at the end of the day. This book alone cannot do that for you. You must have hands-on training with certified and qualified ice diving trainers who understand what is presented here. Best of success and stay safe!

Notes

[1] By Walt Hendrick and Andrea Zaferes with Craig Nelson, 340 pages, Pennwell Publishers 2000.

[2] By Walt Hendrick and Andrea Zaferes, 214 pages, Pennwell Publishers 1999.

Acknowledgments

We are so grateful to all our public safety and recreational students for helping us learn how to run and teach safe ice diving operations. Thank you for working so hard and long in all weather conditions testing procedures and techniques in both training and actual calls. We thank August Zupka for taking hundreds of pages and turning them into the basis of a book. Craig Nelson, thank you for your contributions to text and diagrams and for making the books *Public Safety Diving* and *Homicide by Drowning* realities.

We thank Cameron Jones, Pete Nawrocky, Melinda Keohan, Doug Murray, and Brian Thomas for their photo contributions, Maya Branman for her artwork, Robert McKay for contributions to SOP/SOG writing, Myrna Ross for proofreading, Kathi Norklun for copyediting, Don Hadden for helping take care of our teaching equipment, Carol DiBenedetto for taking care of Sierra while we are on the road, Andy Schmidt and staff for helping keep our computers running, Michael Knorr for his contribution to the section on ice shelters, and Dr. Michael Zorko, M.D. for reviewing the cold stress chapter, and Sgt. Cameron Jones and Lt. David Holland for reviewing the book.

We greatly acknowledge the efforts and support of our ever-hardworking trainers

> Orlando Abreus, Andy and Diane Alwine, Kenneth Balfrey, Pete Dawidowicz, Bryan Duffer, David Harrison, David Holland, Cameron Jones, Patrick Kilbride, Patrick O'Donnell, David McCoy, Mike Mulligan, Brian Nylander, George Safirowski, Greg Scott, Joe Steyer, Walt Szulwach, Cliff Turen, M.D., Jeff Warrick, Jason Yates, and August Zupka.

Special acknowledgement is given to the following friends, family, and colleagues:

- Bob Davis of MARSARS for designing and building the most effective, safe, and innovative ice rescue equipment available today and for reviewing sections of this book.
- Ralph Dodds for whipping out his red pen to proofread and edit all our articles and books year after year, and for always being there for us with his wit and wisdom.
- Nansi Hendrick who helped make Lifeguard Systems possible over so many years with so much patience and caring.
- Martin Nemiroff, M.D. for being the father of cold-water drowning resuscitation and for continuing to make life-saving contributions to the water rescue community.
- Edward Rosacker of Diver's Cove for his equipment contributions to the book and for decades of answering our and our students' equipment questions all hours of the day or night.
- George Safirowski for never-ending support, videography, teaching, and equipment building.
- Brian Thomas for his invaluable help organizing, typing, proofing, problem solving, and scanning during the editing process, and for keeping Lifeguard Systems running so smoothly while we were working on this book and while we are away teaching.
- Tom and Solvejg Zaferes for always being there for us with unquestioning faith.

Chapter 1
Introduction: Ice Diving—An Operation

Some divers ice dive for fun; others do it as a job. Those who ice dive professionally may do so as ice diving instructors, scientific divers, or public safety divers. This book is for the sport diver, dive instructor, non-polar scientific diver, military diver, and public safety diver.[1] Although these diver communities perform very different kinds of ice dives, and therefore have different needs, every community needs safe diving practices.

Unfortunately, many of the standards, practices, and guidelines used today are relatively arbitrary. Except for military divers and for the polar diving scientists who meet in polar diving AAUS (American Academy of Underwater Scientists) conferences to discuss standards, the other ice diving communities use standards presented by their certification agency or their particular public safety department. These standards are typically created by one or two sport diving instructors who are ice divers, or by a public safety diver who learned how to ice dive from a sport instructor. These standards are generally not tested and are not discussed in conferences with panels of experts. Especially for public safety divers, these standards often do not meet the needs of the job.

In 1986, the Ontario Underwater Council stated, "Perhaps the only definite statement that can be made at this time is to recognize the need for a study of ice diving procedures by the various training agencies; with the intention of developing agreed upon procedures that will only add to the safety of divers participating in ice diving."[2] It is now 2002, and after many needless ice diving deaths, we are not much closer to achieving this goal.

One of the primary missions of this book is to demonstrate the need for tested and proven safety, life saving, and training standards and practices that are universally accepted. Standard operating procedures and guidelines that can readily be implemented in public safety sport diving and training programs are also provided.

Although ice diving is technical diving, it is not the recreational technical diving associated with mixed gas and untethered deep diving. Rather, it is technical because it requires advanced specialized training, formal procedures, special equipment, and well-trained surface support. Public safety ice diving operations require even more highly advanced training and equipment.

Photo 1-1 Ice diving operations require specialized equipment, procedures, and highly trained surface support.

In certain parts of the world, for four months a year or more, divers descend below the ice roof. Ice diving is without a doubt severe overhead diving. Due to there being usually only one way in and one way out, ice diving is also confined-space diving. Divers in 10ft (3m) of water cannot push through a 1 1/2in to 2in (4cm to 5cm) piece of ice. They cannot dig their way out, and at depth, the darkness can be euphoric and incredibly disorienting.

Ice diving is highly equipment- and personnel-dependent, and has environmental difficulties that supersede almost every other type of diving. The simplest elements can become major difficulties. Ice itself changes in density and character as continuously and as much as the weather. Unexpected temperature changes are frequent under the ice, as is cold stress for both divers and tenders. Cold hands can prevent divers from performing the most basic tasks. Cold stress can cause irrational thinking, the inability to make decisions, and an incapacity for self-rescue. Cold stress can be compounded by cold and abnormal functioning of equipment and unexpected weak ice. Without sufficient contingency planning, training, and continued practice, such variables can rapidly intensify the severity of small problems and create an escalating chain of events that can lead to injury or death.

When performed correctly, recreational ice diving enables year-round diving and offers new experiences within a unique dimension full of exciting underwater dynamics. Ice diving is exhilarating, akin to how fresh powder is for advanced snow skiers. Ice divers can explore a familiar summer diving site and experience it in a new way. Diving beneath ice brings a new beauty to the underwater world, while fostering improved personal diving skills and knowledge. In the realm of public safety diving, ice diving operations may offer better chances of victim survival due to both cold water and the ability, in most cases, to know exactly where the victim entered the water.

Photo 1-2 Public safety dive teams do not choose dive times or locations.

There are three major differences between recreational ice diving and ice diving that is performed during a rescue mission:

- Recreational divers pick their dive days and sites, whereas in public safety diving, the situation picks the site and time. Recreational divers have time to set up a site that may include shelters, warm-water buckets, and other support tools. Public safety divers responding in a rescue mode do not have access to these advantages. Furthermore, public safety diving is not restricted to 6in (15cm) or more of ice as recreational diving is. Public safety divers typically work on ice that is too thin to support a single standing individual, let alone a group of people with gear.
- In recreational ice diving, two divers are normally in the water at any given time, diving as a buddy-team, whereas the public safety diver is most often diving alone. Solo ice diving is safer and easier to conduct. Hence recreational divers have to learn the skills necessary for ice diving with buddies.
- The mission for recreational divers is simply to have fun, while public safety divers are there to search for a drowning victim or to find evidence.

In the world of public safety diving, there is no "safe" ice. If ice was safe, chances are that public safety divers would never need to be called. If a child fell through the ice, responding personnel will not be able to walk, stand, kneel, or sometimes even crawl on the ice. Rescuers often start falling through the ice long before reaching the victim's point of entry. Lacking sufficient training and equipment, the average ice diving operation begins to fall apart before it even gets started.

Photo 1–3 Dive team members learning how to work in unsupportive ice. Tenders and contingency divers stationed in the water.

Often, drills and training do not prepare public safety divers for the realities of thin ice. Consider the majority of ice training dives you may have observed or in which you may have participated. Divers and tenders probably walked out to a pre-cut hole, and tenders stood on the ice to tend. A bucket of warm water was almost certainly at hand to manage free-flow problems. By comparison, in actual ice dive calls divers and tenders often push themselves to severe states of exhaustion just trying to reach the victim's hole, because the ice is too thin to support them. Frequently, when the team finally reaches the victim's location, the ice is so broken that they are unable to identify the original hole. In addition, tenders and divers, now in the water, are forced to work on self-survival. The *American Heat* video shows such an ice diving incident. It seriously risked the lives of fire personnel because they were not properly trained to work on thin, non-supportive ice.[3] Training should prepare students for the types of situations they may actually face, including the worst ice conditions, too many or too few responding personnel, storms, equipment freeze-ups, nighttime conditions, and more.

Public safety divers will sometimes respond to winter and ice emergencies without proper preparation or equipment. When a team is in the rescue mode, they are more likely to violate safety standards. Although the benefits of quick response may be higher during an ice situation than during warm weather, the risks are nevertheless far greater.

Public safety divers are not alone in being unprepared. It is not uncommon for sport ice diving classes to give students ice diving certification cards without having taught them how to cut holes safely (or at all) or how to tend. Often, students do not understand that a true redundant air source, such as a pony bottle, is mandatory. Furthermore, some divers are not taught how to manage an accidental disconnect or entanglement hazard, how to transport an injured diver to shore, or how to manage other ice-related problems.

In rescue modes, public safety divers must be capable of rapid deployment professional motion that includes the following procedures:

- Dressing in less than three minutes
- Rapidly securing the scene
- Setting up an incident command system
- Performing a risk/benefit analysis
- Creating a plan of action
- Rapidly and safely executing the plan
- Rapidly and gently transporting the victim from the bottom to the shore
- Preparing proper incident documentation
- Cleaning up and maintaining equipment
- Debriefing personnel

Recovery teams should be capable of moving like rescue teams. Even strict recovery-only teams will respond in a rescue mode when the chief's little girl falls through the ice a few blocks from where the team is conducting a drill. Recovery teams need to be capable of professional rapid deployment motion to save their own team members. It is crucial to have strict and stringent safety standards during rescue modes, because that is when noradrenaline levels are high, and greater risks are more likely to be taken for a perceived greater potential benefit. Recovery-only teams are at greater risk than rescue-recovery teams in such operations because they are not trained, nor accustomed, to moving rapidly.

With so much controversy and discussion surrounding recreational technical diving, ice diving is often ignored. Divers, diving instructors, and dive teams need to fully understand all the parameters of safe ice diving. Ice diving, and especially public safety ice diving, is extreme technical diving. The results of the lack of proper training are unfortunately demonstrated every year.

Cave diving is another form of overhead diving. An important difference, though, between cave diving and ice diving is that cave divers have strict, universally accepted, tested, and well thought-out standards. Cave divers understand that they need extensive formal training. Before setting fin in a cave, they obtain cavern certification through a series of advanced cave certification classes.

A basic cave diver in Florida will use similar, or identical, procedures and standards as a cave diver in Mexico. This consistency does not exist in ice diving, as one certified ice diver may use significantly different procedures, equipment, and standards from another ice diver on the same lake. Two ice instructors on that same lake may conduct classes completely differently. One instructor may cut a single round hole prior to the class, exert himself to pull the circle of ice out of the hole and stack it on the ice roof, use octopuses, put divers down on separate lines, and put the lines in the hands of tenders with insufficient training. Another instructor may teach students how to use chainsaws to safely cut three triangular holes with ice screws, straps, and carabiners screwed into each triangle. These triangles are pushed under the ice roof where highly trained tenders will be standing. The latter instructor may also teach the divers and dive students to equip themselves with pony bottles, harnesses, and multiple cutting tools. The divers may also be tethered on a yoked line with harnesses.

When cave deaths have occurred, they have almost always befallen divers who entered caves without cave certification. A certified and experienced cave diver might die in a natural disaster, such as a cave collapse, but the death will not be attributable to poor safety standards or training. Trained cave divers and instructors are rarely found in cave fatality statistics. Moreover, in many ways cave diving has fewer risks than ice diving. Cave diving does not have dangerous free-flow problems that can empty a diver's tanks in seconds; cave divers do not experience the same risks of cold stress and hypothermia; and cave diving does not rely on surface support to the same degree as ice diving.

Sadly, the majority of divers who die under the ice had previous ice training, or, much worse, were in the process of being trained by a certified ice instructor. There are too many headlines reporting "Instructor and Student Die under the Ice." The low standards of ice diving training and diver education that allow those deaths to occur is completely unacceptable!

Does ice diving belong in the advanced technical realm? Yes. Can ice diving be dangerous? Yes. Should any of these deaths have happened? No. In every case, the death was clearly a result of improper procedures and equipment.

The following case histories clearly illustrate that insufficient ice diving training, poorly trained surface support personnel, and improper or missing equipment play roles in these deaths. Let us all learn from these tragedies so that the mistakes are not repeated. The cases can serve as excellent topics of discussion during drills and training programs.

After reading the histories, we recommend that you make a list of each error, broken safety standard, and poor procedure that you recognize. After reading the entire book, return to the case histories, compile a new list, and compare it with your first one.

Case Histories

The first four case histories were selected for review because one or more Lifeguard Systems, Inc. (LGS) instructors participated in the recovery. Consequently, in these cases, we did not rely exclusively on newspaper information or hearsay. No LGS instructors were present at the recovery scenes of the remaining cases, and information was obtained from television, newspaper, and witness reports.

Two divers vanish under the frozen waters of Lake Kiamesha, New York (1994)

A dive instructor and advanced student entered the water through an onshore opening approximately 20ft × 25ft (6m × 8m). Three hundred feet (91m) to the left was openwater, approximately 250ft × 100ft wide (76m × 30m). Four hundred feet (122m) directly out was 5in (13cm) of solid black ice, which from the shore looked like openwater. To the right was approximately 1,000ft × 400ft (305m × 122m) of white ice.

The buddy-pair descended to look for a valve stem with an attached buoy. The divers had a buddy-line approximately 20ft (6m) in length, 68ft^3 (2m^3) tanks, drysuits, and a compass course. There were no alternate air sources, no line to shore, no backup divers, no emergency protocol, no emergency observers to call for help, and no surface personnel of any kind.

The job was seemingly simple: swim the buddy-line until they caught the buoy, then close the valve stem and follow the reciprocal compass course back to shore.

Approximately two hours after their descent, near closing time for the factory that owned the valve stem, someone realized the divers had never returned. A call for help was made. The bodies were retrieved approximately 400ft (122m) offshore and 70ft (21m) to the left. Their knives and masks were missing, and it appeared they had attempted to dig their way through the impossible black ice. The day of the dive was the student's eighteenth birthday.

The recovery effort further demonstrates the need for ice diving standards. When we arrived, most of our time, along with the state police's, was spent trying to keep a multitude of volunteer dive teams from going in, especially after the white ice became somewhat broken up. These teams were going to submerge without lines, harnesses, pony bottles, drysuits, or even a plan. Several of the teams were ice certified. They were little more prepared than the uncertified teams, who saw nothing wrong with going in without ice certification.

This disaster had only one positive outcome, which resulted when restraint prevented others from dying as well. Some typical comments of the day included: "The ice is broken up, why can't we just come up under it?" "We're not stupid. We're not going to come up under the black ice, and if we do, we will just turn right or left to come out."

Two seasoned divers drown in the Oneida River, New York (1994)

A dive instructor and his student attempted to perform a dive in an area of open water with an overhead ice roof a short distance away. They were not tethered, had no buddy-line, no harnesses, no tender, no pony bottles, and no designated area for diving. They had planned to stay in the open water area. They descended for a short period and gently went up under the ice cap to have a look. One or both of the divers caused a disturbance and an instantaneous blackout occurred. Miraculously, the student found his way out from under the ice roof. A divemaster onshore proceeded to descend and search for his instructor and friend. Both the instructor and divemaster died.

The bodies of the divemaster and instructor were removed from under the ice roof. One of the divers was only 10ft (3m) from the open edge of an ice roof and freedom to surface. Incidents like these make us all the more passionate about teaching instructors how to teach and conduct ice diving safely and knowledgeably. Ice divers should survive because of proper procedures, not because of luck.

Our plan is not to die when we perform an ice diving operation. Our plan is not to go under the ice with the attitude of "we're just going to take a look" or "I know this bottom like I know the back of my hand." If you have ever thought that, try putting your hand in your pocket and describe *exactly* what it looks like. Disorganization, lack of planning, lack of air, and sudden panic can create instantaneous chaos. Compound these with the thought that "I must save my friend, my instructor," and another multiple death is only seconds away.

Instructor and divemaster drown under the ice, Pennsylvania (1993)

This next case illustrates not only the lack of safe standards for ice diving and ice diving instruction, but also the critical importance of ice dive tender certification. The instructor and divemaster planned an ice dive and brought along an observer in case something went wrong. A member of the local public safety dive team told us, "These divers were extremely experienced and dove this spot many times." They tied their tether line to the bumper of their vehicle near shore and entered the ice. They had a line, they were tethered, but there was no tender or backup diver, and there was no concept of the maximum amount of line that should be used. They did not have pony bottles, and had no emergency plan. Since the observer left the scene for some unknown reason, there was not even someone to go for help. The water was relatively deep for ice diving and was black. There were rumors that a tender had been present. He perhaps became chilled and tied the tether line to the bumper and left the scene to get warm.

The bumper severed the line. The divers were found several days later, out of air and entangled from end to end in their very long tether line. The recovery dive was so technically difficult and dangerous due to the area, depth, and possible size of the search zone that rescue divers decided to wait a few days, until the ice was gone, before recovering the bodies. The press had a heyday with that delay. We were very pleased that the public safety divers agreed with our advice that public safety divers should never be put at risk unnecessarily for rescues or recoveries.

Ice deaths in Canada

Diver drowns under ice in Canada. At Christmas, two sport scuba divers entered the ice-covered lake—only one went home. From Christmas Day 1986 until New Year's Day 1987, the local fire department dive team and the local RCMP (Royal Canadian Mounted Police) dive team attempted to locate the missing scuba diver. Though numerous holes had been cut throughout the surface of the entire lake and many dives were conducted, both teams were unsuccessful in locating the missing diver. Attempts were made using volunteers walking the surface of the solid lake, in hopes of spotting something through the thick ice. A helicopter search of the lake also produced no results.

Eventually, on New Year's Day, the aid of Canadian Navy clearance divers was requested. Upon arriving at the lake, the Navy divers summoned the surviving diver. The diver admitted that he and his partner had not intended to actually dive under the ice. They had entered where the water was flowing out of the lake and was thus still ice-free. No lines were used, and there were no support personnel onshore. Probably no one knew that they were trapped under the ice. Somehow, the two divers found themselves "lost" under the overhead environment of a frozen lake! For some reason, when they arrived at the shoreline under the ice with no apparent way out, they separated. One diver went to the left; the other went to the right—a better guess. The latter diver was now being interviewed. Apparently, by going left, the other diver proceeded along the shoreline moving farther away from the unfrozen water.

After the interview, a dive location was chosen by the divers who had the necessary training. A full dive team was assembled, complete with all necessary safety equipment and surface support. One triangular-shaped hole was cut and the dive station manned. Only meters from the hole, the second recovery diver found the missing scuba diver. The victim was located wedged against the shoreline and the ice roof with mask off, second-stage regulator out, air still in the single cylinder, and knife missing.

Lake Simcoe: one diver drowns while making a video on ice diving procedures. A diver died after an ice dive in Lake Simcoe, Ontario in 1988. According to the Ontario Underwater Council, "This fatality marked the fourth consecutive year that 'ice diving' had contributed to fatalities in our province and the findings of the Coroner's Jury again seemed to re-emphasize the lack of consistent standards between the training agencies towards preparing divers for this activity."[4] Keep in mind that each year Canada has on average zero to five diving deaths, a high percentage of them due to ice diving. In addition, when one considers how few divers ice dive, the chances of dying under the ice are astronomically higher than dying during other dives.

Six divers met on Lake Simcoe for an ice dive. Three of them were experienced cave and ice divers. The other three divers were there only to make an educational video about ice diving procedures. Five of the divers were certified diving instructors. The victim was a certified divemaster.

They cut a hole large enough for one diver. One diver took a cave-diving reel with 1/8 inch (3mm) nylon line and attached it to a 50lb (23kg) crowbar that he wedged into the bottom just below the hole. The other two divers followed along by holding onto the line. According to one account, "After ten minutes into the dive, and with about 150ft (46m) of line out, the first diver turned around to check on the second diver; but he came face to face with the third diver instead." The two divers returned to the crowbar and found the second diver 10ft to 15ft (3m to 4.5m) from the hole, up against the ice roof hanging face down with his regulator out.

He was pulled out of the hole and onto the ice. The other three divers and a nurse stripped his gear off and began CPR. Emergency medical personnel were told he had had a diving accident, but because he had a bruise on his head, he was sent to a hospital that specialized in head injuries. Hours later he was transported to a hyperbaric center. Police investigation showed that his tank was empty, and his primary regulator was full of ice and contained vomit. After five months in a coma, he was removed from life support and allowed to die.

Unlike the other two divers, the deceased had worn only a single cylinder, no pony bottle, and had an octopus. No buddy-checks were done prior to the dive. Air pressure and other information were not recorded prior to the dive. "Likely the buddy checks were missed because everyone participating was qualified at leadership level. It is easy to fall into the trap of believing that one is above the 'law' as it were."[5] Sadly, ice diving is even less forgiving when the "laws" are broken than almost any other kind of diving.

As you read the case descriptions, you may be saying to yourself, "How could they have done that?" Keep in mind that the red flags are being highlighted here, but were not so obvious when viewed along with everything else that occurred during the day.

Two divers drown in Trout Lake, Canada, during training. Four divers met to conduct an ice dive as part of an advanced diver course. One diver wore a harness with a tether line. The second diver held a buddy-line with the first diver. No pony bottles were used. Two to three minutes into the dive, the line went slack and the tender pulled the line in. No divers were attached. A third diver entered the water with a longer rope and swam a circle, but was unable to locate the divers. Upon his surfacing, a call for help was made. The divers were found after two days of searching. Both divers' tanks were found empty.

"The Inquest found a lack of consistency, regarding standards for ice diving, between certifying scuba agencies."[6] The possibility has been suggested that the divers purposely disconnected because they thought they knew the underwater area well and wanted to go beyond the length of the tether line.

Instructor and student have life-threatening near miss under the ice. Television exploited the "heroics" of a team of ice divers who categorically did just about everything wrong. An ice diving instructor took his student on an ice training dive. The student was tethered, while the untethered instructor planned on diving free near the student. No pony bottles were used. The backup diver was not dressed; his gear was left lying on the ice. The tender did not know anything about tending. The tender was not told a maximum line length to let the divers out, and the line was not marked for distance. The only underwater contingency plan they seemed to have had was that the instructor was going to rely on an untested plan, that buddy-breathing was a feasible skill, and that a first-stage free-flow would not ensue. The untethered instructor probably never even considered what he would do if the student ended up on the bottom, causing a blackout or a whole host of other problems.

The instructor ran out of air. Octopus breathing with the student did not work, so the instructor grabbed the student's primary mouthpiece out of his mouth and put it in his own. The air-starved student pulled himself back on the line causing the tender to give additional line. The more the student pulled, trying to get back to the hole before he drowned, the more line the tender gave.

Meanwhile, the now unconscious instructor became caught in the line and was brought back to the point-of-entry by sheer luck. The hole was too small for a rescue since a rescue had never been planned. The radio topside did not work.

The only thing that went right, most likely by accident, was that the dive instructor's wife, who had medical training, performed basic life support that saved the instructor's life. They still believe they did almost everything right and this incident was just something that could not have been helped. We believe the media often promotes the reoccurrence of such situations by confusing ignorance and heroism.

Photo 1–4 Well-trained tenders, proper tethering, and the right equipment are keys to a safe operation.

In summary, the above fatalities occurred because of a lack of safe, hard-tethering, lack of true alternate air sources, poorly trained or no surface support, and the absence of effective, practiced contingency plans.

Definitions

Overhead environment

Overhead environment applies to any dive in which there is no direct access to the surface due to overhead obstructions along the route between the points of entry and exit. The entry and exit points are the same—the ice hole through which the diver's tether line extends. Overhead obstructions or roofs found in ice, caves, caverns, or wrecks, as well as any obstruction that prevents divers from being able to go directly to the surface qualify as overhead environments. Confined space is defined as an area that has only one way in and one way out.

Advanced diving

An advanced dive is any dive that requires greater than basic dive skills, preparation, planning, and equipment. All advanced diving requires a strong foundation of basic diving skills, including:

- mask clearing
- buoyancy control
- emergency regulator retrieval
- air consumption planning
- contingency planning
- self-rescue skills
- basic gear checks

Divers should not attempt advanced dives without having at least fifteen fully successful basic dives under their belts and without having the necessary training for the particular advanced dives they are attempting.

Tethered diving

Any dive requiring divers to be attached by a line back to shore is a tethered dive. Tethering in ice diving refers to a line that is securely attached to a diver's harness that extends to a trained tender and then back to an ice screw or other non-moving object and often all the way back to shore. Tenders are also often tethered to the ice screw to prevent accidental immersion/submersion under the ice roof. In sport diving, divers are typically tethered on a yoked line, or line that splits at the end into a "Y".

Technical diving

Technical diving requires special planning, training, and equipment. Depths beyond normal recreational diving, penetration, zero visibility, overhead environments, or other concerns such as moving water, rough seas, and ice are conditions requiring technical diving. Special equipment needs may include:

- ropes or lines
- harnesses
- signals or underwater communication tools
- mixed gases
- redundant air sources

Certified shore personnel are often necessary.

There are two primary types of technical diving–recreational and professional. Public safety diving and commercial diving fall under the professional technical diving umbrella. Technical diving is not done for fun. Professional technical diving may have federal standards regulated by the Occupational Safety and Health Administration (OSHA) and national guidelines such as those found under the National Fire Protection Agency (NFPA) Technical Rescue Standards Document 1670.

If you are a public safety diver, it is recommended that you read *Public Safety Diving*[7] prior to reading this book. *Public Safety Diving* is also recommended for ice instructors and serious ice divers. If you are an ice instructor or a public safety diver, it is also recommended that you read *Surface Ice Rescue*[8]. Ice diving instructors must be capable of taking care of everyone, both topside and under the ice. Because someone can accidentally fall through the ice during training, surface ice rescue skills are necessary when teaching ice diving.

Photo 1–5 Public safety personnel and ice instructors should first read *Public Safety Diving*. *Surface Ice Rescue* is recommended complementary reading.

Summary Questions

1. List three major differences between recreational and public safety ice diving.
2. If a public safety team is called to save a drowning victim, the ice is probably ____________________________ since it did not support the victim.
3. List at least eight duties performed during ice diving rescue operations.
4. Ice diving training and operations do not employ universally consistent, proven standards and procedures like those in cave diving. True or False?
5. List four specific, common causes of ice diving fatalities (lack of training or poor procedures are too general).
6. Define overhead environment diving.
7. Define advanced diving.
8. Describe diver tethering in ice diving.

Notes

1 Polar ice dives are often conducted without tethers due to their extremely high visibility. All the diving in this book will be based on tethers. Polar ice divers can gain much valuable information from this book about other ice diving topics.

2 Ontario Underwater Council, *A 10 Year Summary of Sport Diving Fatalities in Ontario,*

(1979-88): 86-5.

3 *American Heat,* 1303 Marsh Lane, Carrollton, TX 75006.

4 Ontario Underwater Council, *A 10 Year Summary of Sport Diving Fatalities in Ontario,* (1979-88): 86-5.

5 Ibid., 88-4.

6 Ibid., 86-4.

7 Hendrick and Zaferes (PennWell Publishing, 2000)

8 Hendrick and Zaferes (PennWell Publishing, 1999)

Chapter 2

Personnel

Whether you are ice diving recreationally or professionally, ice diving is an operation, not just a dive. An ice dive operation is comprised of various personnel, including: an operation supervisor (incident commander) or instructor, divers, tenders, and profilers. Profilers document the diver's exact movement, the search profile, and records all other pertinent information.''

Photo 2-1 Minimum personnel to run one dive (left-to-right): primary diver in water, backup diver, backup tender/profiler, primary tender, and 90%-ready diver.

Minimum Personnel

Ice diving is a labor-intensive undertaking. For a public safety rescue or recovery operation, the following list contains the minimum requirements. (Often one member of the team will function in more than one position.)

- Primary diver and tender
- Backup diver and tender
- 90%-ready diver (and tender if available)

Photo 2–2 The backup tender serves as the profiler/record keeper.

That makes five personnel needed–three divers and two tenders. The senior tender can serve as incident commander and the backup tender can serve as the dive profiler and record keeper. The recommended minimum number of personnel on a public safety dive team is six divers and three tenders. That way, if half of the team shows up, an operation can be conducted.

In recreational ice diving, the following personnel are recommended at minimum:

- Ice instructor or divemaster, who is also a record keeper.
- Diver(s) and tender (at least one tender per diver, unless diving with two divers on a yoke tether).
- Backup diver and tender.
- 90%-ready diver (and tender if available).

To help those who are not trained in water rescue to understand the roles of personnel on a dive team, the roles can be generally classified. The NFPA (National Fire Protection Association) uses the classifications of awareness, operational, and technician levels. These classifications indicate capability and training status levels. The immediate duties of each level for a public safety operation are listed below. Surface support personnel on recreational dive sites should have at least some of the awareness-level training described below, and tenders should have operation-level certification. Recreational divemasters and instructors should have ice diver search and rescue technician-level training, especially if they are diving in remote areas where a public safety response to a lost or injured diver or tender will take longer than ten to fifteen minutes.

In brief, awareness-level personnel have just enough training to cause no harm at a scene and are capable of performing generic cold zone duties. The cold zone is a shore area far enough from the water/ice line that there is no chance of a person accidentally falling on the ice or in water. Awareness-level personnel cannot assist with dressing a diver or physically setting up the warm or hot zones because they do not have the necessary training. Operational-level personnel have the training and certification to directly assist with the operation without being directly exposed to the hazardous atmosphere, which in the case of ice diving is the cold water under the ice.[1] Tenders and the safety officer are the main operational-level personnel in diving operations. Technicians are those personnel who are directly exposed to the hazardous atmosphere (the hot zone) and therefore need the proper training, equipment, and support personnel to function.

In public safety ice diving, tenders sometimes fall into a gray zone between the operational and technician levels because they have the potential of direct exposure to ice water. Tenders can fall through the ice, or, when ice has broken up, be forced to tend from the water. Falling through the ice can occur in both recreational and public safety diving operations, but only properly trained public safety ice divers should ever knowingly dive from ice too thin to stand on. Ice diving tenders therefore must have the skills and appropriate personal protection equipment to survive immersion in ice water.

Awareness level

Anyone who will respond to, or participate in, an ice diving operation should be trained (at minimum) to the awareness level, which involves at least four hours of training. The awareness level is the lowest level of training and has the important main goal of preventing further injuries and problems at the scene. Public safety personnel at the awareness level should be equipped with the proper shore-personnel equipment described in chapter 5.

Photo 2-3 Dive operations often require more surface support personnel than divers

Upon arriving at the scene, awareness level personnel should:

1. Don a personal flotation device (PFD) and other necessary personal protective equipment (PPE).
2. Know where to go to assess the scene and stage the operation, based on pre-determined, documented staging areas and the location of the incident.
3. Perform a rapid scene assessment to determine who needs to be dispatched to the scene and what kind of response is necessary.
   - Identify hazards to life and health, such as slippery embankment areas, holes or rocks in the ground that could cause problems, downed power lines, fuel fumes, a collapsing dock, deep snow, ice, storm conditions, high winds, or potentially contaminated water.
   - Assess the number of victims.
   - Determine whether a rescue or recovery response is appropriate.
   - Know where and how to get needed assistance.
   - Mark a spot onshore in front of each victim if possible.
4. Implement the Incident Management System (IMS).
   - Establish a command post and an incident commander.
   - Establish staging areas, staging officers, safety officers, and other officers and sectors as required by the size and complexity of the incident.
5. Call for the appropriate agencies, including the dive team, Emergency Medical Services (EMS), and law enforcement.
   - Ideally, have an ambulance for each victim, plus one or more for personnel and bystanders who may require medical attention. An extra ambulance provides an effective place for divers to be checked out by EMS when the job is completed. An ambulance is also a good place to secure a very distraught witness or a family member of the victim during the operation. Remember, part of maintaining scene safety is to make sure no one on the scene becomes

hypothermic or injured. Awareness-level personnel may need to notify EMS personnel untrained in ice diving situations that proper exposure clothing will be necessary. These suggestions may sound like common sense, but countless actual incidents show that they are often disregarded. The same suggestions apply to responding police and fire personnel.

6. Secure and manage the scene to maintain scene safety.
   - Determine and mark hot, warm, and cold zones. The hot zone is the water and its ice covering; the warm zone is the area onshore where operation- and technician-level personnel are staged with their equipment; and the cold zone is where all bystanders and personnel with less than operational-level training are staged. The command post is placed in the cold zone with a total view of the scene (if possible).
   - Make sure unauthorized persons do not enter the hot or warm zones. Account for all personnel who enter the hot or warm zones. Photo ID ice diver or ice tender certification cards work well as accountability tags. The cards can be collected when personnel enter the warm zone and returned to them when they make a final exit.
   - Secure witnesses and, if possible, turn them over to law enforcement personnel for interviewing. Law enforcement personnel should determine whether foul play or neglect might be factors in the incident. EMS should evaluate and monitor the mental and physical condition of witnesses and public safety personnel.
   - Determine the wind direction, and direct incoming vehicles to turn sideways to the wind, to form windbreaks for the staging area and thus to protect personnel as much as possible. It may be necessary to set up tarps to prevent wind from coming underneath vehicles as well. Stage arriving ambulances for easy access and immediate egress. Plan and direct a protected staging area for the arriving dive team as well. Plan for a helicopter-landing zone if necessary. In directing vehicles, be sure to maintain access and exit routes for fire, EMS, and other appropriate rescue personnel and vehicles.
7. Draw a profile map of the area and indicate the most likely victim location. Find a safe place from which to deploy divers.
8. If well-trained personnel have not yet arrived, begin interviewing witnesses and document information on the profile map.

Since ice diving incidents are not ordinarily a regular occurrence, it is recommended that someone make an awareness-level incident checklist for awareness-certified members to keep in their vehicles or wallet (see appendix D). The checklist can be printed on ID-sized cards and then laminated.[2] It may also be necessary to keep copies of the pre-planned map of possible staging areas in their vehicles, depending on the size of their jurisdiction.

Operational level: tenders and other shore support

For standard diving operations, operational-level efforts are shore-based operations because the water, or hot zone, is limited to technician-certified personnel only. In ice diving operations, however, as stated earlier, tenders fall into a gray area between operational and technician levels. For that reason, any tender who will work on an ice diving site must have ice tender training and certification appropriate for either recreational or public safety diving response. Public safety tenders who might have to tend from the water, because of non-supportive ice, require tender-technician training.

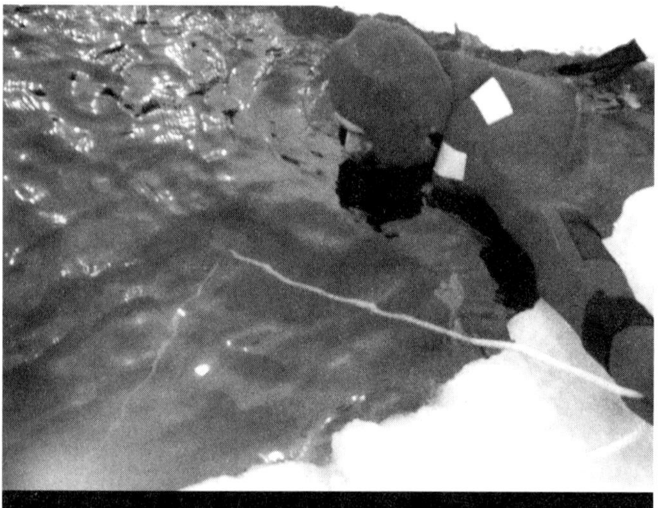

Photo 2-4 Ice tenders fall into a gray zone between operational and technician status because they need the training and equipment to handle accidental or purposeful immersion.

Operational-level personnel have the most important jobs on a site, they essentially direct the diving operation. On the scene, operational-level personnel should:

1. Perform all awareness-level duties, if not already completed.
2. Identify and utilize necessary PPE. Tenders need to be immersion-capable.
3. Conduct risk/benefit analyses, and inform the safety officer of any concerns.
4. Finish implementing the IMS.
5. Create a profile map of the incident with the location of the victim(s), if not already done.
6. Interview witnesses and record information on the profile map, if not already done.
7. Identify available resources and ensure adequate response.
8. Develop a shore-based plan-of-response with available resources and within standard operating procedures and guidelines.
9. Determine the best place to deploy divers into the water. This may require cutting holes.

10. Decide whether the ice is thick enough to stand on for recreational diving. If public safety diving, determine a method for transporting personnel to the area where the divers will be deployed into the water.
11. Determine whether divers are mentally and physically prepared to safely conduct the dive.
12. Assist divers with equipment, preparations, access to the holes, and entry to and exit from the water.
13. Serve as dive tenders, profilers, safety officers, and other shore support.
14. Be capable of carrying out appropriate contingency plans.
15. Be able to throw a rescue rope-throw bag at least 60ft (18m) so that the deployed rope lands in the victim's hands. NFPA document 1670 also requires the ability to throw a coiled line.
16. Perform shore-based ice rescue techniques.
17. Maintain appropriate documentation.

Technician level

Only properly certified and trained ice divers belong in the water performing ice diving. Public safety ice diving training and certification is necessary for public safety divers to perform ice search-and-rescue/recovery operations. Recreational ice training will not meet the needs of public safety divers.

Diver responsibilities

In general, the diver's responsibilities as a team member include the following:

1. Maintain diving skills, including diver-to-diver assistance, swimming, self-rescue, out-of-air procedures, buoyancy, and all other needed abilities.

Photo 2-5 Diver ready? A critical tender job is determining whether a diver is mentally and physically capable of diving from before gear setup time to the point of descent.

2. Maintain good health. If a diver has recently had any type of cold, flu, infection, or injury that may increase the risk of injury during a dive, the diver should not dive. Divers should maintain good physical condition due to the strenuous nature of the work performed.
3. Be familiar with the department-issued equipment and standard operating procedures and guidelines. Divers should learn the signals and emergency procedures before being designated as divers with a team or in an ice diving class.
4. Maintain appropriate department records and personal dive logs.
5. Maintain department equipment. If there is any doubt about the reliability of any piece of equipment, it should be put out of service, and the appropriate officer should be notified.
6. Conduct inspections of assigned equipment at least once a month, and before and after each dive.

On an actual call, divers must strive to reduce the risks to which they and other team members are exposed. Divers should be physically and mentally prepared for the dive. The following is a partial list of the diver's responsibilities at a dive scene:

1. Say "no" to any dive that the diver does not feel capable of performing.
2. Be able to serve as competent tenders.
3. Get dressed quickly and quietly with a tender's assistance. A well-trained public safety diver with a completed buoyancy-control device (BCD)-tank-regulator assembly should be dressed and ready in less than three to four minutes with the assistance of a certified tender.
4. Listen to the briefing, and ask questions if in doubt about the assigned task or the safety of the assigned task.
5. Keep conversations brief and related to the incident/dive.
6. Before entering the water, conduct a proper tender-diver gear check, and review the dive objective, search pattern to be used, line signals, and contingency plans.
7. While in the water, perform the assignment as safely as possible. Additionally, divers should abort the dive if they do not feel right about the dive, experience any signs of cold stress, have cramps, become fatigued, or develop any kind of problem that could compromise their safety or anyone else's safety.
8. Keep the tether line taut at all times to permit line signals to travel, to decrease the chances of entanglement, and to ensure that the search pattern is accurately

kept. A taut tether line increases safety in non-search recreational diving for the same reasons.

9. Use safe in-water skills, including acclimating the face to the water, removing the weightbelt before exiting, and ending each dive with at least 1,000psi (68bar) in the main cylinder, assuming a working pressure of 3,000psi.
10. Stay healthy on the scene. Ensure that a pre- and post-dive blood pressure check is made, maintain hydration, and ask to sit down if left standing too long with gear on.

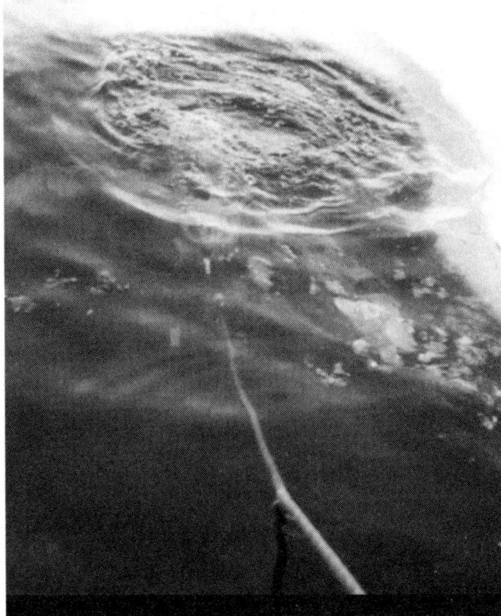

Photo 2-6 It is the diver's job to keep the line taut in tender-directed diving.

If the dive is called off or discontinued, the diver should not question the dive coordinator's decision at the scene. Discussions about the decision should be held out of public view, preferably at the team's headquarters. If an instructor calls off a recreational dive for safety reasons, there should be no dispute.

Divers' responsibilities do not end once they exit the water. They must always:

1. Immediately report any problems or symptoms (including those of cold stress, potential gas bubble injuries, fatigue, equipment problems, or safety concerns) to the dive coordinator and EMS personnel.
2. Properly clean and store equipment until the next operation/dive.
3. Ensure all necessary information is properly recorded.

It is extremely important that divers neither be coerced into making dives that they are uncomfortable with, nor penalized for refusing to make such a dive. No explanation should be requested. In return, divers must recognize their limitations. Divers must learn when to say no, and when to abort diving operations for their own safety. Divers who needlessly endanger themselves also endanger their team members, since the members will try to save them.

Primary diver

In public safety diving, primary divers are the tender's eyes. Primary divers use their hands, bodies, and sometimes eyes, to search at the ends of their tether lines. Divers are responsible for keeping their tether lines taut in order to keep search patterns accurate. These divers are solo-tethered/tender-directed. They do not have a buddy to distract them from the search and with whom they can become entangled. A properly trained solo-tethered/tender-directed diver and tender can search more area, more accurately and more safely than two or more divers down together.

Because the water may be low in visibility or actually black, search divers are often mud dwellers searching on the bottom, where entanglements are a common problem. A tether line provides divers with direct-line access to the surface. It also provides the backup diver direct line access to the diver if he/she becomes entangled on the bottom. The lack of visibility requires that the divers be tender-directed, because they do not know where they are or where they have been. Primary divers rely on the tender to tell them where to search.

Backup diver

In addition to the normal duties of divers, backup divers have the added responsibility of being prepared to assist the primary diver. Backup-diver duties include:

1. Acting as a replacement for the primary diver if the primary cannot dive for some reason.
2. Being prepared to take the primary diver's place in the next rotation of divers, thereby decreasing downtime.
3. Rendering assistance. The primary diver may need help with an entanglement, a leg cramp, or an impaled fish hook. They may also need assistance bringing a body or other object to the surface or extricating underwater vehicles.
4. Rescuing the primary diver. The primary diver may need immediate assistance for out-of-air or injury problems.

The backup diver, also known as a "safety diver", needs to be physically and mentally ready to deploy at any moment in a controlled, practiced, professional rapid-deployment motion. Ready-to-deploy means fully dressed at the hole, tethered, weighted properly, and stationed with a trained backup tender.

A backup diver's regulator or full face mask should never be lying on the ice getting cold. This situation increases the chances of freeze-ups when the diver enters the water. Backup divers are fully dressed except that they are not breathing the air from their cylinder. In recreational or recovery dives with good ice, insulation should be placed between the backup diver and the ice.

Backup divers should monitor the progress of the dive as reported by the tenders so that they will be mentally, as well as physically, prepared to respond if necessary. Backup divers must be constantly alert for any situation that might demand their participation.

Not just any diver can be placed in the backup diver position. Too often, the diver who has the least skill or who cannot equalize that day is placed in the backup diver position, because no one ever really expects the backup diver to be needed. Lifeguard Systems has a saying: "The strong may not survive." This refers to teams or instructors who place strong divers in place as primary divers with weaker divers as their backup divers. The stronger divers are going to go deeper into the weeds and muck than will weaker divers. If a stronger diver cannot manage a problem or get out, what chance does the weaker diver have to manage the problem and get both of them out? Not much chance at all. The stronger diver should be placed in the backup diver position.

Backup divers need to have the necessary training and skills to be able to:

- Deploy with calmness and professional rapid deployment motion.
- Equalize ears effectively while making a safe descent down a primary diver's tether line without pulling on that line in any way to any degree.
- Manage entanglements. This includes feeling and cutting thin fishing line and managing hooks or other entanglement hazards specific to the local environment while wearing appropriate cold-water gloves in silt-out conditions.
- Perform good buoyancy-control skills, including helping another diver make a safe, slow ascent (two seconds per foot) and helping a diver hover in shallow water while managing a problem.
- Show thorough competence with their own and the primary diver's equipment.
- Assist the primary diver in a variety of ways underwater, including, but not limited to:
  - replacing a tank that has slipped out of the BCD tank strap
  - getting rapid breathing under control
  - replacing a loose or lost weightbelt
  - performing out-of-air emergency procedures
  - managing panic
  - finding a lost diver
  - removing leg cramps

Do not place novice ice divers in a rescue-responsible position. When these divers reach the point of being physically and mentally prepared to do the backup job, they should do so with an experienced backup diver serving as the 90%-ready diver.

90%-Ready diver[3]

In recreational and professional ice diving, there are many potential opportunities for the backup diver to fail in an emergency. Equipment freeze-ups, an inability to equalize, the loss of a weightbelt, or any number of other problems can occur. The contingency plan for these events is the presence of a diver in a 90%-ready capability. Typically, for example, it is easier to deploy the 90%-ready diver than it is to make a regulator exchange for a freeze-up problem with the backup diver's regulator. The 90%-ready diver should be ready to take over for a backup diver if, for whatever reason, the backup diver cannot complete the rescue of the primary diver.

In non-ice public safety diving, the 90%-ready diver often sits with fins and exposure suit on, next to fully assembled gear. When a backup diver is alerted or called to go in, all the 90%-ready diver needs to do is don the weightbelt, BCD, and mask—all of which should take fewer than fifteen to twenty seconds. Because the likelihood of a backup diver having equipment problems is greater on ice dives than during warm weather dives, we recommend that the 90%-ready ice diver be fully ready, just as the backup diver is.

If the hole is less than 50ft (15m) from shore, the 90%-ready diver can, if conditions permit, wait comfortably on shore. Conditions permitting means the ability to rapidly transport the 90%-ready diver to the hole, such as with a sled. If the hole is farther from shore, the 90%-ready diver should be set up in a warm and comfortable manner near the hole.

Photo 2-7 The 90%-ready diver is often fully dressed because cold increases the chances of backup diver equipment problems.

Divers placed in 90%-ready diver status should be the strongest divers, since they will only be deployed when the primary diver is in need of help and the backup diver has failed. Stress will be highest at these times, so 90%-ready divers need to be very competent and rightfully confident.

Using weaker divers as primary divers ensures that all divers have backups who are capable of saving them. It also helps build stronger divers and therefore a stronger team. How will the weaker diver gain the experience to become strong if he/she is continually put in the backup or tender position? In such cases, weaker divers typically remain weak, rarely attend drills or actual rescues, and are less involved because they are less likely to be allowed to dive. Weaker divers may also give up, because they know that they will never be asked to be a primary diver on an actual rescue incident. If the stronger diver is placed in the backup position, the weaker diver has the opportunity to dive and become stronger and will feel encouraged to practice and become better.

Tenders

Tenders are the most important people on the site. The tender's role is to ensure that the site is ready, that the diver is ready, and that the diver has a safe dive. The tender is the diver's lifeline back to shore. Too often, we get caught up thinking that the diver is the key figure. Divers are often considered heroes, because they are the ones who go under the cold, often black, and perhaps even fast-moving water, possibly risking death. Although it takes a lot of courage to go into the water in certain conditions, it takes much more than the simple act of going underwater to make an ice dive safe and successful. In search operations, for example, divers are the eyes of the tenders, who direct divers where to search. When a drowning victim is found, the tender is equally, if not more, responsible for the find.

Ice diving tenders are considered specialized tenders, due to the possibility that they may accidentally or purposefully enter the hot zone. They also need specialized, ice-specific skills and knowledge. While all ice tenders do not necessarily have to be capable of performing the specific skill of safely cutting a hole, all tenders must know how to determine whether the holes and site are properly set up. Tenders make sure the diver's blood pressure is checked pre-dive and post-dive and that the diver is mentally and physically ready to make the dive. Tenders dress their divers', perform the first complete equipment check, and then bring the diver to the safety officer for a second complete tender/diver check. Tenders should have at least an estimate of what their diver's breathing and surface air consumption (SAC) rates are, and should monitor the diver for any subtle signs of cold, emotional, or physical stress. Tenders make sure the diver's pressures and times are recorded, and tell the divers when to go in and when to come out. A tender is responsible for the diver's safety prior to, during, and after the dive.

The tender's job also includes communication between upper management and the diver in the water. A good working tender and diver team comes back to surface management with a map showing areas searched and confirms either that "we found it" or "it's not there!" rather than "we couldn't find it...."

A good tender will know within seconds when a diver's tether line is snagged, often before the diver realizes it. Well-trained tenders with good communication systems will recognize the possibility of a problem the moment divers increase the rate or volume of their breathing. Competent tenders will notice changes in the way their divers move by continuously focusing on the tether lines. They will use their knowledge of the diver's breathing and SAC rates to make fairly accurate estimations of how much air the diver has at any point in the dive, and to bring him/her up with at least 1,000psi in his/her main cylinder. *Hence, the diver has little to do or think about other than searching and keeping the tether line taut or simply enjoying a recreational dive.*

It cannot be said enough that trained and certified tenders are the most important personnel at an ice diving site and at all public safety diving sites. The key words are "trained and certified", because a poorly trained tender can cause more harm than good. Tenders do not need to be certified divers, but they must be fully trained in:

- line handling
- line signals
- gear setup
- dive planning
- contingency plans
- emergency equipment
- oxygen and first-aid procedures
- the land supervisory skills of a divemaster

Photo 2-8 Tenders are fully responsible for ensuring their diver's equipment is functioning and perfectly set up.

Tenders should be properly dressed and equipped with appropriate personal protection equipment.

Very importantly, tenders need to know when to call a dive off, because of problems either with a diver or with the situation. Divers may be less likely to call a dive off than tenders for a variety of reasons, ranging from the excitement of wanting to experience diving under an ice roof to the adrenaline stimulated by the desire to save a child lying on the bottom.

It is recommended that dive teams and dive instructors have at least a few dedicated tenders, meaning tenders who are not also divers. These tenders are more likely to take the job of tending very seriously and give it the responsibility and priority it needs. They also save the team money because they do not need dive gear or training. If tenders are trained and certified and are given the responsibilities they are due, they will become long-term, valuable team members. If, on the other hand, they are treated like gophers and are merely handed a rope with the instructions "hold on to this", their lives as tenders will be short-lived, and diver safety will be severely compromised.

Profiler—Record Keeper of Search-Pattern and Dive Profiles

If it's not on paper, it didn't happen

The task of the profiler is to log all necessary dive information. During public safety dive operations, the profiler monitors and records the progress of the search by keeping a map showing exactly where divers have been and what areas were searched. This map is one of the most important tools in public safety diving. Without having the search profile on paper, surface personnel cannot know if and exactly where the diver missed an area in the search.

Although it is not critical for profile maps to be made of recreational diver movements, it will add a measure of safety and tender skill proficiency. In a worst-case scenario—accidental disconnect—an up-to-date profile map will tell the backup tender where to send the backup diver.

The profiler is also responsible for recording information about divers before and during the dive, as well as at the dive's completion. The following information should be recorded on all ice dive sites, both recreational and public safety:

- Diver blood pressure (pre-dive and post-dive) for all public safety sites. Also recommended if someone trained to take blood pressure is present on recreational sites.

- Date, location, surface conditions, and underwater conditions.
  - Surface conditions can include: wind, ice thickness at dive area, snow, etc.
  - Underwater conditions include: visibility, current, and bottom variables such as grass or other debris, etc.
- Names of the:
  - Diver
  - Tender
  - Backup diver
  - Backup tender/ profiler
  - If recreational diving, the buddy
- Cylinder pressures:
  - In
  - Down
  - Out
  - Total used
- Time:
  - In
  - Out
  - Total
- Air consumption rate at depth and the surface air consumption rate.
- Depth.
- Breaths per minute (BPM) recorded every five minutes if a communication system is used, and every ten minutes if only line signals are used.

Public safety dive operation record keeping should also include:

- Incident commander and safety officer names.
- Search pattern used.
- Total number of running feet of search pattern conducted.
- A determination of whether the search area can be secured or should be re-searched.
- A profile map showing every location where the diver moved with marks on any areas where the diver signaled "make a notation".

Profiling a search pattern

The first step in profiling a search pattern is to draw a map showing locations of the tenders, the divers, and landmarks. The next step is to illustrate where the diver searches. It is important to be as accurate as possible. If the diver gives 3ft (1m) of slack, for example (i.e., the diver is 45ft [14m] out and then comes in to 42ft [13m]), for 5ft (1.5m)

1. Know where to go to assess the scene and stage the operation.
2. Don PFD, gloves & other necessary PPE.
3. Perform Rapid Scene Assessment.
4. Identify life & health hazards.
5. Interview witnessess and document information on the profile map.
6. Assess no. and ASAPS status of victims.
7. Establish communication w/ victims.
8. Mark a spot on shore in front of each victim if possible.
9. Implement IMS. Establish command post, Incident Commander, Staging Areas, Staging Officers, Safety Officers, and other officers and modules required by the size and complexity of the incident.
10. Direct positioning of rescue vehicles to form windblocks & exits.
11. Setup demarkation line.
12. Call for appropriate agencies, including the dive team, EMS, and law enforcement.
13. Secure the scene & determine & mark Hot, Warm, and Cold Zones.
14. Secure witnesses for interviewing by law enforcement & check their conditions.
15. Draw a profile map of the most likely victim location. Find a safe place to deploy divers.
16. Reassess ASAPS.

A = Agressive
S = Self-rescue capable
A = Alert
P = Passive
S = Submerging

Divers Alert Network: For Divers Emergencies 919-684-4326

Make sure the diver is physically, mentally, & emotionally prepared for the dive.

SAFETY OFFICER; ______________________

INCIDENT COMMANDER: ______________________

NOTES: ______________________

THE TENDER SHOULD HAVE AND CHECK:

1. Appropriate PFD closed properly.
2. Appropriate footwear and exposure protection.
3. Proper tending gloves.
4. An appropriate time keeper.
5. A water activated flasher.
6. Eye and sun protection.
7. Back tethered harness if working from a steep embankment.
8. If acting as the back up tender, a profile slate and writing utensil.

WATER SITE MUST HAVE:

1. Contingency reg. on full cylinder.
2. Contingency reg. on full pony bottle.
3. At least one rescue throw rope bag per diver-tender pair.
4. BLS, or better, first aid personnel and supplies.

TENDER TO DIVER:

Right: **Diver's**: Left

- 1. Stop/face & tighten line/ you OK?
- 3. go Right (Divers) ⬅
- 4. go Left (Divers) ➡
- 2+2 search immediate area.
- 3+3 Stand by to leave bottom.
- 4+4 Come up.

There is only 1 job you have to do every time. —GO HOME. Team LGS.

TENDER CHECK OF DRIVER:

1. Air on.
2. Hood in place.
3. Mask ready, gloves on.
4. Primary regulator, exhale first, breathe 3 times, check tie-wrap and mouthpiece.
5. Pony reg: same as #4, hose under arm, secured w/ mud mouthpiece cover and holder.
6. Drysuit hose on functional.
7. Diver, press BC inflator button 3 times, while tender checks pressure gauge, gauges secured, all hoses trim to body and secured.
8. Diver, without looking, find 1st shears, 2nd shears, & 3rd cutting tool.
9. Diver, find carabiner at harness tether point, tether line attached.
10. Diver, find weightbelt release– Is it right hand? 10" extra webbing?
11. Fins ready, w/ ankle wts. if needed.
12. Back-ups have contingency lines.

DIVER TO TENDER:

- 1. Diver is okay.
- 2. Tender, make notation.
- 2+2+2 Tangled but okay, alert back-up diver.
- 3+3+3 Okay but needs help from backup-diver.
- 4+4+4 Needs immediate help!
- 6+6 Found the object

Lifeguard Systems
P.O. Box 548 Hurley, NY 12443
Tel/Fax: 845-331-3383
www.teamlgs.com
www.wateroperations.com
www.rip-tide.org

Figure 2–1 Professionally-made profile slates are effective tools to make sure all necessary information is kept, even during inclement weather. This Lifeguard Systems public safety diving slate, for example, prompts profilers to record all the above data, provides a place to draw a profile map, and provides sixteen duties of first responders to secure and begin the operation, tender and diver equipment checklists, and diver-tender line signals. A place to write out the incident command (management) system can also be included on another side of the slate.

of the pattern before realizing it and taking it back up, that should be drawn on the map. Such a 3ft × 5ft (1m × 1.5m) area missed certainly is large enough to have contained a drowned child.

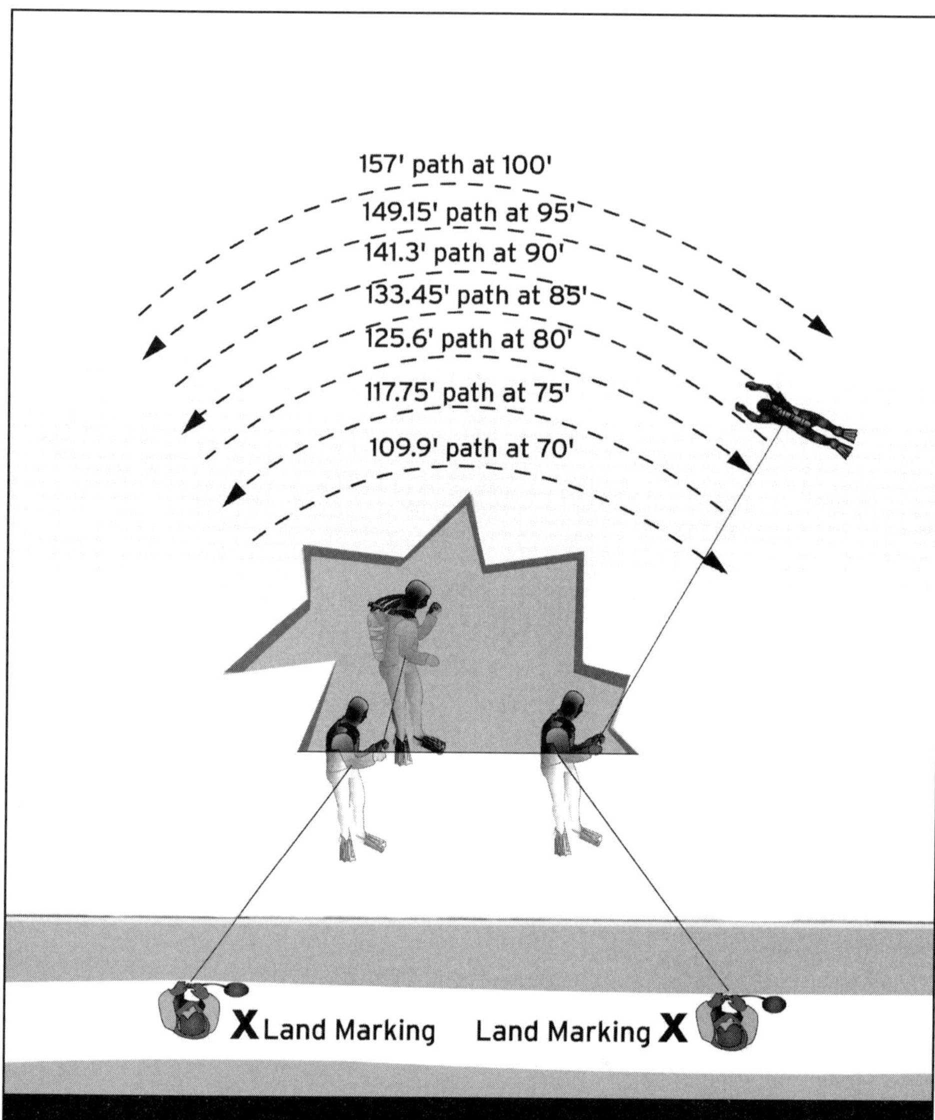

Figure 2-2 A profile map should be drawn so that a person who has never been to the location can pick up the map and walk to exactly where the tender was standing and point out exactly what areas have and have not been searched. Every line is marked and measured. The hole becomes your center. The top of the map should show the opposite shore landmarks in detail.

Who should be the profiler?

The job of profiler is best filled by the backup tender. Profiling increases the backup tender's awareness of the dive situation and diver location at all times; because profilers must map the diver's progress, their eyes are always following the diver. In fact, profilers may even realize that there is a problem at the same time as the primary tender. The increased awareness enables the backup tender to deploy the backup diver that much faster and accurately. Using the backup tender as profiler also enables the team to work with fewer personnel. This is particularly important on ice sites where we want as few personnel as possible on the ice. Primary tenders will call out diver distance and signals given and received. For example, "diver at 52.5ft, given 3, 3 returned". It is the job of the backup diver to look at the tether line and check that the stated distance and direction are correct prior to documenting them. This serves as a double check system for pattern accuracy and diver safety.

Another reason for using the backup tender as the profiler is that when the primary diver is finished with the dive, the backup diver can move to the primary slot if cold stress has not set in. The backup tender then becomes the next primary tender. Backup tenders have the greatest ability to know whether the search area of a primary diver can be secured. This is because backup tenders have recorded exactly where the primary diver searched and where the diver may have missed, how many running feet were covered in what time period, the diver's breathing rate, and other variables indicating whether a good search was done or not. Backup tenders therefore know better than anyone where to put the next primary diver down.

Photo 2-9 The backup tender (left) usually stays with the primary tender to record the primary diver's movements. If the backup diver needs to be deployed, the backup tender relinquishes profiling.

In recreational diving, backup tenders/profilers who become the next primary tenders will know in what areas to keep their divers in order to decrease the chance of snags or other problems. In public safety diving, these tenders will know what patterns need to be used to search the areas with snags so that the snags will not cause further problems.

Shore personnel

Not every ice diving operation has enough extra people to have shore personnel, but if they are available, they can help make everything run smoothly. Shore personnel can help with gear handling and maintaining the area and by providing food and warm drinks in the cold zone. They can help make the shore area safer by moving snow or salting icy areas, and can operate a transport device with a pulley system to move personnel and equipment between the shore and the hole.

Shore support personnel, whether they are recreational or public safety, need to know what they can and cannot do. Shore personnel need to have some degree of ice operation training, or they might do more harm than good. For example, consider a fully certified and well-trained public safety diving tender who is on an ice site with no ice training. The use of a standard gear check rather than an ice check could result in free-flowing equipment. For example, one step in the diver gear check for warmer months is to have the diver hit the BCD power inflator three times without visually looking for it, while the tender looks at the gauge. The functions of this test include showing that the diver can reflexively find the power inflator in blackwater, that the power inflator works, that the diver has a little air in the BCD, and that the pressure gauge needle does not fluctuate downward, which would indicate that the air was not turned on. This important check should not be done on an ice site if the BCD is wet, because it can cause the inflator to start free-flowing. Once an air source, such as a drysuit, BCD inflator, or a regulator has been wet, it should not be used in the air. A tender who is not trained and certified as an "ice" tender would not know this and could inadvertently cause a free-flow while performing nothing more than a standard gear check. Therefore, only shore support with a minimum of operational-level training should be allowed in the warm zone, or "operation zone". Other shore support should remain in the cold zone.

On recreational sites, at least one shore support person should be certified in CPR and first aid and have a strong background in or understanding of hypothermia, cold stress, diving injuries, and drowning. An ambulance with at least one person trained in the specialties of drowning, hypothermia, and dive accident pre-hospital care should be onsite for public safety ice diving incidents.

A leader who makes the final decisions is necessary on all ice diving sites. In recreational diving, this is often an ice instructor. In public safety diving, that person is the incident commander (IC), In commercial dive operations, a supervisor is typically in charge. Sometimes the person in charge is not a certified ice diver or tender. If such is the case, an experienced technician-level advisor should be consulted by the person in charge. This situation most commonly occurs on public safety diving sites, where the IC is a fire or police chief who may have no personal experience on the ice. Such an IC should have, at minimum, ice diving search-and-rescue/recovery awareness-level training, and should allow an officer with sufficient ice experience and certification to make the specific procedural decisions.

Photo 2-10 Incident commander (IC) vests are particularly useful on ice operations because all personnel are so bundled up it may be difficult to recognize individuals and because a traditional command post may not be possible.

Another important person on the site is the safety officer. If only the minimum five personnel are available, then one of the two tenders has to serve as the safety officer, whose duties are described in chapter 9.

A safe diving operation does not need to be a big operation, and an effective team does not need scores of members. Just remember that when running an operation, you should have the right people with the right training and the right plan, so you will all be able to go home at the end of the day.

Summary Questions

1. List the minimum number of personnel to run one ice dive.
2. Define "awareness level".
3. Define "operational level".
4. Define "technician level".
5. List at least eight awareness-level duties.
6. List at least twelve operational-level duties.
7. List at least three post-dive diver responsibilities.
8. In low or no visibility, water divers do not know where they have or haven't searched so search divers are ________-directed.
9. Backup divers should be _______ dressed.
10. What positions should the most experienced ice divers be placed in: primary, back-up, or 90%-ready diver?
11. Explain why 90%-ready divers are necessary.
12. List at least ten tender duties.
13. What records are kept by the profiler?
14. Who serves as the profiler/record keeper?
15. Give three reasons why the above person should perform the profiler/record keeper duties.

Notes

1 The NFPA defines a hazardous atmosphere as that which does not sustain life.

2 Available from Lifeguard Systems, Inc. PO Box 548, Hurley, NY 12443.

3 Term created by author Hendrick in the late 1970s.

Chapter 3
Ice Diving Certification Levels

More divers would participate in ice diving if there were more opportunities to dive under the guidance of trained ice divemasters, and if there was a continuing education training process as found in other types of diving. Currently, most training agencies have only two ice certification levels: ice diver and ice diving instructor. As this chapter will illustrate, these two current certifications do not provide opportunities for advanced or specialized ice operations training.

Too often, divemasters and other leadership personnel believe they are ready to teach or lead ice dives just because they have ice diver certification, which they feel qualified for in a single ice diving course. In reality, they need more training and experience. The progression for cave diving leadership training is a good model to follow for ice diving leadership certifications.

Ice diving instruction and supervision duties are critical for safety. Ice diving instructors should not be certified by recreational diving agencies based on a simple, personal endorsement and claim of experience, without any proof of meeting a minimum standard of knowledge, skills, responsibility, and use of consistent standards. An ice instructor should have more ice training, skills, and knowledge than is taught in a recreational "ice diver" certification course. Like cave diving instruction, ice diving instruction requires special training and certification. A progressive series of instructor and divemaster courses is needed. These should be designed to bring a minimum responsible standard of care to this phase of technical diving. The dive industry has demonstrated its lack of proper training in safe responsible standards in the number of diving and ice diving instructors and divemasters and their students who have died during ice diving excursions.

Some training agencies allow a diver to be certified as an ice diver after minimal training and completion of only two ice dives. An ice diver certification indicates the diver is now capable of ice diving without the supervision of an ice instructor. For that level of competence to be the case, many more than two ice dives must be required. Divers should not only have a minimum of three successful ice dives, each with its own specific skill requirements, but they should also serve as primary tenders and backup tenders; they should be capable of setting up a safe ice diving site; and they should be able to carry out practiced contingency plans. Both dedicated tenders and divers should demonstrate the ability to plan and carry out a safe ice diving operation.

Basic Ice Diver and Tender

Up until the last two decades, newly certified divers were issued a certification called "basic scuba diver" that allowed them to dive under the supervision of a divemaster or higher authority. To be able to dive without supervision, with only a peer, divers had to take a second level of training and become an "openwater scuba diver." Competition among training agencies has lowered certification standards and changed the basic scuba diver certification to openwater diver certification, without increasing the standards of the basic scuba diver program. With this change, newly certified divers have been allowed to dive without any leadership-level supervision. The consequences of this change are still being debated.

It is the belief of the authors that ice diving should follow the original system, and that entry-level, or basic, ice diver certification should allow divers and tenders to ice dive only under the supervision of certified ice divemasters. We also believe that to dive without ice leadership-level supervision, ice divers should need to obtain advanced ice diver/tender certification.

We require that a person serving in an ice diving tender capacity outside of training be ice tender certified. Likewise, for a diver to perform an ice dive, the diver must be tended by a certified ice tender.

To participate in an ice course at the basic ice diver level, students should satisfy the following prerequisites:

- At least be certified divers.
- Be experienced in cold-water diving and comfortable with wearing hoods and appropriate gloves.
- If wearing a drysuit, which is highly recommended, be drysuit certified.
- Be able to calculate surface air consumption rates after every dive and use that information in dive planning.
- Be capable of breathing comfortably underwater without a mask in cold water; achieving neutral buoyancy; and confidently performing basic underwater skills, such as mask doff-don-clear and regulator retrieval with the over-the-shoulder method.

Students seeking only basic ice tender certification should not be required to be certified divers.

During the basic ice diver course divers should complete at least three successful dives, with at least two being tender-directed. In addition, they should complete at least two contingency plans, including diver-disconnect and entanglement management. Students seeking dedicated ice tender certification and those seeking ice diver certification should also serve as primary and backup tenders and profilers. They should use the mandatory safety equipment including quick-release pony bottles and hard tethering.

Basic ice divers and tenders should conduct ice dive operations with leadership level ice divers as part of the operation team. Until they are fully trained to safely cut holes, serve as an ice diving safety officer, and be competent with problem recognition and management, basic ice divers and tenders should not conduct ice dives without some type of qualified supervision.

Advanced Ice Diver and Tender

The advanced ice diver should have additional training in dive planning, site setup, contingency plans, and rescue. The advanced ice diver can make a good safety officer, backup diver, or buddy for less experienced ice divers. The advanced diver course should provide additional supervised opportunities for each student to actually pick a site, plan an operation, cut holes and set up a site. Advanced divers and tenders should be able to demonstrate proper skills in line handling and knot tying, as well as the basic rigging skills needed to set up a safe ice dive operation. Three successful dives in addition to those required for basic ice diver certification should be required along with two underwater contingency exercises, tending experience, and experience using proper

Photo 3-1 At the completion of an advanced or public safety ice class, students must feel, and actually be, capable of running safe ice operations without their instructors.

documentation skills. The underwater contingency exercises should include entanglement management, finding a disconnected diver and out-of-air-entangled diver from both the donor and receiver roles. Both divers and dedicated tender students should serve as tenders during diver emergency drills and should serve as profilers.

Certified advanced ice divers/tenders should be capable of performing ice dives without leadership supervision.

Photo 3-2 The ice divemaster supervises gear setup and donning.

Divemaster

An ice divemaster must be able to safely and effectively run an organized ice diving site, from the point-of-planning and site setup to hole closure and site exit. Before starting a course, the candidates should have the following experience level:

- Be a fully insured divemaster.
- Have completed at least ten ice dives with advanced ice diver certification.
- Have experience with blackwater diving.

To obtain certification, candidates should achieve the following:

- Satisfactorily complete a thirty hour to fifty hour ice diving and ice management divemaster-level course that includes all skills required for advanced ice diver.[1]
- Assist in a minimum of one basic and one advanced ice diving course.
- Properly plan, set up, and supervise fifteen ice dives under the supervision of a certified ice diving instructor. Maintain all proper paperwork for said dives, and debrief these dives with the instructor.
- Demonstrate competence in ice diver contingency exercises, both rescuer and rescuee, including entanglement management, out-of-air, lost diver, free-flow management, unconscious or injured diver transport from water to shore, and distressed diver.
- Participate in at least two full ice diver rescue and evacuation scenarios.
- Conduct/supervise at least two full ice diver rescue and evacuation scenarios.
- Write an emergency standard operating procedure/guideline (SOP/SOG).
- Take an ice divemaster written examination.

Ice diving instructor

The ice diving instructor must be knowledgeable about every aspect of safe ice diving, including ice rescue, ice diving training, and divemaster training. At a minimum, certification should include:

- Complete twenty ice dives supervised by a certified ice diving instructor/ice diving instructor trainer.
- Complete all divemaster requirements.
- Satisfactorily complete a 28-hour or longer ice diving instructor course conducted by a certified ice diving instructor trainer. Demonstrate the ability to teach all skills required for safe ice diving.
- Teach/conduct at least one evaluated ice diver course under the supervision of an ice diver instructor trainer or experienced ice diver instructor.

Public Safety Ice Diving I—
Search and rescue/recovery (SAR) ice diving

Prerequisites:

- At minimum, nationally recognized level-one public safety diver/tender certification. Should include the use of quick-release pony bottles, harnessed tethering, solo-tender-directed diving, surface air consumption calculations and breathing rate calculations, hands-on entanglement management contingency training, blackwater training, profile map/record keeping, and knowledge of cold stress, hypothermia, and dive accident pre-hospital care.
- Same prerequisites as stated above for basic ice diver.

Divers should make a minimum of three successful search dives plus complete diver-disconnect, entanglement, out-of-air, free flow underwater contingency exercises. Divers and tenders should serve in all surface support positions, and as many dedicated tenders as possible should serve in the safety officer and incident command positions. Some team standard operating procedures state that ice must be at least four inches thick to perform an operation. This is a good idea for teams with only Public Safety Ice Diving I training and certification. Diving on thinner, non-supportive ice is actually not difficult or risky if proper, proven procedures are used, hence level II training.

Public Safety Ice Diving II—
Search and rescue/recovery (SAR) ice diving

This training teaches students how to work on ice that may be too thin to lie on without falling through; how to move more rapidly with greater efficiency and safety; and how to work from holes that are closer to shore when searching for objects that have fallen through holes farther away.

Prerequisites:

- Search and Rescue/Recovery Ice Diving I

The same minimum number of dives and surface support standards as the SAR Ice Diving I training should apply to this advanced class. In this class, students are taught how to arrive on-scene, get dressed, check diver-tender pairs, and retrieve evidence. Students are taught to gently transport the victims back to shore. All of these steps must be mastered in less than ten minutes, and all must be based on ice that is too thin to walk on. This situation takes a tremendous amount of team coordination, with each team member practicing their job extensively. Some personnel rapidly set up the ice board pulley system; others dress the divers, profile the site, interview witnesses, and set up the staging area; and the remainder of the team prepares to receive the victims.

Tenders wear ice rescue suits, as they are lying in the water where the ice has broken up. The rigging skills for such an operation are more extensive than for operations with stronger ice. This level of training also teaches divers how to move with more efficient rapid deployment motion, even when the ice is stronger.

In this class, divers are also sent to retrieve victims or evidence from holes that are closer to shore than the holes at which the search objects went down. Divers enter the closer holes, spider-walk under the ice to the farther holes, and then begin their search. The procedures for this type of operation are discussed in chapter 8. The purpose of this type of operation is to greatly reduce the time and effort it takes to transport divers and tenders and to learn how to set up the appropriate rigging necessary to dive from holes farther from shore. Once these skills are accomplished, the team learns how to perform multiple-vicitm operations much further from shore.

Ice rescue diver

Divers may wish to be ice rescue divers. These divers perform all of the skills and requirements of divemaster training, except for the leadership portion of the program. Ice rescue divers are first required to have standard recreational rescue diver and advanced ice diver certifications.

Photo 3-3 Rescue student finds and recovers mock disconnected diver (buoyant mannequin) from under ice and carefully brings it home, sealing the airway from water.

Public safety ice diving divemaster

Prerequisites:

- Search and Rescue/Recovery Ice Diving, I and II.
- Divemaster certification.
- Assist with at least three SAR Ice Diving I and II classes.
- Log at least twenty ice search dives and at least twenty ice search dives as a tender.
- Serve as safety officer on at least two ice operations involving at least six search dives each.

This divemaster course can be conducted as an internship with a public safety diving instructor over a period of time, or it can be conducted in a standard class format.

Public safety ice diving instructor

Prerequisites:

- Public Safety Diving Instructor.
- Public Safety Ice Diving Divemaster.

At minimum, public safety ice diving search and rescue/recovery instructor certification should require at least an additional forty hours of training, with demonstration of the ability to teach all skills associated with safe public safety ice diving. This would include serving as incident commander for an additional ten operational search-and-rescue/recovery dives.

Summary Questions

1. At least one shore support person should be ______ and ______ ______ certified and have a good understanding of how to recognize and provide pre-hospital care for ______, ______, ______, ______.
2. To serve as a tender for ice divers, a person should have the appropriate ______, ______, and equipment.

Notes

1 Offered by Lifeguard Systems Inc., PO Box 548, Hurley, NY 12443 www.teamlgs.com

Chapter 4 Ice Courses and Training

A well-organized ice diving course should excite participants about ice diving and enhance total diving enjoyment and leadership skills. There are hundreds of questions that should be asked during ice diving training, planning, and practice sessions. This chapter provides a short list of what questions need to be asked and what topics examined. The focus of any course should be ice diving. In any ice course, divers need to be trained for, and comfortable with, the gear they will be using.

Choosing an Instructor

Becoming an educated consumer is critical to ice diving training. Any course or training program is only as good as the instructor and staff. Be sure that the instructor and staff meet the following minimum requirements:

- Instructor knows the site and any potential site risks.
- Instructor is certified as an ice diving instructor, and staff is trained and certified to assist in an ice diving program.

- Program follows agency standards and meets either the minimum protocols outlined in this book or the certifying agency guidelines. Students should request to see course guidelines.
- Contingency emergency plans are in place for every dive. Students should ask what these plans are and what emergency equipment will be onsite.

Photo 4–1 Contingency plans. Are divers given hands-on training to manage a disconnected diver incident?

What Should a Good Ice Diving Course Consist of?

The following topics contribute to a good ice diving course:

- Fall pre-planning sessions are not necessary, but certainly may add significantly to the education of students.
- Line signals for both divers and tenders.
- Site management; including safety, neatness, line handling, hole cutting, and area maintenance.
- Safe site set up, and safe use of ice boards or sleds and other transport equipment.
- Site safety equipment, radios, cellular phones, oxygen, and first-aid equipment.
- Safety equipment and procedures (*i.e.*, quick-release pony bottles, hard tethering, record keeping).
- Contingency procedures.
- Dangers of ice diving topics, to include disconnects, entanglement, entrapment, visibility blackout, out-of-air emergencies, free-flows, cold stress, hypothermia, immersion hypothermia, and disorientation.

- Backup-diver training.
- Tender training, including tethered diver techniques, line signals, record keeping, and emergency procedures.
- Safety procedures and contingency training for: nonverbal communication, entanglement, lost diver practice scenarios, self-rescue, free-flowing equipment, flooded mask, breathing without a mask, entanglement management, and underwater self-rescue.
- Practice sessions for exchanging primary second stage with pony second stage.
- Profiling techniques.
- Proper hole entry and exiting, including weightbelt removal prior to exiting.
- Minimum of three successful dives in 15ft to 35ft (5m to 11m) of water for a fifteen to twenty minute duration. Moreover, completion of a minimum of two underwater contingency exercises dealing with entanglement management, out-of-air emergencies, and disconnected diver.
- Securing the scene after the dive operation is completed for the day so that it will not be hazardous to the public.

Ice diving SAR

In addition to the basic course, ice diving search-and-rescue/recovery training should cover at minimum the following skills:

- Site assessments with risk/benefit analysis.
- Calculations of where the body or evidence is most likely to be.
- Ice operation incident management system.
- Planning the safest and most appropriate operation.

Photo 4–2 Whenever possible, look for ways to increase realism during training. Here Hendrick chooses a confined area outside the pool building to teach rapid drysuit dressing to begin preparing students for dressing in tents and trucks.

- Rapid dressing with double safety checks.
- Safe setup of anchors and ice sled pulley systems, and cutting the hole to make it responsible and rescue-ready.

- Operational leadership, tending, profiling, designated-area diving (DAD), and surface personnel procedures.
- Prevention, recognition, and field treatment of cold stress, immersion hypothermia, and drowning.
- Tender-directed/tethered-solo blackwater search diving for maximum search effectiveness and diver safety.
- Entanglement, out-of-air emergency, regulator free-flow, injured diver, unconscious diver, and diver-disconnect emergency management contingency procedures.
- Victim or patient transport from bottom to shore.
- Platform ice rescue operations from boats, docks, kayaks, etc.
- Real-world conditions such as non-supportive ice, steep shore embankments, deep snow, and blackwater.
- Ice diving standard operating procedures/guidelines.
- Minimum personnel operations.
- Blackwater self-rescue and contingency training.
- Proper use of quick-release pony bottle, full face masks, communications, and drysuits.
- Search techniques with single and multiple holes.
- Long distance operations (greater than 300ft from shore).
- In-water tender operations for non-supportive ice operations.
- Poor light and evening operations.
- Poor weather operations.
- EMS pre- and post-dive diver checks.
- Basic surface ice rescue skills.

Photo 4-3 Surface ice rescue training. Public safety ice divers and instructors should have basic surface ice rescue capabilities.

Public Safety Ice Diving Proficiency Training

Public safety divers do not know when or where they will be called out. They lack, for the most part, long-term preplanning, as it pertains to site choice. They often do not know with whom they will dive, nor who the backup will be. Public safety divers often lack proper shelter and may be cold before they start the dive operation.

Because they face these and other unforeseen circumstances, public safety divers need the most comprehensive training possible. They cannot afford to leave any portion of the dive operation to chance. Well-trained divers and tenders alike learn how to work together and how important each of their jobs is to the overall success and safety of the operation.

Public safety divers must keep in mind that personal safety is the first priority. The only mission that must be completed one hundred percent of the time is "Go home when the job is done." Too many teams take shortcuts and regret it. Measured lines, tie-offs, better tending techniques, and a prepared support crew help in every technical dive situation. For the public safety team to maintain proficiency, training must be conducted throughout the year.

Many of the skills practiced have valuable uses in the real world. For example, the recovery technique for a disconnected diver is to send the backup diver spidering along under the ice directed by the tender. Once at the desired distance, the backup performs a circle sweep. In some cases, a drowning victim can become trapped just below the ice roof. While not a common occurrence, this can happen due to clothing and speed of drowning. It has happened with small children. By practicing the lost diver search procedure, the diver and tender increase their sub-surface ice roof searching abilities.

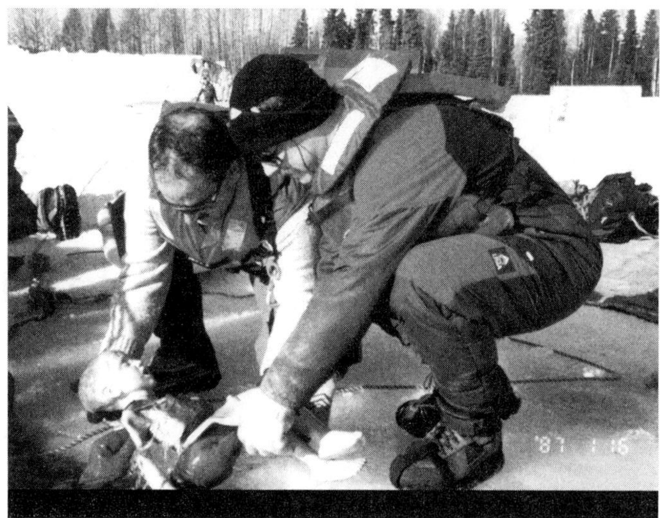

Photo 4-4 Patient handling. Public safety ice dive teams should be taught how to handle victims from the water to the ambulance.

Training tenders and divers to work together

Training tenders can be a very lengthy process. The tender is the diver's lifeline back to the surface, and the two must be able to communicate as if they are one. Sport or public safety diving in an overhead environment such as ice tests the tender's capabilities and his/her awareness of the diver's needs and safety. Sadly, in most of the recreational diving world, tending is not considered an important role. Often the only qualification is that tenders show up and be able to hold a line.

Public safety diving can place not only major mental stress, but also physical stress on the tender. Often, just getting to the victim point of entry (VPOE) can be exhausting, requiring strength, redundant training, and practiced skills. With advanced training, the public safety diver may be able to reach the VPOE from under the ice roof. This is far more efficient when the ice is weak, but takes even greater coordination between divers and tenders.

Photo 4-5 Land drills. Make sure tenders and blacked-out divers can effectively work lines and signals prior to diving. Here a backup diver takes the primary diver's hand (front) and puts it where his pony bottle is for a pass-off.

Dry land tender training

Start by having divers and tenders tape the lines at distance marks of 5ft, 10ft, 25ft (1.5m, 3m, 8m), etc. This gives them the opportunity from the beginning to understand the line markings.

Make up signal reference cards in plastic holders and give them to divers and tenders. Conduct a dry-land, blacked-out diver training session, so that divers and tenders learn to work with and trust each other. Without visual input, the diver must trust the tender's directions. Have the tenders and divers set up as if this were a water session by wearing harnesses, blacked-out masks, personal flotation devices, and gloves. Once the divers are connected to the tether line, there should be no further verbal communication, since realism is an important part of the training process. Go over each of the tender-to-diver and diver-to-tender signals, and have the tender-diver pairs practice them.

This ensures that both sides truly understand what the signals are and how to send and receive them. Pre-dive review of signal and contingency plans is imperative even when simple signal systems are used. A well-trained instructor will catch and correct a variety of mistakes during this drill, which will make the ensuing real dives much smoother and safer.

Next, place small objects in slightly obscure places about 60ft to 80ft (18m to 24m) from the tender. Have tenders direct their divers to the objects.

To help tenders understand the need to be extremely attentive, place items in the way of the tether line. As the line entangles in the obstruction, the tenders should notice it happening and respond to the problem. First, they will notice that the line is no longer moving freely with the diver. Second, they will begin to recognize the feeling of a line whose movement is restricted. These are skills the tenders need to understand.

If tenders will be tending two divers for a recreational dive, have them practice with one diver first, then the other, and then both divers on the yoke or double-line system. The yoke system is often preferred over two separate lines. Yoked lines reduce the number of tenders required to perform safe dive operations. Two separate tether lines are more likely to entangle unless advanced techniques are used.

Confined water training

Note: *Confined water training can be used to start the ice training process in the fall.*

If conducting confined water exercises, use the pool as if it were an open water environment. Divers and tenders should go through the entire dive process, including: dressing, equipment checklists found in this chapter, safety review, and proposed site setup. During this training session, use all of the gear normally used during actual ice diving. This creates a full hands-on exercise before reaching open water and the actual ice dive environment, plus it allows divers to become neutrally weighted with the equipment in which they will be diving.

To help provide more realism for diving and tending in ice conditions, you can use a tarp to create an overhead environment in the pool.

Have the divers work with blacked-out masks, thereby augmenting the trust relationship between divers and tenders. Designate an area of the pool as off limits to the divers. The tenders must learn how to comfortably control the divers, while keeping

Photo 4-6 Students learn how to maneuver with fully inflated suits during drysuit training. Divers need drysuit certification to dive dry in open water, and especially under ice.

them out of the no-dive zone. In the beginning of the exercise, have the divers move along the bottom. Once the tenders and divers demonstrate that they can communicate, have the divers remove their duct tape and swim up freely from the bottom. This exercise gives the tender an enhanced perception of what tended diving is all about. The idea is to create as much realism as possible during the confined-water training.

Note: If you are going to do blacked-out pool sessions with divers lacking zero visibility diving experience, it is a good idea to have a safety diver underwater who can see clearly.

Photo 4-7 All ice divers, whether recreational or public safety, need blacked-out mask training to be able to dive in unexpected siltouts.

For extended confined-water practice, create the following game:

1. Build a cutting station, using a milk box-type of crate. Tie sections of fishing line, plastic tie-wraps, old electrical wire, chicken wire, rope (anything that a diver could find and possibly become entangled in during the dive) all over the box. Have the blacked-out diver go under, cut one or more of each piece, and tie them together.
2. Send another diver in from a different part of the pool and create a lost-diver scenario.

3. Using blacked-out masks, have the primary diver practice the blackwater contingency. As the primary diver sends the emergency signals, have the backup diver and tender follow the emergency plan, including full deployment of the blacked-out backup diver. A second cutting station is a great training tool at this point, allowing the backup diver the opportunity to simulate a full-cutaway rescue.
4. Using blacked-out masks, practice search techniques by placing twenty pennies and two nickels on the bottom of the pool. Have the divers retrieve only the nickels with gloved hands and tended signals.
5. Place either a diver or an object underwater so that the tether line becomes entangled.

Summary Questions

1. A diver should be permitted to dive in a drysuit without having drysuit training and certification if the diver is a divemaster or better. True or False?
2. List at least five prerequisites for taking an ice diver course.

Chapter 5
Equipment

Specialty diving often requires techniques and equipment configurations different from those of normal diving. Some standard equipment, such as regulators, requires special rules of use when used during ice diving. Proper maintenance of equipment becomes imperative during the ice diving season. Ice diving is heavily equipment-dependent and Murphy's Law is the norm: *If it* ***can*** *go wrong, it* ***will*** *go wrong.*

The following chapter discusses ice-specific equipment, rules of use specific to ice diving, equipment maintenance, and configuration of equipment for divers and tenders. Site equipment will be covered in chapter 11, and tethering equipment will be discussed in detail in chapter 6.

Diver Equipment

Consistent equipment for public safety diving teams increases in importance when divers enter zero-visibility water or water that has potential entanglement risks. A blacked-out backup diver, while assisting a primary diver, needs to mentally picture exactly where the primary diver's equipment is and what the diver is touching. Recreational divers need to have this skill as well, in water that could silt up when the bottom is stirred. Entanglement risks increase the chance that a backup-diver assist will be needed. If consistent gear is not possible, at minimum, place the important items in the same place on each diver, and make sure that the system for releasing the pony bottle is the same for each diver. For example, the pony mouthpiece can be secured on a neck strap just below the diver's chin and a pair of shears can be mounted on the left shoulder strap of the harness. Gauges can be placed under the arm, through the BCD armhole, and secured with a Fastex™ quick-release buckle to the BCD chest area. A consistent weight-ditching system should be a high priority. Consistent equipment also makes it easier and quicker for surface personnel to perform gear assembly and gear checks.

Scuba versus surface supply

Surface-supplied diving requires additional equipment, diver training, and surface support training. It is therefore not recommended for recreational divers. Public safety divers will find it cumbersome and more time consuming to set up than scuba; therefore, it is not as good for teams in rescue modes. Lacking strong ice, surface supply systems require umbilicals long enough to reach the diver's point of entry from the shore or a platform on the ice. Surface supply diving should be used when water is contaminated enough to require helmets rather than full face masks, and when dive depths or times are extended. This holds true for ice diving operations as well. Superlite™ helmets are lighter than some helmets, but 28lb or more of dry weight is still a factor on thin ice.

The air management and contingency procedures presented in this book are designed to make scuba air sources sufficient for recreational and public safety ice operations. Notice that the word "safe" was not used, because safe means without risk, and that can never apply to scuba or surface supply diving.

Photo 5–1 A well-organized surface supply air station on the ice with helmets kept high off the ice to decrease the chances of free-flows.

Regulators

Free-flow causes and prevention. "When diving in water colder than 5°C (41°F), freezing of the regulator may represent a serious safety problem [that] is especially dominating as the temperature approaches 0°C (32°F) or colder."[1] The U.S. Navy has special planning guidelines for ice diving or diving in water at or below 37°F (3°C).

"The water temperature of 37°F (3°C) was set as a limit as a result of Naval Experimental Diving Unit's regulator freeze-up testing."[2]

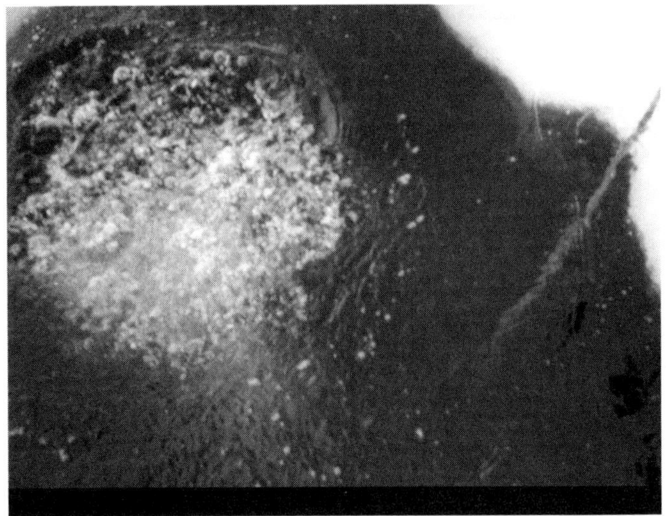

Photo 5-2 A free-flowing regulator can be a serious problem for unprepared divers.

Why regulators free-flow. Charles' Law essentially states that as the pressure of a gas increases, its temperature also increases; and when the pressure of a gas decreases, the temperature of the gas decreases as well. Anyone who has let air out of a tire has experienced this first hand. As gas goes from the cylinder (high pressure) through a narrow metal orifice into the regulator (low pressure), it is cooled. The gas coming from the cylinder is at high pressure, for example, 3,000psi (204bar). The regulator first stage drops that pressure typically to 130psi to 150psi (9bar to 10bar). At a small location in the first stage, the high-pressure end of the regulator where the pressure drops, the temperature can cool to as low as -70°F (-57°C). Because the first stage is constructed primarily of metal, the rest of the regulator experiences a drop in temperature. The part we are mainly concerned with is the other end of the first stage, namely, the spring chamber.

The diaphragm in the spring chamber is affected by ambient pressure, which typically is in the form of water pressure. The pressure exerted by the spring on the diaphragm is increased when water pressure increases in the spring chamber. As pressure increases on the diaphragm, the high-pressure poppet is lifted, allowing air to flow from the scuba cylinder into the intermediate-pressure hose and onto the second-stage regulator. In piston regulators, a piston replaces the diaphragm.

In warm-water diving, the water raises the temperature of the regulator that was cooled because of the air passing from high to low pressure. In ice-cold water, however, the water may not be able to warm the regulator enough, and ice crystals may form in the spring chamber. The ice exerts greater pressure on the diaphragm, because ice takes up more volume than water. When this happens, more gas is allowed through the orifice, and subsequently, a further drop in temperature occurs. This condition could continue to the point that enough ice forms in the chamber to allow gas to escape from the first-stage regulator into the second-stage regulator in great volumes. This is a free-flow. Piston regulators free-flow in a similar manner.

Some regulators have a rubber diaphragm separating the spring chamber from the rest of the regulator. The rubber provides more insulation to protect the chamber from the cold high-pressure end of the first stage than a piston regulator, which has only metal separating the chambers.

Another potential cause of a free-flowing regulator is water in the first stage itself, other than in the pressure-sensing chamber. Ice may form and prevent the high-pressure seat from forming a positive seal. The result in either case is a first stage free-flow. The free-flow may start with a low amount of gas passing through, but that gas causes more cooling and more freezing, and small free-flows often become larger ones in a relatively short time.

A first stage free-flow can empty a cylinder in less time than it may take an ice diver to get back to the surface. Therefore, a true redundant air source with a redundant first and second stage, such as a pony bottle, is an absolute necessity for ice diving. The possibility of free-flow is also one of the reasons the distance diver's travel from the hole should be limited and why entanglement prevention and management skills are critical.

Free-flows occur in the second stage most commonly in one of two ways. A first stage free-flow can cause a second stage free-flow. Think of how snow is made on ski slopes. High-pressure air is blown through water. Imagine 2,500psi (170bar) blowing through a second stage that is filled with water. Ice crystals are likely to form, and if they block the action of the second-stage seat, it will free-flow. A second stage free-flow can also occur independently of a first stage free-flow when divers allow the second stage to become wet, either from immersing it in water or from laying it down in wet snow. If the diver breathes on this regulator in air, the remaining water or moisture in the second stage can freeze, causing a second stage free-flow. Therefore, an important rule of ice diving is: once the regulator is wet, only breathe on it while it is underwater.

Full face masks present an exception to this rule, because if divers lay their masks in the water to don them, the upside-down position could cause a free-flow, especially if the mask is a positive-pressure mask. Divers using full face masks should don their masks prior to immersion, and need to make long slow exhalations to help decrease the likelihood of freezing. Should a diver surface and need to speak with tenders, a communication system allows divers to keep their faces in the water while speaking. If surface breathing vents are used on positive-pressure masks, divers should have the positive-pressure lever in the down position before pulling open the vent.

With two modifications, most modern regulators can be used for ice diving. One modification provides environmental protection. Many regulators use water to transmit ambient pressure to the first-stage piston or diaphragm, as described above. Environmental protection is obtained by sealing off the spring chamber and filling it with silicone gel or liquid. The silicone, rather than water, transmits pressure changes. If you know you are going to go ice diving, purchase regulators that are already protected; otherwise, check whether the regulator you already have can have an environmental protection package installed after production. Modern environmental packages are great improvements on the old method of placing condoms filled with petroleum jelly over the first stage. Some regulators use alcohol instead of silicone, but because alcohol can have some water in it, the regulator is more likely to corrode. An increasing number of modern regulators are built with only air in the spring or piston chamber for environmental protection. This is also a good option.

Intermediate pressures of 130psi to 150psi (9bar to 10bar) are designed for easy breathing rates. While this is good for summer and deep-dive operations, this range may allow a regulator to free-flow too easily in ice diving. The second modification that helps prevent free-flow occurrences is to tune the regulator down a little for winter diving. This can be done by a certified regulator repair technician. In summertime, if your second stage free-flows when the mouthpiece is upside-down in water, it is too sensitive for cold-water diving. Do not tune it down so that it is difficult to breathe, because comfortable breathing is very important. However, if it free-flows just because it flips upside-down underwater, the chances of a free-flow while ice diving are much increased. The pony regulator should be tuned down as well.

A third preventive measure is regulator care and maintenance. Jeff Bozanic and Jim Mastro conducted 1,191 ice dives in Antarctica from 1989-91 and reported their findings about free-flowing regulator failures. They found that the failure rate reached almost 100% during initial dives for those divers with no prior polar diving experience. The type of regulator did not alter this outcome. Further instruction on proper care of the regulators solved the problem.

Often compounding this problem was the manner in which the regulators were transported to the dive site. Equipment was generally transported in the back of open vehicles such as tractors or snowmobiles. "When exposed to these conditions, the regulators were pre-chilled and had a higher incidence of failure."[3] Whenever possible, keep regulators in warm, dry environments prior to use in ice dives.

Photo 5–3 Notice the second-stage regulator on the ice (right). Make sure to keep all regulators and other air sources up off the ice.

Another important factor in regulator failure in ice environments is moisture in the regulator mechanism. After cleaning a regulator in a warm environment, blow it dry by forcing air through it from a cylinder. Then, dry the external parts of the regulator before storing it in a warm place (if possible). Sport divers may be able to store their regulators in a warm environment long enough for the equipment to dry, but public safety divers never know when they may have to respond again. Moreover, some dive teams keep their equipment in vehicles that are housed outdoors, rather than in a warm building.

Once the possible sources are understood, moisture in a first stage can be more easily prevented. The following list describes some of the ways to keep moisture from occurring:

1. Be sure to dry the dust cap before replacing it over the first-stage filter. Moisture can appear on the filter and be blown into the regulator if it is too soon placed on a cylinder for use. At an incident, keep something available to dry the dust cap, in case a regulator needs to be removed from a cylinder, Chamois are great because they fit easily into a PFD pocket and can be used effectively all day.
2. Following a dive, some divers attempt to blow their first stage dust caps dry by blasting them with air from a cylinder that has no regulator. This is not a recommended practice, because it places the uncapped first-stage filter in the line of fire of water droplets that may be blown into the first-stage mechanism. Instead, dry the dust cap thoroughly with a dry towel or chamois.
3. Make sure the tank valve orifice is dry before placing the regulator on it. Icy rain or wet hands can inadvertently place water droplets at the orifice which will then be blown into the first stage.
4. When washing regulators, keep the first stage higher than the second stage. If a wet or immersed second stage has its purge button depressed when it is no longer pressurized by a cylinder, water can enter and run backward to the first stage. Do not depress the purge button when cleaning the regulator.
5. Do not wash a regulator without the dried dust cap firmly in place.
6. Make sure all plugs and hoses are properly and securely attached to the first stage, as an improper seal may allow water in if the regulator is not pressurized. Avoid changing hoses or removing plugs in wet environments or with wet hands.
7. Make sure tank and other o-rings are in good condition. Damaged o-rings can allow water leakage when the air is turned off, because the regulator is not pressurized.

Water can enter the second stage in several ways, too. Good maintenance and preseason servicing by a repair technician can discover and fix such problems as leaks in the diaphragm, an improperly seating diaphragm, and faulty exhaust-valve seals. As with first stage maintenance, improper rinsing or any material that normally would cause a free-flow should be avoided, such as bottom sediment in the second stage. Some modern second stages are designed to better trap and hold the heat from a diver's exhalation. Such second stages should be considered when purchasing regulators for ice dives.

Diver exhalations contain moisture. Although divers cannot avoid exhaling into their regulators, we have found that long, gentle exhalations seem to cause fewer problems than rapid quick ones. The former method of breathing is safer for a variety of reasons, in addition to potentially decreasing the chance of free-flows. Divers who find themselves breathing rapidly should stop and work on gentle, slow, three-second exhalations, followed by normal inhalations. They should practice exhaling a normal to slightly larger breath volume in a longer period of time. Also, work towards relaxing enough to have a respiration rate between nine and thirteen breaths per minute during a non-exerting dive.

Divers should also avoid doing anything that will cause a rapid flow of air out of a regulator. If there is moisture in the tank, a rapid decrease in pressure may cause that moisture to condense and freeze. One of the rules for filling lift bags is: never use a breathing air source, such as a diver's main or pony cylinder; instead, take down surface-supplied air or use a cylinder that will not be used for breathing. In recreational ice diving, divers sometimes have the desire to purge their main or pony regulators (or, God forbid, an octopus) directly under the ice roof to play with the air bubbles, which behave and look similar to mercury as they run along the ice. This is not a good idea. Play with the bubbles that are already there from normal breathing.

Other preventive measures involve the dive itself. Once the second stage is wet, it should remain wet. Divers should put it in and out of their mouths with their faces in the water, always keeping the second stage under the surface. They should try not to push the purge button at any time. They should keep regulators off the snow and ice. Divers should wrap the second stage in a small, insulated picnic-type bag or a wool blanket, if there are no other means of protecting it from the weather.

Even though the weather may not be cold enough during training to cause equipment to freeze up, divers should train as if the weather was as cold and windy as the most extreme they will encounter. They should make it reflexive during the winter to breathe from the regulator only when it is immersed, if it has already become wet.

Photo 5–4 Twin 80ft^3 cylinders with two regulator first stages provide much air, but are more cumbersome and do not afford the ability to pass air off.

Chapter 9 covers in detail what to do on a scene when free-flows occur. In recreational ice diving and in recovery diving, it is usually possible to keep a container of warm water on the scene; therefore, you may want to add this item to the equipment list for ice diving. Public safety dive teams, however, rarely have the luxury of warm-water buckets during actual calls; consequently, they need to train without them.

Alternate air supplies

Divers, whether recreational or professional, do not belong under the ice without a true alternate air supply. This can be a pony bottle or other alternate cylinder with redundant, environmentally protected first and second stages, tuned down a touch to reduce the chances of free-flowing.

Quick-release pony bottles. An octopus is *unacceptable* for any public safety diving and for all ice diving. The concept of the octopus was designed in the early 1960s by Dave Woodward and Walt Hendrick, Sr., for high-visibility, shallow, mid-water, buddy-based sport diving, to replace the task loading contingency plan of buddy breathing. Public safety diving, whether under the ice or not, often involves zero-visibility, bottom-dwelling, solo diving. An octopus becomes useless if the wearer runs out of air or experiences a first stage failure, because the octopus is connected only to the main air source. An octopus is simply another mouthpiece and nothing more. In addition, if a buddy passes off an octopus to an entangled diver, the two divers are stuck together until the buddy runs out of air and leaves.

An octopus under the ice can be a hazard. When an octopus is passed off to a diver in need of air, two things happen. First, the donor's first stage supports twice the air demand because two people are breathing off one first stage. Secondly, out-of-air stress can cause the first stage to be faced with ten or more times the demand for air. The risk of free-flow then becomes significant. If a first stage free-flow occurs in these circumstances, the likelihood that both divers will run out of air before reaching the surface is great.

Some divers incorrectly think that they should carry an octopus in case their primary second stage fails. Today's divers use downstream regulators, which free-flow rather than stop giving gas if the regulator fails. Divers should *not* use an octopus if their primary regulator second stage free-flows, because the air in the cylinder will be used up at an even greater rate. Also, switching to an octopus when a primary second stage free-flows can cause the first stage to free-flow, because of the increased demand for air. First-stage regulator free-flows can empty a cylinder far more quickly than a second stage free-flow—especially if there are two mouthpieces from which gas can escape. These problems hold true for all diving, not just ice diving.

As a safer option, ice divers should use quick-release pony bottles. These small, additional cylinders allow divers to have an air reserve even when their primary tanks have been emptied. In addition, pony bottles mounted to the main tank with a quick-release harness provide back-up divers with an air supply that can be passed off to entangled or entrapped primary divers. Backup divers can then leave and return with still more air. Pony bottles, therefore, are mandatory for responsible, safe ice diving.

Ideally, a pony bottle should contain at least a 19ft^3 (.5m^3) tank and must have an independent regulator. For depths of 40ft (12m) or more under ice, divers should consider using 30ft^3 (.85m^3) pony bottles. The goal is a good ten-minute window, provided by the pony bottle. As divers increase their depth, they increase the amount of air they consume. In addition, it takes a backup diver longer to reach a primary diver and go back to the surface for more air when the primary diver is at a deeper depth or farther away from the ice hole.

Because they have very low volumes of air, Spare Air™ units are not an acceptable option. They can be easily dropped during use or pass-off and require different operating procedures than a standard regulator and bottle. Spare Air™ units require a great deal of continuous practice to make their use reflexive.

If there is no risk to the diver of entanglement or entrapment between a shallow bottom and the ice roof, the issue of a quick-release alternate air source is not as important. The alternate air source can be fixed to the main cylinder. Some divers prefer to wear double main cylinders, such as two 80ft^3 (2.3m^3) tanks with independent regulators, instead of a main cylinder and a pony bottle. For most ice dives, this is unnecessary, because most ice dives have shallow depths and relatively short dive times. In addition, the added weight of two main cylinders may be undesirable if the ice is thin and the air source cannot be passed off to an entrapped/entangled diver.

Some dive teams have made it their policy to require surface-supplied air for ice diving. This usually happens after the team experiences an accident or "close call" on scuba. Surface-supplied air certainly adds an additional measure of safety, but it can pose problems for public safety dive teams, who often have to work on thin ice or who may have to respond in rescue modes as described earlier. This takes more time and effort. Surface supply requires additional training of both divers and surface support, and it involves greater costs. Divers have to contend with an umbilical air-supply hose that is heavier and thicker than standard tether lines. The additional bulk makes it more difficult to keep accurate patterns when searching for small items. Surface-supply divers typically walk or crawl along the bottom rather than move in the swim position as scuba divers do. It is more difficult to conduct accurate searches for small objects such as small handguns or bullets when using surface supply.

Some divers feel that wearing a pony bottle or other second tank makes them more susceptible to entanglements. If entanglement is a concern, then common sense and liability concerns actually demand that a pony bottle be worn. Entanglement management includes a true redundant air source for self-rescue and the ability to have diver-to-diver exchange of redundant air sources. If worn properly, a pony bottle does little or nothing to increase the risk of entanglement. If there are many entanglement risks, such as wires or cables from an airplane crash, then a section of inner tube or other similar item can be secured over the main and pony regulator first stages to prevent branches or other debris from becoming stuck between the two cylinders. Should the pony cylinder need to be passed off, the tubing can be easily pulled off first.

One important note about pony bottles: having a pony bottle is not all that divers need to be safer. In addition to carrying a pony bottle, divers must practice using them. Practice should include transferring to the pony bottle and releasing and receiving a pony bottle from a backup diver. When training with pony bottles, be aware that aluminum pony bottles, like standard aluminum cylinders, may float. When a pony bottle is passed off underwater, the diver should be careful not to let go of it, to prevent it from floating up and hitting the donor or receiver in the face. Divers may need to adjust the placement of weight on their belts to counter positively or negatively buoyant pony bottles. Divers should routinely surface on their pony air during training dives to gain practice with this lifesaving skill.

Photo 5-5 If a metal pony bracket system is chosen, make sure it is quick-release capable.

Mounting pony bottles

Generally, pony bottles are mounted on the right side of the tank, so that the regulator second stage is properly set up for the diver to use. Whether your team mounts the pony bottle on the right or left side, be sure that every diver sets it up the same way, so that team members are able to locate and retrieve every other diver's pony bottle, even in blackwater. If an Interspiro Divator™ full face mask is used with an RSV-1 block, the pony needs to be mounted on the left.

Photo 5–6 The simplest quick donning and release system for pony bottles is the Lifeguard Systems BCD pony pocket™. It also ensures that the bottle is in its correct position up against the BCD backpack.

Wherever you mount your pony bottle, be sure that it has some type of quick-release system. Mounts can either consist of a webbing harness from which the bottle can be pulled, or be constructed of two metal brackets, one mounted to the pony bottle and one mounted to either the main tank or the BCD strap. Metal, bracket-type pony mounts are good, because they hold a pony bottle very securely. The only downside to metal bracket systems is the cost, which is much more than that of webbing systems. When evaluating metal bracket systems, make sure the system is quick-release-capable, and that the receiving diver can make the disconnect easily. Another option is a BCD with built-in pony bottle pockets, such as Lifeguard Systems' public safety diving BCD.

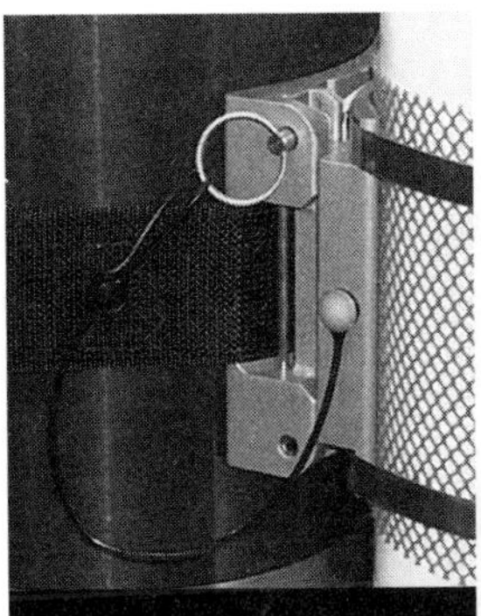

Photo 5–7 Tiger gear mount. We feel this is the best metal quick-release bracket available.

Photo 5-8 An inexpensive, simple to use, quick-release pony bottle harness is the LGS webbing system. The disadvantage is that the bottle will not hold as stationary as when a metal bracket system is used.

Contingency pony and main cylinder

If entanglement or entrapment is a possibility, the dive site should have a contingency pony bottle and main cylinder. After a backup diver passes off his pony bottle to an entangled/entrapped primary diver, the backup diver needs to return to the surface. The backup tender should replace the backup diver's pony bottle. The backup diver then descends with a contingency main cylinder to ensure plenty of air for the primary diver while the backup works to free him/her.

In more than thirty years of teaching public safety and ice diving to many thousands of students, we have never needed to use these cylinders, but that does not make them any less important on a dive site, particularly an ice dive site. A contingency cylinder usually contains at least 72ft^3 (2m^3) of air and has some type of harness that enables a contingency diver to bring it down to an entangled or entrapped safety diver. The harness system should also allow the receiving diver to don the contingency tank if the diver is forced to ditch the BCD assembly.

Photo 5-9 This 30ft^3 contingency cylinder has a contingency bottle harness system that allows the backup diver to easily bring it to an entangled primary diver who needs air. The harness can be worn by the diver if necessary and has attached cutting tools, flashers, a standard second-stage regulator and a quick-release hose to attach to the RSV-1™ block. 80ft^3 cylinders are preferred.

Tools can be mounted on the harness contingency cylinder, perhaps extra cutting tools, a flasher, or an extra contingency strap. These tools should be mounted securely so they do not dangle and become entanglement hazards. The contingency bottle should be weighted so that it is -1lb to -2lb buoyant. If an aluminum cylinder is used, small weights may need to be duct-taped to the lower end of the tank, maybe right above the tank boot if one is used.

If divers are using a redundant breathing supply block, the contingency cylinder regulator, like the pony regulator, should have both a quick-release hose to connect to the block and a standard second stage.

Pony bottle and contingency cylinder regulators

Pony bottle and contingency cylinder regulators should have the exhalation ports removed or cut back to reduce the risks of mud collecting or ending up in the mouth of a receiving diver. The pony regulator intermediate-pressure hose should be longer than the primary regulator hose and should have a bright-colored wrap that looks and feels different from any other hose. The hose wrap and removal of the exhaust port also enables tenders to immediately see, and then stop, divers from descending while accidentally breathing on the pony regulator.

Pony regulators need to be environmentally protected, and their first stage pressure can be tuned down a little lower than those of primary regulators. A pressure gauge is not necessary on the alternate air supply, as long as the pressure is checked before the bottle is mounted onto the BCD tank assembly. Port plug pressure gauges work well for pony regulators, because the diver does not have to carry an extra hose. Pony regulators can be unbalanced, so that they become harder to breathe as the bottle pressure decreases. Unbalanced regulators cost less than balanced regulators, which may be a factor for some dive teams.

The placement of the alternate air source's second stage is very important. It should be in the same place and position on every dive, so that the diver can learn to reflexively grab for it. This is especially important during a potential state of panic.

Pony regulator second stages can be secured to a strap or tubing worn around the diver's neck. This placement helps keep them out of the mud and snow and in a position where divers can place the mouthpieces into their mouths without searching for them.

Do not place the second stage in a BCD pocket. It is too difficult and task loading to remove the second stage from a pocket during an emergency. The second stage should be worn so that one quick hand movement releases it to pass it on. The ideal location, enabled by using a neck strap, is just under the chin, where the diver can reach the mouthpiece with his/her mouth simply by bending the head forward and down. Secure the pony hose under the wearer's arm. Or mount the mouthpiece on a snorkel keeper placed somewhere else in the golden triangle area.

Photo 5-10 The tender secures a pony second-stage neck strap on the diver. The diver should be able to reach the mouthpiece by tipping his/her chin down.

Blocks—redundant supply valves

When only a standard pony regulator second stage is used with a full face mask, the diver is forced to remove the mask in order to switch to the pony mouthpiece. With this setup, divers should carry a standard facemask in their BCD pockets to don when breathing on a pony regulator. This standard facemask should be required for ice dives. Due to the cold water, ice divers are more at risk for problems stemming from breathing without a mask than other types of divers, such as an unplanned or uncontrolled bolt to the surface which has even greater consequences than in warmer water.

To minimize these risks, divers can wear a transfer block, which allows them to switch easily from the main tank to the pony tank without removing the full face mask. This option for transferring to pony bottles also has the advantage of not exposing divers to potentially contaminated mouthpieces, contaminated water, or ice-cold water.

Standard commercial blocks enable divers to keep regulators, full face masks, or helmets in place when they switch to an alternate gas source. They are typically designed for surface-supplied commercial divers. Because commercial blocks typically cost more than $250, some divers choose to build them at home. Home-built blocks, though, open the owners up to liability problems should the system fail or should some type of injury or accident occur during use.

Photo 5-11 Diver wearing a positive pressure AGA™ mask, with an RSV-1™ block, and a surface breathing vent™ in the communications port.

One problem with commercial diver blocks is that, while they do allow a diver to change gas sources, they do not usually allow one diver to pass a self-contained, redundant gas source to another diver while underwater. Such a system will not work for public safety scuba diving, because a backup diver must be able to pass a pony bottle or other cylinder to an entangled primary diver.

Some full face masks, such as the Ocean Edge™ mask, have ports for two second stages. The front port is for the primary mouthpiece, and the second port is for a pony bottle mouthpiece. With this system, divers can switch to a second air source without removing their masks. Again, though, donors are unable to pass off or receive a pony bottle or other contingency bottle.

Breathing Blocks

Because of concerns about divers removing full face masks to transfer to pony bottles, author Butch Hendrick developed the concept of a regulator-mounted transfer block. This block enables divers to transfer their gas source quickly and safely, while still permitting a contingency diver to pass-off a pony bottle to a primary diver without mask removal. Hendrick met with Sartek Industries, which designed and now manufactures the RSV-1™ Second Breath block.

An important and unique feature of the RSV-1™ Second Breath block is its mounting on the mask itself. This position was based on Hendrick's observations that most panicked divers in almost any underwater emergency will remove their masks during ascent. The RSV-1™ has been placed in a position to combine the emergency panic reflexes of mask removal with the physical activation of the emergency pony bottle transfer. It is not only kinetically consistent with the mask handling, but also incorporates panic motions into a self-rescue technique. Because the block is in a place consistent with mask removal, it can disrupt the panic cycle by providing air before the mask is removed.

In any discussion of transfer blocks, it should be noted that the system must include some type of overpressure-relief valve mounted in the first stage of the primary regulator (if it is dead-ended). That relief valve is necessary in the case of a first stage free-flow on the primary cylinder. If the first stage free-flows, and the diver switches to a redundant gas source, the vent for the free-flow coming from the primary regulator is shut off. That shut-off could result in over-pressurization of the intermediate-pressure hose on the first stage and cause catastrophic failure. In a first stage free-flow, the high-pressure hose could blow if a first stage blow-off valve is not installed. Therefore, with any transfer block system, an overpressure-relief valve must be mounted in the regulator first stage.

Safety Note: *Block users should still wear a pony regulator second stage to manage the situation of a full face mask failure. If, for example, the regulator second stage of the mask freezes up and free flows, then switching the block to access pony air will quickly empty the pony bottle. In this case the contingency plan would be to doff the mask, switch to the pony second stage regulator, don a standard face mask, and abort the dive.*

Whatever options and equipment your team selects for dealing with out-of-air emergencies, one factor remains critical: every diver must practice handling these problems until they become reflexive! Divers should practice being the one that runs out of air, as well as the backup diver who provides more air. Only consistent, frequent rehearsal will ensure that a diver performs adequately and safely in an out-of-air situation. Having the right equipment in the right location does little good if the diver does not think to use it, or is incapable of using it, during a state of high stress or panic.

Surface breathing vents for full face mask

When full face mask divers are on the surface, they may want to wear their masks to keep their faces warm or to keep snow or rain out of their faces. If public safety divers are working on ice so thin that divers and tenders are immersed in water, then backup and 90%-ready divers may want to keep their masks on. In order to do this without using gas from their main cylinders, these divers need a way of breathing outside air while still wearing their masks. At the time of writing this book, there are two systems available enabling divers to do this: the Gill™ and the Surface Breathing Vent™ (SBV™).

The Gill™, which was the first unit on the market, is a pullout vent affixed to a full face mask faceplate. Users replace the faceplates on their masks with the Gill™ faceplate. The SBV™ takes the place of the communication port in most full face masks. The advantage of the SBV™ is that it sits at the bottom of the mask's oro-nasal cavity and thereby provides a larger volume of air with significantly less inhalation effort than with the Gill™, which sits at the diver's temple. The SBV™ therefore also creates less dead-air space in the mask than the Gill™. This decreases the danger of becoming "out of breath" when using the system.

Photo 5–12 The Gill™ was the original piece of equipment to allow full face mask divers to breathe outside air. Its main downside is the location at the diver's temple, which significantly increases inhalation effort and dead-air space.

Another advantage of the SBV™ is that it can slowly be opened underwater to allow the excess gas from a full face mask free-flow to escape. This increases the probability that divers will be able to maintain their masks on their faces during a free-flow event. A disadvantage of the SBV™ is the added cost of retrofitting a communication system microphone into the SBV™. If divers have not already purchased either an SBV™ or a communication system, then they should purchase the SBV™ with a communication microphone already built in.

Another advantage of the SBV™ is that it does not stick out from the side of the mask like the Gill™; therefore, it does not increase the risk of entanglement snags on the full face mask.

Buoyancy-control device (BCD)

To make an effective BCD purchase, you must determine which features you need and do not need. If you are a public safety diver, you need front flotation to keep you from falling facedown into the evidence or the body you are holding at the surface. If you need to carry a marker buoy, a standard facemask, a window punch, shears, a knife, or other tools, you need plenty of usable front pockets. Some BCDs come with several D-rings on the shoulder straps for attaching tools. That is not how tools should be carried, for entanglement- and buoyancy-control reasons. For good buoyancy-control, divers need to be "one with their gear". Equipment that dangles will not move at the same moment or in the same direction as the diver and will therefore affect the diver's position and movement in the water. Good buoyancy-control capabilities are especially important in ice diving, to prevent silt-outs and to decrease fatigue. They also enable a diver to operate in the shallow depths at which ice divers tend to dive. Tools should be carried inside BCD pockets or mounted on a harness worn under the BCD.

Another BCD feature that relates to good buoyancy-control is a hard backpack. An increasing number of BCDs are made with soft backpacks so they can be easily carried in a suitcase for dive trips. That is where soft backpack BCDs belong, in suitcases, not on divers. Soft backpacks allow the tank to move in whatever direction the diver moves, so that if a diver leans to the right, the tank falls to the right. More dangerously, when divers need to retrieve their regulator mouthpieces, the top of the cylinder is often leaning back away from their head so that they cannot reach the first stage of their regulators. To compensate, divers must reach down, grab the bottom of their cylinders with their left hands and push

Photo 5–13 BCD features to look for include quick-release pony pockets, front flotation, large buckles, front cutting tool, standard facemask pockets, bright color, hard backpack, and a pull-down purge.

the bottom away and up, to bring the regulator first stage up and forward, hopefully within reach. Soft backpacks were never designed with pony bottle use in mind. The pony bottle will pull the main cylinder to an uncomfortable angle that will continually tilt the diver in that direction. Lastly, soft pack BCDs are more buoyant than hard pack BCDs, requiring divers to wear more lead. Doing so, they will have greater buoyancy change with depth change.

A good feature for ice diving is a pull-down purge, which allows divers to empty the BCD rapidly in case of a free-flowing self-inflation. Pull-down purges help prevent BCD power inflator free-flows, because the BCD can be emptied at the surface without having to raise the power inflator out of the water.

Bottom exhaust valves, for upside-down BCD deflation, are not a priority feature. If a drysuit diver is inverted and therefore has air in the legs and feet, purging the BCD is not the correct plan of action. Doing so would keep the torso heavier and inverted.

Shoulder quick-release buckles are a good idea, because without them drysuit exhaust valves on the suit shoulder can get in the way of removing BCDs. Because the hands of divers and tenders can become cold during ice diving and because thicker than usual gloves are worn, extra-large quick-release buckles are a good feature to look for. And finally, before purchasing a BCD, try it on with full gear to make sure it will not obstruct the drysuit inflator valve, weightbelt, or harness tether point.

Diver exposure suits

Most noncommercial divers wear passive-exposure suits, which are suits that do not provide a heat source to the wearer, but rather, provide insulation between the wearer and the environment. Active-exposure suits are more commonly found on commercial and military diving sites and typically involve hot-water suits. Public safety divers, and even sport divers, can manage with passive suits because their dive times should be much shorter than those of commercial or military divers. Passive suits are self-contained and far less expensive than the actively insulated suits, which typically burden divers with an umbilical and require highly trained surface support.

There are, however, battery-powered heating units on the market designed to be worn under drysuits for recreational, public safety, or research diving. These heated units are particularly advantageous for divers who move very slowly searching for small objects in low-visibility water, divers who stay in one place for long periods of time conducting research, or divers who will sit through the 90%-ready and backup diver positions before acting as a primary diver. The heated units are also helpful for divers who feel the effects of cold quickly, such as some women and people with very little body fat.

The choice of passive suit is very important. Although thousands of divers have survived ice dives while wearing wetsuits, wetsuits are not recommended for such extreme temperature operations. "The wetsuit is only a marginally effective thermal protective measure, and its use exposes the diver to hypothermia."[4] Drysuits, also known as variable volume suits, should be worn to increase the effectiveness of the operation and, most importantly, to help keep the diver safe. Drysuits are also important for sport divers, who typically lack the surface support system that should maintain highly watchful eyes on the public safety diver.

A good drysuit, with appropriate winter-weight undergarments and a layer of air, will keep the diver's skin and core temperatures at normal levels for the duration of an average ice dive. What is often problematic is keeping divers' hands, feet, and faces warm.

Everyone reacts differently to cold exposure and should be dressed accordingly. What is good for one team member may be insufficient for another. A Navy study of the effects of immersion found that a higher level of body fat reduced the rate of rectal temperature loss.[5] A 110lb (50kg) person losing heat at the same rate as a 250lb (113kg) person will most likely have a lower core temperature than the larger person. A study of eight Navy combat swimmers in 6°C (43°F) water showed that "each diver appears to have a particular core temperature decline profile with respect to cold-water exposure."[6]

For example, one diver may be perfectly comfortable using wetsuit gloves, while another will feel pain after fifteen minutes of wearing an insulated dry glove. A previous cold injury can increase susceptibility to cold problems, even if the original injury took place many years earlier. Individual differences between ability to withstand hand immersion in cold water can be great and should not be ignored, especially in the "grin and bear it" atmosphere of law enforcement and fire personnel. Divers and surface support should be observant of individual cold-stress susceptibility. Some divers may feel fine in summer-weight undergarments, while others may need winter-weight undergarments with shortened dive times. Divers should dress to be comfortable submerged for twice as long as expected in order to be prepared for unexpected problems such as entanglements.

Hoods

If your environment becomes cold enough to have ice in the winter, then you can wear a hood year round. Typically, divers ditch their hoods as soon as the water warms up because they are not fully comfortable with them. Then when the water cools, they don head protection that they are not fully comfortable with. Divers should not dive under the ice without being fully comfortable with all their gear. Make sure the hood fits properly and then dive with it as much as possible before winter to gain that necessary second-nature comfort.

Wet hoods may be somewhat adequate for short-duration ice dives if the diver does not have to later sit with that same hood after immersion as a contingency diver. If the weather includes wet snow or rain, wet hoods are not a good idea, because divers will lose evaporative heat when wearing the hoods out of the water. Wear a

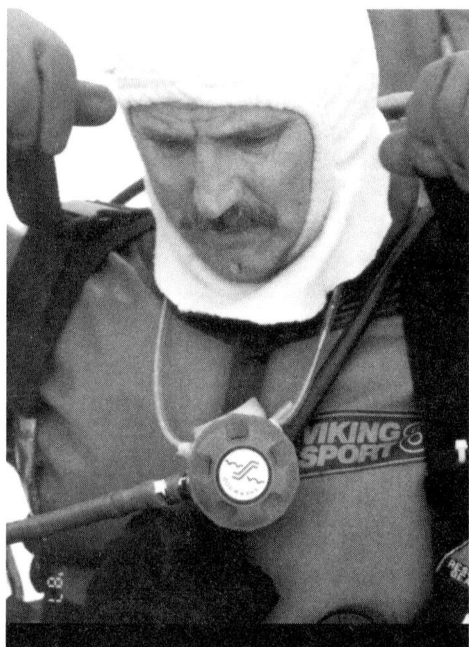

Photo 5–14 Nomex™ hoods make excellent drysuit hood liners.

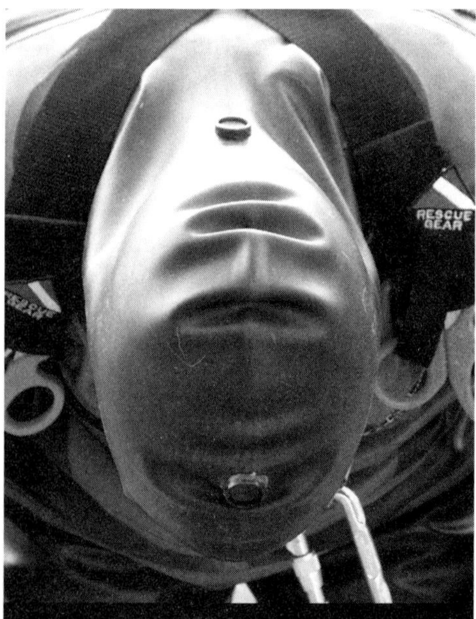

Photo 5–15 Hoods can fill with air, especially when full face masks are used. This can cause hood leaking, buoyancy changes, and discomfort. Install a one-way valve hood vent at the top of the head for vertical postures, and one at the back of the head for horizontal positions out of the way of full face mask straps.

thin-material hood under the wet hood to cover the ears and prevent water movement if the hood does not have a snug fit. A firefighter's Nomex™-type hood with the neck portion cut back is great for this purpose. This inner liner keeps the head warmer and keeps the neoprene hood away from the ears. One advantage of wet hoods is that most of them enable divers to wear their standard face mask straps under their hoods.

Dry hoods, especially latex, need some sort of layer, such as a skullcap, between the ear and hood. This layer will help prevent ear problems and equalization difficulties. Nomex™ hoods also work very well for this. Most dry hoods require that the mask be worn over the hood to keep the diver's head dry.

Commercial hood vents for both wet and dry hoods are a good idea for non-hazardous-material dives, because they let in less water than vents made by cutting a notch or a cross into neoprene hoods. Place one vent at the top of the hood for vertical diving, and one vent at the back of the hood for horizontal postures. Full face mask divers typically have a higher incidence of air leakage into hoods; therefore, they should have hood vents installed when they purchase the suit.

Protective helmets

Divers may wish to consider wearing a helmet over their hoods to protect themselves from a head injury should they have an uncontrolled ascent under the ice roof. One of the Canadian fatalities described in chapter 1 involved a diver with a bruised head. Cave diving helmets can be worn, or swift-water helmets can be used if the foam is removed.

Lights can be mounted onto helmets, thereby allowing the diver to have both hands free. Lights can be a helpful option, since the ice roof can greatly reduce the amount of ambient light underwater, especially if there is a deep layer of snow on the ice. If the water has low visibility due to turbidity (particles in the water), lights are not helpful and should not be used.

Drysuits

Drysuits, which can be made of several types of material, encapsulate all parts of a diver's body, except the face. Surrounded by a layer of air in the suit, the diver stays totally dry and is able to wear warm, insulating undergarments inside the suit. To compensate for the compression of undergarments and air in the suit as the diver descends, drysuits have a power inflator, much like that of a BCD. Because of the air within the suit and the additional power inflator, drysuits are a unique piece of equipment that, unlike wetsuits, require special training and certification. Beyond the additional cost of training, a good drysuit will cost more than a good wetsuit. Still, drysuits offer many advantages to recreational, research, and public safety divers, including greater thermal protection, rapid dressing, ease of storage, and protection from hazardous materials. A good quality drysuit should also last longer than a wetsuit.

Heat loss in water is largely due to conduction. A layer of heat-trapping, dry insulation in the form of drysuit undergarment greatly reduces heat loss. Because the required amount of thermal protection can be provided by what divers wear under them, drysuits can be used for diving in any range of temperatures, from icy water to sun-heated ponds in August. Consequently, the same piece of gear has applications for all diving seasons. If divers have drysuits, they should dive with them year-round so they can stay comfortable with them. They should not wait for the more extreme conditions of diving in the winter.

Drysuits, with appropriate undergarments, will enable a diver to stay warmer for longer lengths of time, both in and out of the water, than wetsuits. Bear in mind, though, that a vulcanized rubber or trilaminate drysuit alone has no thermal protection, so some type of undergarment is always necessary. Crushed neoprene likewise has little inherent insulation.

The reason drysuits will keep the diver warmer, especially at depth, is that according to Boyle's Law, the volume of a gas decreases as pressure increases. Wetsuits are filled with insulating gas bubbles, which are crushed flat with depth, thereby canceling their thermal protection. At approximately 4ata (4bar), a wetsuit will be half its 1ata (1bar) thickness. A drysuit diver, however, can simply add air to the suit to maintain full thermal protection.

Another advantage of drysuits over wetsuits is rapid dressing capability. Wetsuits must fit snugly in order to function and, therefore, take a little longer to put on. Drysuits, however, have a looser fit. Well-trained public safety divers with good tenders can have their drysuits on before wetsuit divers have even begun putting on the tops of their suits.

Protection from hazardous materials is another advantage of drysuits. Members of the public will forever find the concept of walking on water irresistible, and the cleanliness of frozen water does not seem to concern them. As a result, public safety ice divers are sometimes required to dive in water filled with petrochemicals, pesticides, sewage, heavy metals, or other hazards. In these situations, drysuits offer a tremendous safety advantage over wetsuits, because they keep water-borne contaminants away from a diver's skin. For greater protection against hazardous materials, drysuits should be used with dry gloves, a built-in dry hood, and a full facemask or diving helmet properly mated to the suit.

Drysuits can be easier to store than wetsuits. Wetsuits and neoprene, bilaminate, or trilaminate drysuits must be hung to dry after use. In addition, because neoprene will crease, neoprene suits are stored on large hangers and occupy a lot of space. A vulcanized rubber drysuit, on the other hand, never needs to be hung. To store a vulcanized rubber suit, simply towel dry it, powder its seals, and paraffin-wax the zipper. A drysuit can be rolled snugly into a small duffel bag that takes up very little room—an important consideration when gear must be packed neatly, quickly, and efficiently into an apparatus truck. As with all exposure suits, do not lay heavy items on top of vulcanized rubber suits during storage.

The myths of drysuit diving. Many divers and dive teams have at least a few mistaken beliefs about drysuits which prevent them from taking advantage of the suits. These apprehensions can increase safety risks. Below are the most common myths, followed by the reasons they are untrue.

Myth #1: Drysuits require more weight than five millimeter or seven millimeter wetsuits. In reality, a properly weighted vulcanized rubber drysuit with warm undergarments may require the same or even less weight than a thick wetsuit. The key to weighting a drysuit is to vent excess air from the suit after it is donned. That way, the suit will have very little air in it when a diver enters the water, and the diver can descend with less weight.

Myth #2: Drysuits are more restrictive to movement than wetsuits. A properly fitting drysuit should allow a full range of motion. Additionally, because they do not need to fit as tightly as wetsuits, drysuits are often more comfortable and less restricting. A key to freedom of movement in a drysuit is to stretch the arms fully up, forward, and then in a large circle before entering the water. This will put the suit in the right place on the body. Repeat this procedure after entering the water. Make sure to add enough air into the suit to keep the suit comfortable.

Myth #3: To vent a drysuit that is malfunctioning and self-inflating, pull the neck or wrist seals out to allow air to escape. While this procedure certainly sounds good, reaching through a hood or wrist seal while wearing gloves is extremely difficult at best, and to do it before hitting the surface may be impossible. Instead, if your inflator is stuck, disconnect the hose, then purge air through the arm-mounted exhaust valve as you flare out to slow your ascent. Do not forget to continue breathing normally.

Myth #4: Drysuit divers do not need buoyancy compensators (BCDs). While forgoing BCDs was commonplace years ago, this practice has since been discarded. To be as safe as possible, drysuit divers should wear BCDs to give them good flotation on the surface and to supplement the buoyancy of the suit underwater, if necessary. Drysuits were not designed to take the place of BCDs. Unless the drysuit is constructed of full thickness neoprene, a flooded drysuit offers little or no positive buoyancy.

Myth #5: Add air to the drysuit and not the BCD for buoyancy control. The assumption behind this dangerous practice is that divers cannot properly vent or control both a BCD and a drysuit during ascents, particularly uncontrolled ascents. One of several hazards of adding air solely to the drysuit is that it could cause an uncontrolled ascent if the diver is over-weighted, as the majority of divers are. Too much air added to a BCD can rapidly, easily, and reflexively be purged. The same is not true when too much air is added to a drysuit. A drysuit can hold far greater volumes of air than a BCD, so mistakes can also be much greater. Air in a drysuit can become trapped in lower extremities and can invert a diver and cause the loss of fins. This cannot happen with a BCD.

Pressure on a diver's neck caused by too much air in a drysuit can cause a carotid sinus reflex and possible unconsciousness. An over-inflated BCD cannot do this, and a BCD has its own over-inflation self-purge valve. During ice diving, the more a valve is used, the greater the chance for free-flow. A free-flowing drysuit inflator poses greater risks than a free-flowing BCD inflator. The safest procedure is to add just enough air to the drysuit for warmth and freedom of movement, and use the BCD to adjust for buoyancy.

Myth #6: A flooded drysuit will cause a diver to sink. Water weighs the same inside and outside a suit. The only way water can cause negative buoyancy is if the diver is very over-weighted and is using the drysuit (instead of the BCD) to compensate, and if so much water entered the suit that it replaced the air. Author Hendrick overweighed himself by 4lb to 5lb (2kg) and then removed the inflator valve of a trilaminate suit at a depth of 100ft (30m). He discovered that the undergarments became stuck in the hole and kept the water out. He then fully unzipped the suit and waved the opening in and outwards in an attempt to fill the suit. After filling the suit, he made a slow, controlled ascent to the surface.[7] The real dangers of a flooded drysuit are contamination and hypothermia. Getting out of the water onto the ice roof may be a bit more difficult, though, if your legs are full of water. Should that occur, use a sled or tender assistance.

Myth #7: Divers do not need a drysuit course and drysuit certification to dive in a drysuit. Drysuits are a unique piece of equipment that require training and practice to use safely. Otherwise, divers risk drowning, carotid sinus reflex, decompression sickness, or lung over-expansion injury.

The following topics regarding drysuits should be covered in an ice diving course:

- It is imperative that divers reflexively learn reality-based, effective self-rescue procedures for self-inflating/free-flowing drysuit inflator valves and for inversion events.

- Students should experience swimming without fins on the surface with a fully inflated suit.
- They need to learn how to properly weight themselves for different thickness undergarments and varying amounts of air in the suit.
- A good course will teach them how to set the dump valve for different types of dives, and how and why to adjust it during a dive.
- Without good training, divers are likely to use poor drysuit dressing procedures, which often result in suit flooding, damaged zippers, torn seals, restricted wrist movement and circulation, and lengthy dress times. Divers need to learn how to properly burp the suit prior to gearing up, how to dump air at the surface, and how to stretch to make sure the suit sits on the body properly.
- Divers need to learn how to let enough air into the suit for warmth and comfort and then to make necessary buoyancy adjustments with the BCD. Students should learn how to adjust air in the suit and the BCD with depth changes.
- A knowledgeable instructor will direct students to the drysuit and features that best meet their needs and will then teach them maintenance and repair procedures to increase suit longevity.
- A good course will teach students how to properly cut neck, wrist, and face seals for increased seal longevity, increased diver comfort, and increased safety. Drysuit divers need to be capable of immediately recognizing signs of carotid sinus reflex in other divers and themselves.

Sadly, many of today's drysuit certification courses are somewhat inadequate. That is why many divers believe that they do not need to become drysuit certified before diving dry. A good program is imperative. It is irresponsible for a diver to dive dry without good training, not only for the diver's own safety, but for the safety of anyone diving with him/her. Take the time to find good training; do not just take the first course that comes along.

A department or instructor who allows divers to dive in drysuits without drysuit certification is taking on a significant liability should a diver become injured or killed. Even more negligent is permitting uncertified drysuit divers to dive under the ice.

What to look for in a drysuit

Fit. There are numerous drysuits on the market, but a key factor with all of them is fit. Drysuits that are too big require more air internally to fill the excess space. This increases the amount of weight needed to dive and decreases overall mobility and diver safety. Mobility is based on a proper fit without compression in areas such as elbows, knees, shoulders, and spine to the point of severe restriction. If the suit legs are too long, the diver could lose one or both fins if accidental suit inflation caused body inversion.

Try on the suit with the thickest undergarments that may ever be worn, then bend down with hands reaching for the floor to make sure the torso is long enough. Stand straight with arms raised to the sky and make sure the crotch is not too low. That would mean that the torso is too long and possibly the legs too short. With arms raised forward, ensure that the end of the suit material at the wrist does not extend beyond the wrist bone. Squat down to make sure the legs are long enough. Lay supine on the floor and have someone pull on the boots. If the boots come right off your feet then the legs are too long. With the warmest undergarments on, the suit should fit well to every part of your body. If it is necessary, suit customization is worth the extra cost.

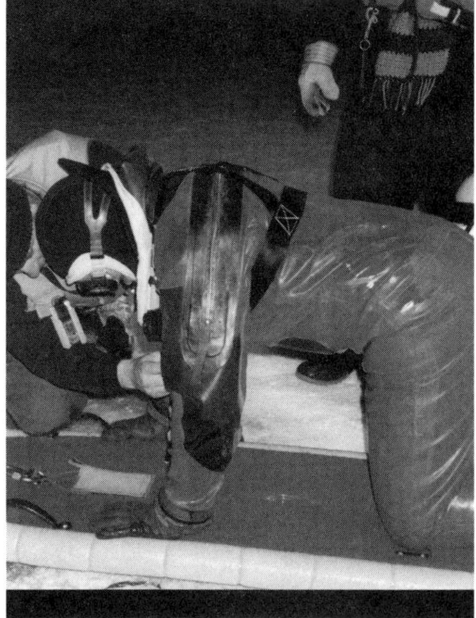

Photo 5–16 This diver wears a hazardous materials (haz/mat)-tested (Viking™) vulcanized rubber drysuit with a dry glove ring system, back-mounted heavy duty zipper, and a wet hood with a Nomex™ liner.

Materials. Full thickness neoprene offers the most intrinsic insulation, but it has drawbacks. It requires a significantly greater amount of weight and significantly more time to dry. It is difficult to repair on the scene, bulkier to store, unacceptable for contaminated water, and more difficult to clean. Crushed neoprene suits have much less intrinsic insulation than full thickness neoprene suits, but they are often more comfortable, easier to store, require less weight, and take a little less time to dry. They can be costly, though, and have the same vulnerability to contamination. They also cause evaporative heat loss in air after immersion and are difficult to repair in the field.

Photo 5–17 When used for surface duties, all drysuits except full-thickness neoprene drysuits, require the wearing of PFDs. Notice cutting tool on PFD.

Bilaminate or trilaminate suits require less weight, take less time to dry, are relatively easy to repair on-scene before immersion, and take little room to store; however, they offer no intrinsic insulation. The durability of trilaminate suits and their ability to protect a diver from contamination varies widely depending on the manufacturer and type of suit. Be wary about low-priced suits because you will often get just what you pay for.

Vulcanized ethylene propylene diene methylene terpolymer (EPDM) rubber suits require the least amount of weight because they are intrinsically negatively buoyant. Less weight means less compensation for depth with the BCD, better buoyancy control, less risk of uncontrolled ascents, and safer diving. Rubber suits can be immediately repaired on-scene even after immersion. Simply dry the area with a towel, sand a little, apply glue, wait a little bit, apply a little more glue, apply a patch, roll with a soda can, wait a few minutes, and the suit is ready to dive.

Vulcanized rubber suits can be simply toweled off and then rolled for storage. These suits do not require time to be hung to dry. This is a real plus for public safety divers, who can simply replace the suits in their proper place after a dive and be ready for the next call. The ability to towel the suits dry is a great advantage for ice divers, because the suits won't freeze between dives if there is no shelter, and divers can keep the suits on after a dive in wet weather since there will be no evaporative heat loss. If EPDM suits are shared between divers, the next diver can don an externally and internally dry drysuit. These advantages are particularly important for dive teams who do not have the time or the ability to set up a heated shelter on the site.

Viking Trelleborg™ manufactured the first rigorously tested and approved vulcanized EPDM rubber suits. The suits were exposed to a multitude of contaminants using the National Fire Protection Association (NFPA) 1991/1994 standard for vapor protective suits for hazardous chemical emergencies. They also applied the American Society for Testing and Materials (ASTM) F1001 1989 guide for evaluating protective clothing materials. "It also should be mentioned that the only approval mechanism in place in the world today [that] deal[s] with standards of a diving suit, is CE95 0403. These are standards to which Viking is certified after passing through examination procedures in accordance with the Personal Protective Equipment Directive 89/686/EEC (PPE), class 3, which is the highest class with the toughest demands of the European Union."[8] If contaminated water is a potential or realized concern, we recommend reading *Diving in Contaminated Water*, which includes the most comprehensive test data available on any drysuit for diving in contaminated water.[9] Before purchasing a drysuit, ask to see data on testing for contaminated water.

Because of their impermeability to contaminants, vulcanized EPDM rubber suits are the easiest suits to clean. In terms of durability, some high-quality trilaminate suits may be more resistant to punctures and abrasion than vulcanized rubber, and much more resistant to punctures and abrasion than neoprene or crushed neoprene suits. Make sure the suit comes with a good repair kit.

Zippers. Not all zippers are of the same quality. Again, you often get what you pay for. Ask what grade of zipper is installed in the suit. Use a heavier weight zipper (often referred to as heavy-duty). Heavy-duty zippers seal better and last longer than lesser quality zippers. They have larger teeth, some almost twice the size of medium-quality zippers. Heavy-duty zipper teeth are crimped and riveted into their rubber-coated canvas base, whereas medium-quality zipper teeth are only crimped. The zipper will last longer if it is mounted on a rubber cloth material rather than a vinyl composite.[10]

Photo 5-18 Full-thickness neoprene suits provide the most inherent insulation, their own positive buoyancy, and generally are inexpensive. They do, however, require the most weight, take the longest to dry, and are not acceptable for contaminated water.

A never-ending controversy revolves around the question: Which is better, front- or back-entry zippers? Back-entry suits are usually more comfortable and easier to don. Since tenders should close all zippers, whether back or front, the advantage of a self-donning suit to public safety or ice divers is irrelevant. Front-entry suits often require a flap of material that is buckled down over the abdomen. This means one more step for donning and more suit material to fill with air. Front-entry zippers require a protective flap over them, since they are more exposed to bottom and water debris than back zippers.

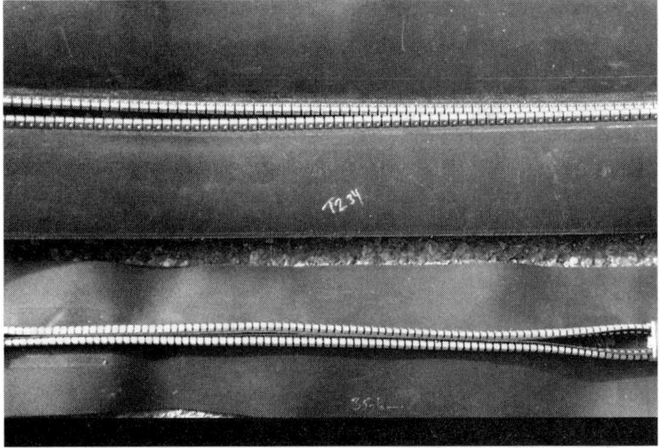

Photo 5-19 To decrease the chances of zipper failure and possible hypothermia or aborted dives, make sure to purchase drysuits with heavy-duty zippers. Notice the larger teeth on the heavy-duty zipper top.

Inflator and exhaust valves. Exhaust valves with a domed top are preferable to flat-topped valves because the former can be purged by raising the left arm up (if the exhaust valves are on the left arm) and pressing the dome against the left side of the diver's head. This enables divers to disconnect a self-inflating drysuit hose with the right hand while simultaneously flaring and purging the suit. Again, check for valve quality because not all valves are created equal. A swivel feature in an inflator valve is advantageous, because it allows the diver to position the inflator hose for the most effective and rapid disconnect motion possible. The inflator valve should be pointed downward at a 45° angle for a center-chest mounted valve. This is the most natural position to reach to disconnect the hose if the valve starts to self-inflate. Make sure the valve placement will not be covered by the BCD cummerbund.

Suspenders and urination zippers. These are nice features for ice dive instructors and divemasters, who may be in their suits for extended periods. Suspenders help hold the crotch of the suit in the proper position to allow full leg movement.

Cuffs and seals. Ice divers may want to have heavy-duty latex cuffs and seals installed on their suits, as cold increases the risk of latex tearing. Latex will tear more easily than neoprene, but it can be toweled dry and is better for contaminated water diving. Cuff ring systems are recommended because a torn wrist seal can be replaced in the field in minutes. If drysuits are shared, there are two options for individualizing neck seals. The less expensive option is to cut the neck seal for the diver with the largest neck, then have the other diver wear neoprene neck collars under the seal. Each collar would be of the appropriate thickness to make the suit neck seal fit properly. A second option is to purchase neck ring systems. Each diver then has an individual-sized neck seal that secures into the suit's neck ring. Cuff ring systems also work well for multiple-diver suit use. Each diver has appropriately-sized dry gloves that are secured onto the cuff ring. The wrist seals are cut to fit the largest diver. The dry gloves keep the water away from the wrist seal.

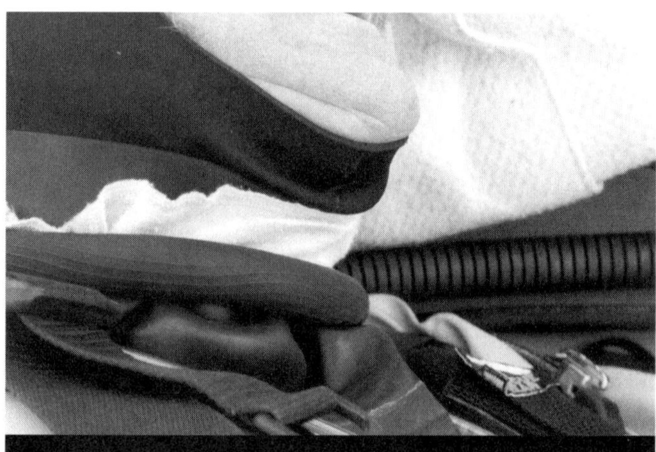

Photo 5–20 A plastic tube neck ring is useful to keep pressure off a 90%-ready diver's neck. The ring goes over the neck seal, and the top of the seal is laid back over the ring.

Drysuit undergarments. Always be prepared to stay comfortable for longer than the planned dive time. Especially for public safety diving operations, divers may need to remain as backup or 90%-ready divers for times longer than expected.

Make sure the undergarment fits. If it is too tight, it can restrict movement and blood flow. Even an undergarment that is too big can cause movement restrictions and decreased circulation.

For example, if the torso is too long, the crotch area may end up too low, making it difficult to kick well; or, if the legs are too long, they can bunch up at the ankles, causing decreased circulation to the feet.

Photo 5-21 If only one set of undergarments are purchased, make it the warmest, such as this Arctic Plus™ underwear because fleece or wool sweatpants or sweatshirts will suffice for warmer months. Arctic Plus™ underwear can also be worn during surface support duties.

Avoid undergarments with a great deal of compressibility, which will result in unnecessary and possibly hazardous buoyancy changes if your diving will occur deeper than 33ft (10m). Pile-type undergarments, for example, will greatly compress with depth, making the diver heavier. The diver will respond by adding air to the BCD and drysuit, which makes buoyancy control more difficult and the chances of rapid ascents more likely. Remember, rapid or uncontrolled ascents under ice are far more hazardous due to the overhead environment.

Choose undergarment material that will continue to insulate even when wet, so that divers will not become chilled if they exert themselves and sweat before they dive, or in case the suit leaks for some reason. Polar fleece and wool work very well. An Acrilian pile undergarment was tested under an immersion coverall during immersion for one hour in -2.5°C (-36°F) water. Subjects without the undergarments lost 13.1°C (55.6°F) of mean skin temperature, whereas the protected subjects only had a loss of 9.9°C (49.8°F).[11] Avoid cotton, because once wet or damp, it loses its insulation value and the wearer will quickly chill. One study tested three types of undergarments—cotton, polyester, and teviron—worn under a drysuit in 10°C (50°F), 7°C (44.6°F), and 5°C (41°F) temperatures. Today's more popular undergarment materials include Thinsulate™, for wicking away moisture, and Moleskin™. A wetsuit was also tested. Rectal temperature, skin temperature, exhalation volume, and decrease in body weight were the variables investigated. "Teviron undergarments provided the best heat insulation and the wetsuit was the worst."[12]

Drysuit maintenance

If the water has any level of salt or contamination, the suit should be washed down with freshwater, or with water and Joy™ liquid soap, and then rinsed. Ask the suit manufacturer if there are any cleaning fluid recommendations specific to the suit that you have. If dirt has settled in the zipper, brush it out gently with a soft toothbrush.

If a suit becomes wet on the inside, due to either diver perspiration or suit leakage, the inside must be dried before the suit can be stored. If the suit has not been dried inside and mold has begun to grow there, sponge the inside down with a 5% bleach solution and rinse with freshwater before drying. Suits that are naturally heavy, or that become heavy when wet, such as 1/4in (6mm) neoprene drysuits, may develop stressed horizontal seams if hung to dry. Such suits may last longer if they are laid to dry across a fire department hose rack, a horizontal ladder, or even a few chairs. A fan may be used to speed up the drying time.

Once the suit is dry, generously powder the inside and outside of all the seals with pure talcum powder. The best source for talc is a store that specializes in large farm machinery. They sell large containers of talc at a fraction of the cost paid elsewhere. The talc can be placed in socks or cloth sacks for ease of use. Avoid inhaling the talc when applying it. Talcum powder will make the seals slip onto the diver much more easily with less strain on both the seals and the diver, and will help prevent dry rot during storage. Avoid powder containing scents or oils.[13]

Photo 5-22 Zippers can be paraffined on or off the diver. This should be done after every dive.

Next, lay the suit zipper-side up on a flat surface with the zipper closed. Rub paraffin wax into the sides of the teeth and open and close the zipper several times. Repeat this process until the zipper closes smoothly and effortlessly. Do not wax the inside of an open zipper, because that wax will not wash away during a dive and will accumulate. If the zipper has been poorly maintained, do not force it closed. Carefully wax the sides of the teeth with the zipper open and gently close it. The zipper is the most expensive part of the suit, and replacement costs can be as much as one-third of the suit's value. Consequently, proper maintenance is important.

Never use beeswax on the zipper. The wax will attract dirt that can cause the teeth to bend when the zipper is opened. If beeswax has been used on the suit, clean it out with a toothbrush and perhaps a solution of Joy™ liquid dish soap. Some manufacturers recommend commercially available zipper lubricants other than beeswax. Experience with thousands of ice and warmer water dives in which zippers were lubed solely with paraffin has demonstrated that paraffin wax works very well. We have had vulcanized rubber suits dry rot after twenty years of use, without ever needing to replace a zipper. Purchase a block of paraffin wax at a supermarket or hardware store, cut it into slices, place each slice in a Ziploc™ bag and store one with each drysuit.

An effective procedure to ensure that all divers have properly waxed suits is to wax the zippers after a dive, before the diver removes the suit. Keep the zipper closed, wax it, then remove the suit.

If the zipper develops loose threads that protrude from the stitching, carefully cut the threads without pulling on them. If there are many loose threads, send the suit in for repair.

After the suit is dried, powdered, and paraffin-waxed, it is ready to be rolled and stored. Lay the suit flat, zipper-side down, zipper open. Put the boots together toe-to-heel and loosely roll the suit up to the hood. Fold the arms over themselves, flip the hood over the arms, and put it into its storage bag. Store it with the zipper open to prevent unnecessary zipper strain that will decrease the life of the zipper.

Argon gas

Some divers use argon instead of air in their drysuits because it provides better insulation. Argon is a colorless, odorless, inert gas. Unlike diatomic nitrogen and oxygen (N_2, O_2), argon (Ar) is monatomic and has a low specific heat per molecule. Let us look at what this means.

Our body heat is transferred from the gas in our drysuits to the outside cold water through a process of gas molecules colliding with one another and with the drysuit. "Warm" molecules collide with "cold" molecules and thus transfer heat energy to them. As the "last" molecules in this chain of collisions collide into the drysuit shell, heat is transferred to the shell and then finally to the water. Because molecule collisions occur in all directions, heat goes back and forth between our body and the drysuit shell, with the net heat transfer being from our body to the shell and water. Molecules store heat in some form of motion. Simple molecules like the monatomic argon can only store heat by moving. More complex molecules, such as those found in air, can also store heat by rotating and vibrating, and can therefore steal more of our body heat than monatomic molecules such as argon.

Argon may be useful for divers who have to be at depths longer than the fifteen or twenty minutes allotted to public safety and recreational divers. Scientific divers or exploration divers may find that the benefits of argon outweigh the cost of argon bottles, mounting systems, and fills.

Wetsuits

Wetsuits enable a layer of water to enter the suit. The diver's body warms this water, which is supposed to then serve as an insulating barrier. This means that divers lose heat at the moment of immersion, as well as later in the dive if cold water moves in and warmed water moves out of the suit. Movement of water within the suit also can cause temperature loss, because areas such as the armpits, lateral thorax, upper chest, and groin lose heat at greater rates than other parts of the body, such as the extremities. Thermal gradients of the body's surface are increased during cold-water immersion.[14]

As you will learn in chapter 7 in the heat loss section, water rips the heat out of a person twenty-five times faster than air; therefore, heat loss in ice water can be significant, even when a wetsuit is worn. Research comparing dry survival suits and survival suits filled with 27lb to 28lb (12kg to 13kg) of -46°F (-8°C) water showed that drysuit wearers lost heat at 0.46°C per hour while those in the wetsuits lost heat at 0.9°C per hour, or almost twice as fast.[15]

Only wetsuits that fit snugly, without restriction of circulation, should be considered for ice diving. A set of long undergarments or leotards worn under a less than perfectly fitting suit helps to reduce movement of water in and out of the suit. For multiple dives, have extra sets of long undergarments, so that a dry set can be worn for each dive. Having an extra wetsuit is even better. Wool socks under the booties also help. Suits customized with spine pads to decrease water movement in the suit are preferred, as the shoulder and spine areas are often those where cold is felt first when wetsuits are worn.[16]

Remove the wetsuit between dives, as wearing a wetsuit in air can cause significant evaporative heat loss. Covering up with a big coat is not enough to prevent heat loss during cold ambient air temperatures. Between dives, if possible, dry wetsuits in a dryer on the "air fluff" setting for twenty to thirty minutes. Wring out the suits before putting them in a dryer. An extra shore support person might be able to get this done at the local fire department or laundromat. Divers who serve as backups or 90%-ready divers can wear wool hats or balaclavas (head socks) over their hoods and can wrap themselves in blankets to decrease heat loss while in air.

Warm water flushes have been a topic of discussion for years. Adding warm water to a wetsuit just before the diver descends makes the suit feel much more comfortable and certainly makes the diver feel warmer. Warm water also vasodilates peripheral blood vessels, bringing warm core blood to the surface. However, cold water then replaces the warm water, the result could be a greater heat loss than if the warm flush was not used. During actual rescue and recovery operations, warm water may not be available, so do not use it during public safety training. If warm flushes are used, consider duct-taping ankles and wrists to prevent water movement in and out of the suit. Make sure the water is lukewarm, not hot.

Remember that as depth increases, the insulation value of wetsuits decreases, because the gas bubbles inside the neoprene decrease in volume. Wear an appropriate thickness for the dive profile. Suits with step-in tops are preferred over tops with beaver tails, because the former provides more insulation for the groin area, which is an important heat loss area. One-piece suits with attached hoods that enable a vest to be worn underneath may be warmer than the two-piece suits, if the former allows less water in and out of the suit. Terpolymer, more commonly known as Rubatex™ neoprene, has been shown to be warmer, with better retention of flexibility in cold temperatures, than other neoprene rubbers.[17]

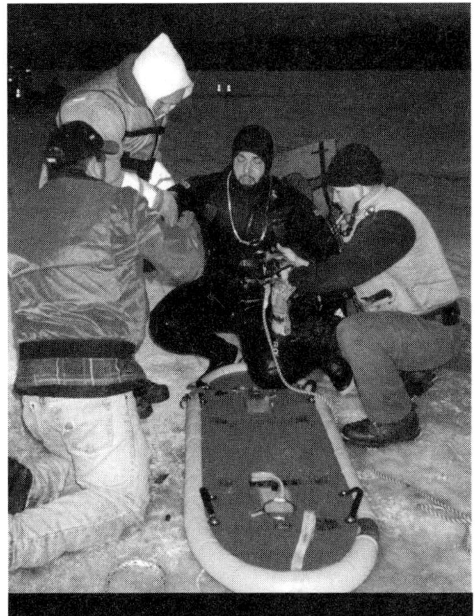

Photo 5-23 Wetsuit divers may require shorter dive times than drysuit divers.

Hand and foot protection

No matter how warm the rest of the body is, the hands can become dangerously cold; therefore, hand insulation is critical on ice dives. If hands are not properly insulated, the resulting vasoconstriction will severely restrict blood flow to the hands, and hand function will be seriously compromised. However, fully insulating the hands leaves divers with little ability to use them. Divers wearing 1/4in (6mm) neoprene wetsuits or drysuits with neoprene mittens or gloves were tested for grip strength at 2ata and 3ata depths.[18] Results showed that there were significant grip-strength decrements at 2ata or less. At 3ata, the neoprene compressed enough so that grip strength was not compromised. Most ice dives, however, are shallower than 3ata.

During unprotected cold immersion in -18°C (-64°F) water, finger blood-flow has been reported to drop to 5% of maximum blood-flow in warm water. This percentage was unchanged during exercise, which indicates that your hands can become useless from cold even though your body feels warm from increased heat production during exercise.[19] The same study found forearm blood-flow was restricted to 17% to 21% of maximum blood-flow during cold-water immersion.

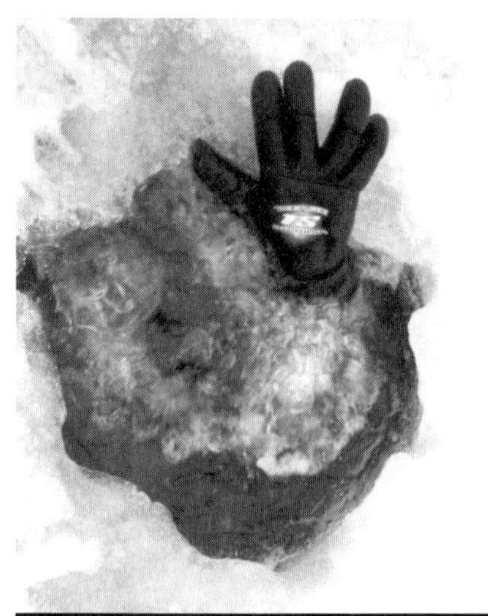

Photo 5-24 Wetsuit gloves are not the best option for warmth. Duct-taping wrists can help to decrease water flow.

Photo 5-25 Drysuit gloves with appropriate liners keep hands warm and allow needed fine mobile dexterity.

The US Navy contends that divers should not let their hands become colder than 15°C (59°F). In a polar-diving workshop proceeding, Bob Stinton noted that hands begin to hurt in water at 10°C (50°F). Your knuckles will feel like they have been hit with an ice pick. Hands that are in 8°C (46°F) temperatures for thirty minutes may develop a nonfreezing cold injury. "To recognize a nonfreezing cold injury, get out of the water and warm up. Your fingers will tingle and be numb, and your joints will not work properly for days. If they [your hands] are still tingly or numb the next day, you have got it [nonfreezing cold injury], and once you have it, you cannot get rid of it."[20] Prolonged exposure will result in an affliction much like trench foot.

Individual differences do occur, as noted by Lieutenant (Navy) Retired David Holland, SBStJ, CD, BMASc, Canadian Forces: "While conducting three hour experimental dives in 4°C fresh water, while wearing a drysuit, custom made undergarments and a hot water heating vest (core only), I found that my fingers remained completely functional despite the skin surface temperature being almost 6°C."[21]

The inability of divers to use their hands, either because of cold or thick insulation, puts them and their teammates at risk for injury or even death. Therefore, divers need to wear enough hand protection to keep their hands functional with regard to insulation from the cold water, while not wearing insulation that is so thick that it significantly restricts dexterity. Dry gloves are definitely recommended over wet gloves. Separate liners for dry gloves are useful because the liners can be changed should they become wet. Dry gloves with built-in liners that become wet for whatever reason typically take hours to dry before they can be used again.

Dry mittens or lobster claw gloves are warmer than five-fingered dry gloves. But, we only recommend the use of mittens or lobster claws for divers with plenty of cold-water diving experience who have demonstrated that they can easily perform such skills as mask donning and clearing, BCD adjustments, drysuit inflator rapid disconnect, primary-to-pony regulator switch, equalization, carabiner use, and weightbelt ditching and donning, while wearing these hand-protection options.

Ice divers should have drysuits that include built-in boots, so that the feet share the same insulating air as the rest of the body. Wear large enough fins and ankle weights to prevent blood-flow restriction.

Do what it takes to prevent feet, hands, and the rest of the body from becoming cold before diving. If divers are first serving as tenders, they need to be especially careful about preventing heat loss. The hands of a tender may experience multiple immersions while assisting divers. Those hands need to be warm enough to record necessary information, work tether lines, and assist divers.

Avoid anything that will constrict blood flow to the hands or feet, such as: thumb loops on undergarment wrist sleeves, watches, jewelry, cutting-tool straps (which should not be worn on the extremities anyway), or tight socks. Avoid hanging vertically in the water for any length of time while diving dry, because the air will be pushed up to the torso, causing constriction and loss of insulation in the lower extremities.

Members of the Allentown Fire Department in Pennsylvania use Dr. Scholl's™ inserts in the bottom of their drysuit boots. The inserts definitely keep feet warmer and protected, especially when feet encounter ice or snow when topside. Use these inserts with wool or polar fleece socks and booties and your feet will be warm for hours. If a diver's feet have a tendency to sweat, the diver should generously apply powder before donning socks. A little cayenne pepper mixed in with the powder will add warmth sensations to tenders and divers who commonly get cold feet (physically, not mentally).

Ankle weights

Many of today's fins are positively buoyant. In conjunction with neoprene or drysuit boots, the fins force diver's feet upward, often causing knees to bend upward. This causes lower back strain as divers work to keep their legs horizontal, and it often causes over weighting on the belt as divers attempt to compensate for legs that keep pulling them up. This in turn compromises the diver's hydrodynamics, resulting in far more energy being used to swim through the water. Of greater concern, though, is that drysuit divers can lose their fins if their feet rise higher than their body. More air is allowed into the boots, which makes the feet even more positively buoyant. The diver can end up fully upside-down. If the suit legs are too long, the boots can then extend beyond the feet, with the result that the fins pop off.

Divers should adjust ankle weights as needed to keep their feet in the proper position without being too heavy. The diver's legs should be just about neutral at a depth of about 15ft (4.5m). Ankle weights not only enable feet to be neutral, but they help keep boots on feet and decrease the amount of air going to the feet.

In addition, public safety divers should wear ankle weights in low-visibility water to enable them to keep their feet on the bottom, so that they can search with their entire bodies.

When purchasing ankle weights, try them on with whatever you will be wearing on the coldest dives to ensure that they do not restrict ankle movement or blood-flow. Divers wearing heavy rubber fins may need to remove some of the lead from their ankle weights or may not need them at all. If the latter is the case, a strip of duct tape or a Velcro™ strap can be worn to prevent the loss of a boot during an inversion.

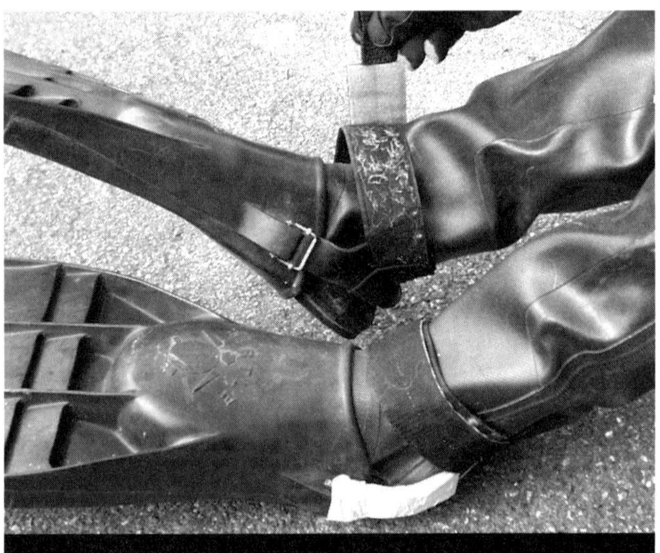

Photo 5-26 MARSARS™ fin keepers prevent fin loss underwater.

It is always good to have an extra pair of ankle weights readily available. If a diver discovers that he is too light when already in the water, an ankle weight can be placed around the tank valve or diver's harness.

Dive masks

Standard masks. Masks must fit well. It is particularly important to make sure they do not leak on ice dives, when cold water can shock a face and cause a gasp reflex. Seals are also less forgiving in cold, so they need to fit the face well. To check for fit, gently lay the mask against the face while breathing through the mouth. Have someone look all around the mask and make sure the mask skirt is touching the face everywhere. The most common method of checking for mask fit is to have the diver inhale through the nose to see if the mask will stay on the face without holding it. This is not a good technique, because if you inhale hard enough, you can cause any mask to stick to your face, no matter how badly it fits. Inhaling hides the areas where the skirt does not meet your face.

A neoprene mask strap is easier to use than silicone or rubber straps. Neoprene straps do not become stiff when cold, they are easier to adjust with gloves on, they do not break as easily, and they do not grab the hood and pull it back when the strap is donned over the head.

With regard to strap placement, make sure it is not over the ears, as that could decrease equalization capability. If divers are wearing wet hoods, they can wear their mask straps under their hoods to prevent an accidental full mask removal.

Masks are more likely to fog in cold water. Saliva is not an effective cold-water anti-fogger. If there is water visibility, properly use an anti-fog solution.

Full face masks. A full face mask with a drysuit hood will greatly decrease the amount of heat loss from the face, and it will prevent that uncomfortable ache sometimes experienced when the forehead is left exposed between a wetsuit hood and standard mask. A drysuit hood and full face mask will serve to insulate the head's heat radiators: carotid arteries, lips, nose, and ears. Keeping the mouth warm is important, as the lips and mouth muscles need to be able to hold the regulator. Full face mask divers do not have to be concerned about maintaining mouthpieces in their mouths; all they have to do is breathe normally with their mouth in the oro-nasal cavity of the mask. Divers using standard masks can use hoods that are built like a head sock, with holes for the eyes, mouth, and nose. Become comfortable wearing these hoods in confined water before using them in ice diving.

Another important function of full face masks is helping to protect divers from contaminated water. All public safety teams can be exposed to contamination, even if their local water areas are clean. A vehicle immersion, for example, will contaminate the cleanest of lakes. A body also presents a potential biohazard. Full face masks that are positive-pressure masks also serve to keep the airway of an unconscious diver dry, which can make the difference between life and death. If contamination is a real concern, the diver may want to consider a positive-pressure mask. Otherwise, demand masks are preferable for ice diving, because they are less likely to free-flow.

There are several different full face masks available to the ice diver. When cared for properly, they all work well. If not fitted properly, though, they have a tendency to leak or free-flow. Full face masks normally fit over the top of the hood skirt; hence, hoods with a smooth rubber area along the outer skirt, where the mask meets the hood, usually are more forgiving to leakage. Proper fit and placement are extremely important, as are hood vents, because air from the mask can leak into the hood.

Divers require training by experienced, full face mask instructors and practice in warmer water before diving with full face masks under ice. The need for strong full face

Photo 5-27 Full face masks keep divers warmer than standard half masks and allow the use of communication systems. They also provide protection from contamination.

mask skills significantly increases for ice diving, where the problems of accidental mask dislodgment, cold water on the face, and freeze flows are more likely and have more potentially serious consequences than in warm-water, non-overhead diving.

Spare masks. Full face mask divers should carry spare masks in case they have to switch to a pony regulator, without a block. A situation in which a diver needs to switch to a pony bottle and use a spare mask can cause stress. To help prevent the elevation of stress, and possible evolvement into panic, use the easiest techniques possible. A neoprene mask strap is recommended for easy donning. A low-volume mask is also recommended, not only for easy carrying, but for easy clearing.

The best place to carry the mask is in a BCD pocket. If that is not available, the mask can be worn on the back of the diver's neck, but this is not as comfortable. Some drysuits have thigh pockets that may fit a low-volume mask.

Cutting tools

Entanglement is a serious concern, and the most common entanglement encountered while ice diving is with fishing line. Such entanglements can entrap divers on the bottom. Public safety divers in low- or no-visibility water are particularly at risk for entanglements, because they have to search by feeling the bottom with their bodies. Because the dangers of entanglement are so great, divers should be well trained and practiced in managing a variety of entanglement problems common to their local area. One important aspect of managing entanglement problems is having the right tools to deal with them.

One cutting tool is not enough for cold or low-visibility water. Cold hands can easily drop a cutting tool. Low visibility also increases the risk of dropping and losing tools. If the water is just cold, then two cutting tools may be sufficient, but add the possibility of no visibility and three cutting tools becomes the minimum. Bottom-dwelling divers should also make three cutting tools the minimum. These tools should be secured in the golden triangle area as described below.

At least one of the tools should be mounted on the harness, rather than the BCD. That way, if divers need to remove their BCDs to deal with entanglements, they will still have their cutting tools in a known, easy-to-reach location. Additional tools can be mounted on the BCD or other places in the chest area within easy reach. Cutting tools should not be worn on a leg, where they are at the farthest point from the diver's hands, and where the tool itself can cause entanglement. Because legs are what kick up the fishing line, a knife on the leg is a prime cause of entanglement. In addition, knives worn on legs can catch a purposely or accidentally ditched weightbelt.

Photo 5–28 Cutting tools on harness with fin taping. Mount at least one set of shears upside-down on the harness. Tape fin straps to reduce entanglement occurrences.

While most divers carry dive knives, as they were taught to do in their scuba courses, there is a better solution. Rather than carrying multiple knives, we strongly recommend that divers carry one knife and at least two pairs of shears. Paramedic shears are one of the most effective tools for handling entanglement problems, particularly if fishing line is involved. To cut monofilament and other types of line with a knife, the line must be looped-and-pulled and then cut with a pulling motion. Not only does that motion require two hands, but the loop in the line can easily slip from a diver's grasp, as fishing line becomes slimy after remaining only a brief time in water. Even worse, some materials, such as barbed wire or some new synthetic fishing lines, cannot be cut with a knife. Shears, however, can cut through wire, fishhooks, and quarters. They can chew up line and other entanglements with little problem and often require the use of only one hand. The other hand can be used to hold a fishhook in place to prevent it from being pulled farther into the diver's suit or equipment, for example.

Knives have additional drawbacks, especially in low-visibility or cold water. Divers can accidentally stab or cut themselves, another diver, a regulator hose or other equipment, or a drowning victim. Shears, on the other hand, have blunt tips and a protective flange. Even if you accidentally hit yourself or your equipment with the shears, no damage will be done. In low-visibility or cold water, knives should be dropped after use, to avoid possible injuries that can occur during re-sheathing. Backup divers should use only shears when managing a primary diver's entanglement problem, not knives.

If you use knives, use only blunt-tipped knives. If you already have knives with pointed tips, grind the tips to a rounded or flat point. That way, if you are forced to use a knife underwater, you are less likely to stab or puncture yourself, another diver, a victim, or equipment.

Besides shears, divers may also wish to carry additional cutting tools if entanglements in their areas include items that shears cannot cut. Sturdy wire cutters can be useful underwater.

Maintenance tip: *While most cutting tools are made of stainless steel, good steel with a cutting edge will rust, especially if exposed to salt water. To remove rust, simply rub it off with steel wool or a copper penny. Also, to prevent rust from forming, keep shears and knives coated with paraffin wax.*

Dive lights and flashers

Visibility underwater depends on surface conditions and turbidity (particles in the water). Wind can inhibit visibility in two ways. First, wind causes the water's surface to be choppy or wavy, which causes more sunlight to be reflected off the surface and less to penetrate the water. Secondly, wind-created wave movement can stir up the bottom and cause turbidity.

Ice-covered water is not directly affected by wind. If the ice is black (clear), divers may actually experience significantly better visibility under the ice than during warmer months. Thick ice with a foot or more of snow can cause a noticeable loss of light underwater. Dive lights come in handy on such dives as long as turbidity is not present. The best place for dive lights for solo divers is on a helmet or mask. This keeps both hands free for good blood circulation and to perform tasks.

Photo 5-29 If low visibility is due to loss of light, not turbidity, attach a flashlight to the mask. (Note: The drysuit inflator hose is too short.)

Caution: *Buddies with mask- or hood-lamps can temporarily blind each other if not careful.*

It is not necessary to hang a light or flasher in the hole, as divers are tethered, and therefore, always know where home is. Hanging lights can cause entanglement; and, in the case of a diver-disconnect, divers should not try to search for the hole! If the hole has a hanging light, a disconnected diver is more likely to search. Hanging lights have no real benefit and can actually do some harm.

Snorkels

Ice divers and public safety divers should not wear snorkels. In ice diving, because the diver enters and exits a single hole, there are no surface swims, so snorkels are not useful. When divers are on the surface, tenders assist them by taking tension on their tether lines to keep their heads out of the water; therefore, divers never need to switch to a snorkel.

Snorkels can actually be hazardous in ice and tethered diving. The snorkel can find its way around the diver's tether line and can cause the mask to dislodge. When an ice diver is near or against the ice roof, the snorkel can wedge between uneven ice, causing the mask to leak. Mask leaking and dislodgment under ice should be prevented as much as possible.

Weightbelts and harnesses

Recreational weightbelts should be set up so that when worn, there is 6in to 8in (15cm to 20cm) of extra webbing beyond the buckle. Public safety divers, who are often bottom-dwellers, are the most likely divers to have their belt buckles accidentally opened. Public safety divers should therefore have 10in to 11in (25cm to 28cm) of extra webbing, to allow them to catch the webbing before it pulls through the buckle. Short belts cannot be quickly caught or held if the buckle opens. Accidental weightbelt losses under ice can be disastrous. Metal buckles are preferred over plastic ones, because they grip the webbing better and are less likely to break. Expander buckles are always a good idea, to compensate for suit compression and re-expansion; and, if the diver is well trained, double buckles work well. Since ice diving is in an overhead environment, proper weightbelt placement is extremely important. Make sure the webbing is perfectly straight as it exits the buckle. Crooked webbing will prevent the buckle from fully closing. Make sure the webbing is correctly threaded in the buckle.

If a diver forgets to don the weightbelt before the BCD, or if the belt accidentally slips off, it is important to first remove the BCD before donning the belt. If the belt is put on after the BCD, there is a greater risk that it could slip off during the dive. Do not shortcut. Shortcuts can lead to serious problems, particularly in ice diving.

Harnesses reduce the chances of weights falling off divers with pear-shaped bodies or those with minimal hips. One advantage of a weight harness over a weightbelt is that the harness is worn less snugly than a belt. Subsequently, a harness allows air to move more freely between the torso and lower extremities. Weights need to be removable at the end of the dive before the diver exits the hole. Not all weight harnesses allow this, nor do all weight-integrated BCDs.

We do not advocate the use of BCD integrated weights.

- Unless an ice board is used to remove divers from the ice hole, weights should be ditched prior to exiting the water. This allows divers to practice that very important life-saving, self-rescue skill at the end of each dive.
- If divers need to ditch their BCD tank assembly underwater due to severe entanglement, they will shoot up like a torpedo if wearing integrated weights.
- BCD weight pockets give far less options on putting the weight where each individual diver needs it.
- Integrated weight is not visible, so dive leaders cannot see how much weight a diver is wearing. We find that integrated weight wearers are less likely to know how much weight they have on, and are less likely to make adjustments.
- Some integrated weight release mechanisms include webbing with a small plastic unit for the diver's hand to grip and pull. We have observed incidents where the plastic unit disconnected from the strap, leaving the plastic piece in the astonished diver's hand and the weights still in the BCD. We have also observed incidents where weights did not release because the diver did not thread them in properly or did not maintain the system properly. This is a serious concern. Weightbelt release systems are far more infallible.

Fins

With fins, fit is important. Tight fins can result in cold feet. When purchasing fins for a drysuit, try them on with winter booties and the drysuit boot. Fins should be matched to the strength and kicking ability of the diver. We prefer a heavy rubber fin that enables divers to use both the downward and upward stroke for propulsion. Furthermore, heavy rubber fins counteract the positive buoyancy of drysuit or wetsuit booties and allow straight-legged, effective kicks that will enable the use of advanced buoyancy techniques.

A common entanglement location is a fin strap, because legs and fins kick up fishing line and other entanglements. Tape fin buckles to prevent them from snagging fishing line. If fins need adjustment before each use, tape the outside buckle only, so the diver will have a left and right fin. Because the human leg bends inward at the knee, the inside edge is easy to reach to clear an entanglement. Attempting to reach an entanglement on an outside buckle, on the other hand, can result in leg cramps.

If fins are too loose, they can fall off. Gently tape the fin to the diver's foot or use fin straps to reduce the chances of loss.

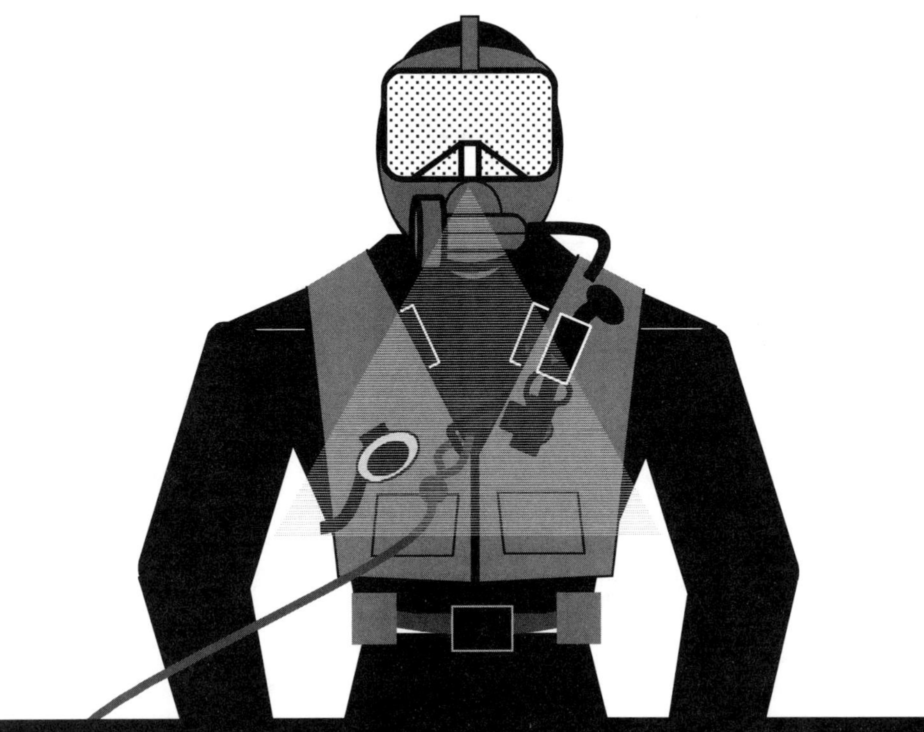

Figure 5–1 Golden triangle area. This imaginary boundary on a diver's body runs from the mouth to the bottom of the rib cage on each side of the diver.

Golden triangle area

All gauges, knives, shears, and pony regulator mouthpieces should be secured in the golden triangle, so named by Dr. Glen Egstrom.[22] The points of the triangle are the navel and the shoulders; it is golden because it provides simple access at all times. Equipment should never be allowed to dangle from the diver, as that will increase the risk of entanglement and inhibit buoyancy control. Place the submersible pressure gauge and drysuit inflator hose through the armpit between the BCD shoulder strap and the body under the arm.

Contingency strap

Each backup diver should have a contingency strap that snaps into the primary diver's tether line when the primary diver asks for assistance. The contingency strap enables contingency divers to descend to the primary diver without pulling on the primary diver's tether line. This is important because pulling on the line could increase the severity of an entanglement problem or could drive a fishhook deeper. The contingency strap also enables the contingency diver to have two hands free without losing the primary diver's tether line.

Photo 5-30 Contingency divers wear contingency straps to connect to the primary diver's line when they need to be deployed. The strap frees both hands without the possibility of losing the line, as well as several other important advantages.

The contingency strap should have enough length to allow some maneuverability around the primary diver, but not enough to allow it to reach the primary diver's throat should the contingency diver attempt to rise and reach behind the primary diver. The strap has a quick-release buckle in its center that enables the contingency diver to disconnect from the primary diver's tether line and work behind the primary diver. The carabiner on the contingency strap should feel different than the carabiner that attaches the tether line to the diving harness.

Diver Equipment List

- ❑ Equipment is standardized as much as possible
- ❑ Primary regulator with gauges
- ❑ BCD gauge clip
- ❑ Pony regulator
- ❑ Standard mask with neoprene strap
- ❑ Full face mask
- ❑ Surface Breathing Vent/Gill
- ❑ Redundant Supply Valve (Second Breath Block)
- ❑ Main Cylinder
- ❑ Pony Cylinder
- ❑ Pony quick-release system
- ❑ Pony neck strap
- ❑ Dry or wet suit
- ❑ Hood with vents
- ❑ Skull cap or Nomex™ hood
- ❑ Gloves/liners
- ❑ Undergarments
- ❑ Boots/liners/inserts
- ❑ BCD
- ❑ Weight belt with 10 inches extra webbing
- ❑ Fins/fin keeper/taped straps and buckles
- ❑ Ankle weights
- ❑ Diver harness with shears
- ❑ Locking carabiner/duct tape
- ❑ Two extra cutting tools
- ❑ Pre-measured and marked tether line in bag
- ❑ Wrist-mounted ice awls
- ❑ Helmet
- ❑ Dive lights
- ❑ Hard-wire communication system
- ❑ Contingency strap
- ❑ Paraffin wax and talcum powder

The contingency strap is also used to connect the primary diver to the backup diver should the primary diver's tether line need to be ditched because it is fouled. The backup diver attaches the contingency strap's carabiner to the primary diver's harness tether point and then cuts or disconnects the primary diver's tether line.

Tender Equipment

Tender personal protection equipment (PPE)

No one belongs on the ice without proper clothing, including:

- gloves
- eye protection from sun or snow blown from a chainsaw
- personal flotation device (PFD)
- proper footwear. Since public safety diving tenders need to be prepared to go into the water, they have additional equipment needs.

Exposure suits

If they expect to be immersed, public safety diving tenders should wear ice rescue suits or drysuits. A good option for tenders on stronger ice is flotation suits and coats. These will insulate well against heat loss, protect from wind, rain, and snow, and keep the wearer afloat should immersion occur.

Photo 5-31 Author in full-flotation suit that provides excellent protection from cold, wind, and accidental immersion, while being comfortable for long-term wearing.

Exposure suits should extend beyond the groin to insulate that area of significant heat loss and to prevent rain or snow from entering the pants. An attached hood is recommended to protect the neck and to keep snow and rain out of the jacket. Some hoods have built-in visors to help keep rain or snow out of the face. Make sure the pants are also wind and waterproof, and that they extend over bootlaces to help keep water out of the boots. Gaiters are good to wear if there are areas of deep snow or if pant legs are not long enough to cover the boots.

Photo 5-32 Tenders who test the ice should wear immersion-capable suits. Neoprene surface ice rescue suits may not be good for long-term wearing as rubber works better for insulation in water than out of water.

Mask and Snorkel

Snorkels can be useful for in-water tenders who hang vertically in the water at the surface and who deploy their divers under the ice roof toward the Victim-Point-of-Descent (VPOD). These tenders can wear a mask and snorkel to keep their face in the water in order to more easily monitor their diver's tether line.

Fins

Be sure to size fins to fit over exposure-suit boots. Tenders who plan to tend in the water need training in the proper use of fins.

Ankle weights

Ankle weights can be a great help for tenders who are tending in the water and are therefore wearing surface-ice rescue suits or drysuits. Heavy rubber fins can, however, often do the job of keeping the legs down and thereby eliminate the need for ankle weights.

Photo 5-33 Ice cleats should be worn by tenders at the hole as well as other surface personnel as is appropriate.

Ice cleats

Everyone going on the ice should have some sort of ice cleat on their feet if they intend on walking or doing anything that requires standing. Cleats typically affix to the bottom of the boot.

Divers who plan on walking around on the ice, especially those with wetsuit booties, may want to look into acquiring a pair of calf-length crab fisherman boots. Place a liner inside and fit the bottom with short screws. The boots provide traction and keep the feet warm. Most drysuit boots will not fit in these boots.

Recreational divers sometimes sprinkle dirt (not salt) on the ice. This provides reasonably free movement to and from the ice hole.

Ice picks or awls

Ice picks or awls are 4in to 6in (10cm to 15cm) hand-held units with a nail-like point on one end. They are for digging into the ice. Generally, picks and awls have a lanyard to secure them to their holster. Commercially made awls are positively-buoyant and may have a retractable head that, when pushed into the ice, exposes the nail-like point. Carry ice awls on a harness, ice rescue transport device, PFD, one wrist, or on both wrists. The last works best for divers. The diver's wrist tethers should be no more than 8in (20cm) in length to reduce the risk of entanglement. Surface support personnel should adjust the lanyards on their picks when they first get them. If using a single-wrist pouch, reset the lanyard so that one side is 6in to 8in (15cm to 20cm) long and the other takes up the rest of the cord. This allows for no wasted lanyard and better ease of use.

Photo 5-34 Wrist tools. Author wearing flotation coat with 2-way radio, stop watch, ice awls, and cutting tool on his wrist for easy access.

Photo 5-35 Divers and tenders use ice awls to get out of holes. Surface technicians can slide them to self-rescue-capable victims in holes, and divers use them to spider under the ice roof.

Modify wrist cases for instructors, divemasters, and command personnel by attaching a small waterproof radio and stopwatch.

The use of ice picks or awls in case a diver or tender has fallen through the ice cap makes them one of the most valuable tools surface support crews and divers can carry. In fact, everyone on the ice should have a set.

Gloves

Hands can lose heat not only in water, but in air as well. One study tested twenty subjects with bare hands in air at 60°F (16°C), 20°F (-7°C), 0°F (-18°C), and -20° F (-29°C). In attempting block-stringing, knot-tying, and screw-tightening, significant performance decrements were found in the lower ambient temperatures.[23]

Gloves should be mandatory for all divers and tenders. Rubberized gloves are a good idea with a pair of wool liners. Contamination is always a concern, and tenders are usually the least protected, yet most vulnerable personnel because they are constantly touching wet lines and equipment. Every sort of bottom sediment ends up on the tender, and tenders often remove their gloves to perform functions that they were not trained to do with gloves. EMS gloves as an underglove may not hurt; however, they have a tendency to cause the hands to get cold faster due to sweating and possible circulation restriction. Divers should not use EMS liners. Cold wet hands do not make for a safe dive operation. Double-layered wool mittens with liners work extremely well, even when soaked or ice covered. The mittens can be removed when fine mobile dexterity is required, and the liner will provide the tender some protection.

Photo 5-36 Balaclavas (head socks) protect the head's heat radiators: the ears, neck, nose, and mouth. Wool mittens keep hands warm even when soaking wet.A PFD coat with mid-layer garments protects the torso.

Footwear

Boots are necessary even on warm days. Wear proper winter, waterproof boots. Wool or polypropylene socks for divers and tenders provide the best protection on cold days. Layering is also a good idea. Boots should not be so tight as to cut off circulation or restrict breathing, and the toes should be able to move freely. Bringing a change of socks is a good plan for those whose feet perspire a lot.

Cold, wet feet do not make for a safe dive operation.

Headwear

It is no secret that a significant portion of our body heat can be lost through our heads. Be sure to cover up the heat radiators, namely the ears, mouth, nose, and neck areas. Fleece or wool balaclavas work very well. Headwear should be both waterproof and windproof.

Clothing under exposure suits

Clothing can be divided into three layers. Inner layers are the first items donned, middle layers are the next set of clothing, and the shell, or exposure suit, is worn on top. All clothing should be loose to enable layers of warmed air to add to the insulation of the clothing materials. Loose clothing also does not restrict movement or circulation. To decrease evaporative heat loss, make sure that inner-layer materials wick moisture away quickly. Wool, polypropylene, and silk are recommended for inner-layer materials. Do not wear cotton. Cotton has little insulation value when wet, and it takes a long time to dry. Some departments have a standard maximum of 30% cotton for any material worn under the exposure suit. This is true for socks and underwear as well as other layers.

Mid layers should be looser than inner layers because their primary purpose is to provide insulation, not wick away moisture. If tenders know they will be exerting themselves, at least some mid layers should be designed to allow water vapor from sweat to pass through to the shell layer. Wear at least two mid layers, so one can be removed to decrease perspiration if the weather warms up. Fleece vests work well as one of the mid layers because they provide insulation to the core without restricting arm movement.

Outdoor, army-navy, and ski stores usually have good options for clothing, and end-of-season sales often provide good prices. Ski shops may also have battery-operated sock heaters for tenders who have difficulty keeping their feet warm. If these are used, make sure they can be turned low enough to prevent perspiration and can be kept running for the entire duration that the tender is outside.

Eye and skin protection

Tenders should wear good eye protection to prevent sun glare off snow and ice from causing tearing eyes, or worse, retina damage. Tenders need to keep constant visual attention on their diver's line; therefore, having to look away from the sun's glare is not an option. If it is snowing or raining and there is a wind, ski goggles should be worn. Use defogging wipes or fluids on the insides of goggles to keep tenders from having to stop what they are doing to clean out the inside of their goggles to see. Lip balm and sunscreen should also be taken to ice dive sites and used, even if the day is cloudy, because harmful sunrays still come through the clouds. For the same reason, some type of eye protection should still be worn on cloudy days when eye pupils are more dilated and let in more rays. Even clear glass will provide a degree of eye protection.

Turnout gear

Turnout gear is a controversial item of tender equipment. It provides warmth and protection from wind, rain, and snow; it is readily available to fire personnel; and some fire departments mandate that personnel respond in it. There is a downside if the turnout

Photo 5–37 PFDs need to have enough flotation to keep afloat the turnout gear and tender.

Photo 5–38 Drown-proofing turnout gear training is recommended for any fire personnel who will respond to ice or water calls in turnout gear.

Photo 5–39 If there aren't enough PFDs with sufficient buoyancy, a large-capacity BCD can be used.

gear is in poor condition or the wearer does not have drown-proofing turnout gear training. In the latter case, accidental immersion can quickly become submersion and drowning. If turnout gear is worn by ice diving surface support personnel, the following requirements apply: the gear needs to be in good enough condition to hold trapped air; a PFD with a proven sufficient amount of flotation must be worn; ideally, the wearer should have drown-proofing turnout gear training.

Since most PFDs will not fit properly over turnout gear, they can be worn under the coat. This also allows the wearer to ditch the coat should that become necessary during accidental immersion.

Ice Equipment

Safety is a prime concern at all times. Cutting the hole is no occasion to take safety for granted or to forget proper ice cutting training.

Chainsaws

Based on personal experience over the past twenty years and on hundreds of ice holes, the authors find that STIHL™ is an excellent choice for chainsaws for use on ice. They work extremely well when wet and do not freeze up as often as many other brands. These chainsaws are for the most part an excellent weight for ice use, and when used properly, they start in freezing cold weather and after becoming wet. The 039 or equivalent sized saw or larger is usually the best, as smaller saws are a little too light and the bar width is a little small. Cuts made by the narrower blades have a tendency to re-freeze if they are not back-bladed or widened several times in extreme cold weather.

Remove all of the bar's lube oil to make it environmentally safe. This oil functions to keep the bar cool and lubricated. The cold water and melting ice will lube and cool the bar. Do not put gasoline in the bar's lube section to clean it; simply empty it, and run the saw for a few seconds. Whenever possible, make sure the chainsaw is at least 2in to 4in (5cm to 10cm)—6in (15cm) is preferable—longer than the depth of the ice. There are few things more annoying than arriving at a site to find that the chainsaw is shorter than the ice depth.

Always read the manufacturer's safety manual.

Ice auger

An ice auger is a piece of steel with a flat cutting end that is picked up and driven into the ice. The auger may seem like an obsolete and inefficient tool, but in the hands of an expert, it can be used to make holes faster than most people can make with chainsaws.

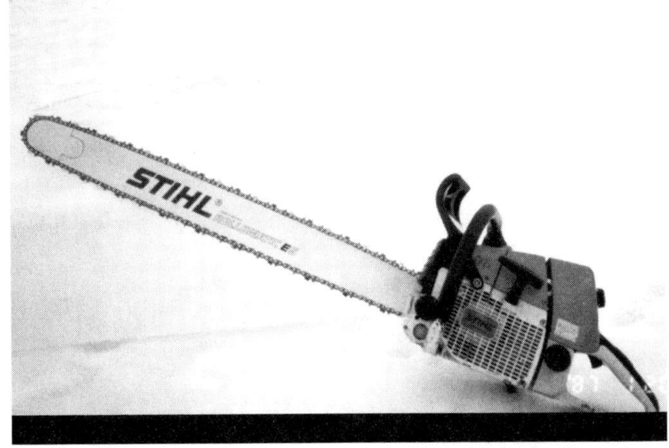

Photo 5-40 Whenever possible, use a chainsaw blade a few inches longer than the ice thickness.

Hand saw

There is an incredible hand saw made in Minnesota. Starting the hole is difficult because the blade is a little too flexible. An initial starting point hole is needed to get the blade in. Do this with an ice auger, chainsaw, or slam tool. Once the initial hole is made, the hand saw moves almost as quickly as a chainsaw, even through ice thicker than 18in. It is safer than a chainsaw and does not require much practice to be safe and effective. The hand saw cuts a hole wide enough so that a second or back cut is not necessary. Refreezing is not a big problem.

The saw is not weather dependent, because it cannot freeze up. It opens and closes like a giant jackknife, so it is safe to transport across the ice in the closed position. Use it at a slight angle of 35° to 45° with the handle toward you and use a smooth up and down motion. The hand saw is the tool of choice for safety and effectiveness. In the hands of an expert, a chainsaw will be as safe and even faster than the hand saw, but in the average person's hands this hand saw is preferable.

Photo 5-41 Do not skimp when purchasing ice screws. A good screw will screw into black ice with no effort and will hold in place all day in as little as two inches of ice with a heavily loaded ice board pulled against it dozens of times.

Ice screws

Features to look for in ice screws include:

- Parabolic-shaped threads for high pull-out strength
- Material made of chromium molybdenum
- Nickel-plating for durability and to minimize friction
- Hangers large enough to be used with mittens
- Plastic end-caps to keep screws sharp during storage and to protect other equipment stored in the same area as the screws

Do not use standard steel screws; they are too dull and will have to be banged in. Ice anchor kits that come with a good quality screw, webbing, and a black non-locking carabiner are useful. A good quality screw will easily enter ice with nothing more than finger strength.

Ground mats

Sections of military rubber ground mats work well. They are light, inexpensive, non-absorbent for the most part, great insulators from the cold, and they float. Small 5ft (1.5m) sections of mat work well for contingency divers to sit or lay on while the primary diver is diving.

Cocoa matting also serves as a good ground mat. It is made of rope and is very durable and excellent for ice diving. The bottom sticks to the ice, but the top remains pliable and graspable. At the end of the day, you peel it off the ice and hang it so the water drains out. It freeze-dries overnight, ready for use the next morning. The matting can also serve as carpeting in an ice shelter to keep temperatures from getting too cold.

Photo 5-42 Covering the ice with cocoa matting works well especially for long operations. Notice how divers keep regulators out of the water to prevent free-flows.

Blankets

Wool blankets are excellent, unless someone has a wool allergy. Space blankets are not bad for the wind, but do not provide insulation or warmth for a cold diver. When using a space blanket, put it on over the wool blanket. Fleece and wool blankets also work well when wet. Blankets should be wrapped around contingency divers' heads, upper body and hands, as well as feet.

Photo 5-43 Make sure contingency divers are insulated and comfortable. Notice the foam and blocks under his feet, and the tether-line bag supporting the weight of the cylinder.

Hole barrier

Barriers consist of four stakes that are approximately 5ft (1.5m) tall for each planned hole. Sharpen one end of each stake and place it in the ice surrounding the area where the hole is to be cut. String fire or police tape or other weatherproof, wide, international orange or yellow strips between the stakes at the end of the operation to make sure the hole is easily visible.

Photo 5-44 The MARSARS™ ice board is the most versatile transport device; it moves easily from ice, snow, openwater, and back onto ice. In addition to other advantages, it has low wind susceptibility, can serve as a backboard, and is the most effective surface ice rescue transport device.

Photo 5-45 The diver lies on the board while surface support puts his fins on to make him ready to go to the hole. When the line tender (right) pulls the line, the diver is moved to the hole.

Ice board

The most useful ice transport device is the ice board.[24] It is the most effective tool for standard surface ice rescue operations, for getting a diver out of an ice hole, and for transporting personnel and equipment back and forth between shore and holes. The ice board should meet the following criteria:

- Can be easily carried by one person across difficult shore areas.
- Is easily stored in a vehicle without needing assembly on the scene.
- A tender or technician can easily cross over snow, ice, open water, and back onto weak ice with the device.
- Is not very susceptible to wind.
- Makes an effective transport device for divers and tenders across weak ice as part of a pulley system operated by shore personnel.
- Two persons plus a technician can be transported to shore.
- Can serve as a backboard for patients with head, spine, or other injuries.
- Is not made of metal, which would conduct heat from riders, and has padded or rounded edges.

- Cannot injure anyone if it collapses or is not put together properly.
- Can be used to extract divers from ice holes.

Photo 5-46 Transport devices need to be able to cross over weak ice, snow, open water, and back onto ice.

If the users may be called upon to perform surface ice rescue operations, in addition to the above criteria, the following apply as well:

- Allows a technician to establish independent-victim-positive-buoyancy in less than ten seconds from the time of the technician contacting the victim.
- Aggressive or multiple victims cannot easily tip over the device, and if tipped over, can quickly be righted by one technician.
- Can be used for patients in statuses ranging from aggressive to no breathing or heartbeat.
- Removes patient horizontally and very gently from the water.
- Can transport severely injured victims.

If an ice board is not in the budget, use some other type of sled.

Photo 5-48 Surface ice rescue teams can save a victim 200ft (61m) from shore across very weak ice in less than six minutes from time-on-scene until the victim is back-to-shore if an ice board is used.

Photo 5-47 Ice boards can transport victims safely, gently, and horizontally to shore.

Photo 5-49 Bob Davis, creator of the ice board, ice rescue flotation sling, and other important ice-related tools, holds the MARSARS™ ice pole with a shepherd's crook attachment. Commercial poles offer the advantage of better durability and a variety of useful attachments.

Ice poles

Ice poles should be approximately 6ft to 7ft (2m) in length. Stair railing works well, but do not use pine since it splinters on hard ice. A nail should be counter-set into the nose of the pole and exposed 1in to 1 1/2in (2.5cm to 4cm). The nail will dig into the ice when the pole is thrust in a forward, downward motion. Ice poles have a variety of uses. The first personnel who step on the ice to test it or to prepare the ice diving area should bang the pole in front of them to make sure the pole does not go through and to listen to the sound of the ice. The pole serves as a quick testing device for a large area of ice. If the person punctures the ice, the pole can quickly be positioned horizontally to prevent full submergence. The pole can be laid across the edge of the ice hole and used as a self-extraction aid for getting out of the hole. The pole can be used to assist someone else out of a hole.[25] A pole can be held down into a weep hole during training programs to simulate a lost diver for a backup diver to find.

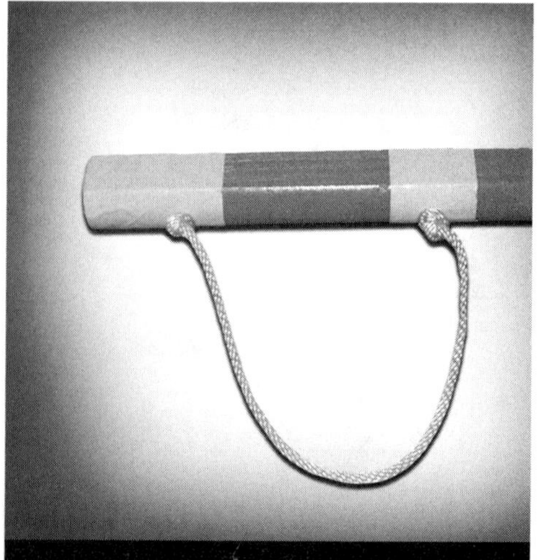

Photo 5-50 A stiff line loop is placed at one end to reach out and secure the wrist of a weak victim during surface ice rescue before the rescuer approaches close enough to possibly break the ice supporting the victim.

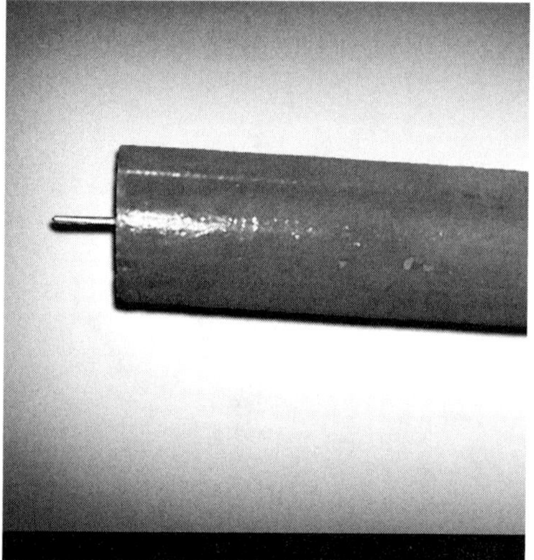

Photo 5-51 A nail with its head cut off is secured to the other end of the pole to provide stability for the technician who is crossing the ice and to make it easier to push the ice block under when setting up a diving hole.

Tender Equipment

- ❑ Appropriate headwear
- ❑ Eye and skin protection
- ❑ Exposure suit/drysuit
- ❑ Undergarments
- ❑ Appropriate gloves
- ❑ Appropriate footwear
- ❑ Personal flotation
- ❑ Ice cleats
- ❑ Ankle weights/fins & keepers
- ❑ Ice awls
- ❑ Time keeper
- ❑ Dive profile slate and pencil
- ❑ Diver-tender equipment checklist card
- ❑ Harness
- ❑ Locking carabiner
- ❑ Tether strap
- ❑ Ice pole
- ❑ Hole cutting tools
- ❑ Mask/snorkel
- ❑ Backpack for hard-wire communication system
- ❑ Communication system headset
- ❑ Ice screws, carabiners, and straps
- ❑ Ice transport device and pulley system

Ice platforms (boats)

Boats and other divable platforms can be extremely helpful on bad ice or during extended dive operations. Light rubber type boats (without wooden transoms) or inflatable platforms work best.

Inflatable boats and other transom-fitted vessels are hard to pull over the ice or through snow and ice. The transom drags across the ice, increasing the workload. If using an inflatable vessel, 12ft to 13ft (4m) long is best; anything larger is very difficult to move, especially if using a human piton. A small overland soft hull unit is great on the ice and can be moved by a couple of people.

Photo 5-52 An inflatable platform is secured to the weak ice and stays in one place during the operation. Equipment and personnel are transported between the platform and shore by an ice board.

Photo 5-53 Inflatable ramps come in lengths varying from 10ft to 50ft (3m to 15m). They can make useful ice platforms. Carry a small pail as they are not self-bailing. They are not as versatile year-round as are inflatable boats.

Photo 5-54 An ice board can be used as a one-person platform to spread weight out across thin ice, to provide insulation from the ice, or as an effective place to stage a 90%-ready diver. Notice the gauge hose running through the BCD armhole under the diver's arm.

Photo 5-55 Class-5 inflatable, self-bailing kayaks such as this Aire™ can be used as ice platforms and are effective swift-water rescue tools. They do not provide the carrying capacity of an inflatable boat, but are far less costly.

Photo 5-56 In very weak ice operations, a second inflatable (left) can be used across openwater as a transport device between the shore and the platform (right).

Metal vessels, duck boat types, V- and flat-bottom boats are not good for ice work. They can freeze to the ice in extreme weather conditions, do not balance well on ice, and are difficult to transport from ice to soft water to ice. In addition, metal increases the potential for hypothermia.

Safety equipment checklist

- ❏ First on scene duty checklist
- ❏ Fire/Police tape to delineate scene zones and demarcation line
- ❏ IC and safety officer vests
- ❏ Personnel accountability system (*e.g.*, photo ID certification operations or technician cards)
- ❏ Tarps
- ❏ Wind shelters
- ❏ Propane heaters
- ❏ Personal flotation devices
- ❏ Extra tender and diver PPE
- ❏ Ice cleats
- ❏ A fathom line for the first tender out, to check the depth of the water to assure safe dive operations
- ❏ Ice board or backboard/litter to transport a patient
- ❏ Field neurological diver evaluation slate
- ❏ Blankets
- ❏ Thermal recovery capsule
- ❏ Warm, non-caffeinated, non-alcoholic fluids and food
- ❏ Oxygen and first-aid kit
- ❏ Blood pressure cuff, stethoscope, and documentation paperwork
- ❏ Heat packs for warming divers, oxygen system, warming first and second stage regulators, and loosening locks on carabiners
- ❏ Sunscreen, lip balm, and eye protection
- ❏ Extra carabiners, webbing, and line
- ❏ Duct tape
- ❏ Radios or cellular phone to contact 911 for assistance
- ❏ Radios for personnel to communicate between shore and the ice (advisable, but not mandatory)
- ❏ Many extra batteries
- ❏ Flashlights and dive lights
- ❏ Chamois for drying off anything, especially the human body
- ❏ Extra gloves, hats, socks, and undergarments
- ❏ Dive slates and pencils
- ❏ Contingency bottle harness
- ❏ Contingency bottle (at least 72ft^3)
- ❏ Contingency pony bottle with regulator
- ❏ Marked lines and non-locking carabiner
- ❏ Bucket of warm water
- ❏ Save-a-dive kit and extra dive gear
- ❏ Four wood stakes and fire/PD tape
- ❏ Hole-cutting tools
- ❏ Ice screws and webbing
- ❏ Fire pipe poles
- ❏ Shovels
- ❏ Sand, cat litter, or salt
- ❏ Jumper cables

Summary Questions

1. Public safety divers or low-visibility, recreational divers should have _________ equipment so that divers in blackwater will be able to assist each other.
2. Sealing off the spring or piston chamber with silicone gel will help prevent the regulator first stage from _____________.
3. List at least eight ways to decrease the chances of regulator free-flow problems.
4. Give at least three reasons why an octopus is not an acceptable option for ice diving.
5. To manage out-of-air-entanglement emergencies, divers should wear ______-______ pony bottles.
6. A _____ main cylinder and a ____ pony cylinder should be on every dive site in case an entrapped/entangled diver is in need of air.
7. The _____ ______ should be removed from the second stage of the pony regulator, and a ______ _______ should be put on the pony regulator hose to enable divers to feel which hose it is in blackwater.
8. A _______ attached to the full face mask allows a diver to switch from the main cylinder to the pony cylinder without removing the mask.
9. List at least three items divers may want to carry in their BCD pockets.
10. A _____ backpack is preferable in a BCD over a ____ backpack because it holds the cylinder more securely in one place on the diver.
11. Divers should dress to be comfortable enough to be submerged for _____ as long as the planned dive time.
12. List at least four advantages of drysuits over wetsuits.
13. Drysuit divers do not need to wear buoyancy compensators. True or False?
14. An overly tight neck seal can cause a _____ _____ ______, which can result in unconsciousness.
15. A flooded drysuit will cause a properly weighted diver to sink. True or False?
16. Drysuit training and certification are required to dive dry. True of False?
17. List three disadvantages of EPDM drysuits.
18. List at least nine advantages of vulcanized rubber EPDM drysuits.
19. List three advantages of back drysuit zippers.
20. Cotton is a good material to wear under a drysuit. True or False?

21. _____ wetsuits between dives to decrease evaporative heat loss.

22. Neoprene drysuits or wetsuits lose insulation value with increased _____.

23. Why should divers avoid holding flashlights, wearing anything tight on their arms or legs, hanging vertically if drysuit diving, or wearing drysuit seals over their wrist bones?

24. List two functions of ankle weights.

25. To check for facemask proper fit, divers should inhale through their nose to see if the mask will stick on their face. True or False?

26. Full face masks can decrease facial heat loss, can protect divers from contaminants, and may help protect the airway of an unconscious diver. True or False?

27. An important part of full face mask training is for the diver to go to the _____ _____ instead of surfacing when unable to fully clear the mask.

28. Ice and public safety divers should carry a minimum of ____ cutting tools, with at least one of those tools being _____.

29. Cutting tools should be mounted in the _____ _______ area, not on the ______.

30. Snorkels should not be worn when ice diving. True or False?

31. Ice and public safety divers should have ______ inches of extra webbing beyond their weightbelt buckle.

32. Fin straps should be _____ to decrease the risk of entanglement.

33. A backup diver uses a ____________ ______ to attach to the primary diver's line when performing an assist.

34. List at least three functions for the item in question 33.

35. If the ice is potentially weak enough for accidental tender immersion, tenders should wear ____________ for full-body personal protection.

36. ______ ______ are worn over boots to decrease the chances of the wearer slipping on the ice.

37. List the three main functions of ice awls.

38. When choosing tender gloves, what are the three main issues to consider besides protection from cold?

39. List the four main heat radiators in the head and neck region.

40. When choosing a chainsaw to cut an ice hole, a bar width of 039 or __________ (larger, smaller) is recommended.

41. _____ blankets should be available to wrap around contingency divers stationed on the ice.

42. List at least six functions to look for in a surface ice transport device.

43. List at least three functions of an ice pole.

44. All personnel in the warm and hot zones should be wearing a ________ ________ ________ or the equivalent.

Notes

[1] A.J. Arntzen and D.M. Furevik, "Pilot Investigation to Prevent Freezing of Scuba Regulators by Use of Catalytic Hydrogen Combustion," Ytre Laksevag, Norway: Norwegian Underwater Technology Center (1982) NUTEC rep 40-82.

[2] U.S. Navy Manual: 11-2.1.

[3] J. Bozanic and J. Mastro, "Regulator Function in the Antarctic," (San Diego: AAUS Polar Diving Workshop [May 1991] eds. Lang and Stewart): 22.

[4] U.S. Navy Manual:11-2.9.1.

[5] W.R. Keatinge, "The Effect of Work and Clothing on the Maintenance of the Body Temperature in Water," *Quart J Exper Phsiol* (1961) 46: 69-82.

[6] M. Strauss and W. Vaughan, Jr., "Effects on Core Temperatures of Suited Divers Exposed to 6° Centigrated Water for 4 and 6 Hour Durations," *Undersea Biomed Res* (1978) 5 (Suppl): 31.

[7] Walt Hendrick, *Drysuit Flooding Video* (Delphin Productions 1983).

[8] V. DiGiglio, "Diving in Contaminated Water, Just How Safe are You?," *SORTIE* 1, issue 3 (1999).

[9] Available at no charge from Trelleborg Viking™.

[10] Trelleborg Viking™.

[11] P. Marcus and S. Richards, "Effect of Clothing Insulation beneath Immersion Coverall on the Rate of Body Cooling in Cold Water," *Aviat Space Environ Med* (1978) 49: 480-483.

[12] K. Tomiyasu, M. Nakamura, M. Yamada and T. Murai, "Study of Body Heat Loss and Heat Insulation or Heating on a Diver: Effects on Heat Insulation of Undergarment for Dry Diving Suit," *JAMSTEC* (1980) 5: 187-199.

[13] As per manufacturer such as Trelleborg Viking ™.

[14] J. Hayward, M. Collis and J. Eckerson, "Thermographic Evaluation of Relative Heat Loss Areas of Man during Cold Water Immersion," *Aerosp Med* (1973) 44: 708-711.

[15] S. Gorden and A. Pasche, "Thermal Evaluation of HH-E351 Survival Suit," Ytre Laksevag, Norway: Norwegian Underwater Technology Center (1983) NUTEC rep 43-83.

[16] R.R. Given, V. Pilmanis, B. Bassett and A. Pilmanis, "A Feasibility Study of a Temperate Shallow-Water Habitat off Santa Catalina Island, California, *Hydrolab J* (1975) 3(1): 115-124.

[17] M. Snyderman, "Parkway's New Encounter," *Skin Diver* (1983) 32(11): 47.

[18] D. Brennan and J. Weelihan, "Thermal Protection: Effect on Static Grip Strength at Various Depths," in Proceedings of the Eighth Conference of Underwater Education, San Diego NAUI (1976): 64-70.

[19] D.M. Fothergill, W.F. Taylor and D.E. Hude, "Physiologic and Perceptual Responses to Hypercarbia during Warm-and Cold-Water Immersion," *Undersea and Hyperbaric Medicine* (1998) 25: 1-12.

[20] B. Stinton, "Thermal and Hand Protection," AAUS Polar Diving Workshop, eds. Lang and Stewart (San Diego [May 1991]): 1-5.

[21] Personal correspondence.

[22] University of California at Los Angeles.

[23] J. Lockhart and H. Kiess, "Auxiliary Heating of the Hands during Cold Exposure and Manual Performance," *Hum Factors* (1971) 13(5): 457-465.

[24] Manufactured by MARSARS™.

[24] Walt Hendrick and Andrea Zaferes, *Surface Ice Rescue*, 1999

Chapter 6

Tethering

Why Tether?

The inability of divers to find their way back to their exit hole means certain death. Ice divers are essentially confined-space divers, with one way in and one way out. Along with the possibility of blackouts from stirred-up bottoms and entanglements left over from ice fishing, ice divers need to be tethered for the simple reason that they have to be able to get back to the hole. Polar ice divers often go untethered because of the tremendous visibility. The rest of us do not have that luxury. Move three feet from the hole and it could be as hard to find as a needle in a haystack.

The result of a backup or 90%-ready contingency diver not reaching an entrapped, severely entangled, or unconscious primary diver is also death. The ability to follow a diver's bubbles down to an untethered diver is a myth in non-overhead, zero visibility diving and even more ludicrous in ice diving. Even if that myth were a reality, the diver might no longer be breathing or sending up air bubbles. The contingency plan of sending a marker buoy up as a distress signal works almost as poorly. Clearly and simply, tethering provides direct line access between contingency divers and a primary diver in need.

This sounds so logical and seems like common sense, but consider the fatalities described in chapter 1. None of those divers understood the need for effective tethering. In fact, the double fatality that was related to a disconnect is thought to have occurred because the divers disconnected intentionally. It is believed that because the divers often dove in that location, they thought they knew the site well enough to come back to where they left the ends of their tether lines or perhaps to the hole. The double fatality with the instructor and student who thought they knew the area well and were compass experts again proves the necessity of proper tethering. The instructor-student double fatality in Pennsylvania demonstrates that proper tethering involves more than simply how the line is affixed to the diver. It also involves who holds the line topside and how much line is available. The instructor near-drowning incident demonstrates that if two divers enter a hole, they both need proper tethering. The incident also reinforces the importance of limited tether lengths and trained tenders. The divemaster who died going after his instructor friend was an untethered contingency diver looking for a lost untethered primary diver. Far too many ice divers have needlessly lost their lives due to poor, or nonexistent, tethering techniques.

Tethering is a common practice in public safety diving because of low to no visibility and entanglement risks. Besides giving direct line access between primary and contingency divers, tethering gives tenders the ability to direct divers in searching and to know exactly where the divers have and have not searched. Recreational divers rarely use tethering, however, since they typically dive mid-water and dive with some degree of visibility.

To make ice diving as safe as it is enjoyable, divers and tenders need to be tethered properly. The tethering techniques used by some public safety dive teams, such as hand-held loops or ropes around the waist, are less than safe for warm weather diving and completely unacceptable for ice diving. This chapter covers tethering techniques that meet ice diving safety standards and that also work extremely well for dive teams in warm weather diving. Whenever possible, choose techniques that can be used all year so that they can become reflexive.

When all the rules pertaining to tethered overhead environment diving are followed, the lost diver procedure becomes mostly academic. However, a disconnect is still a possibility, albeit remote, and therefore needs to be practiced in training. This topic will be addressed in chapter 9.

Rule number one of ice dive tethering is *never* disconnect a tether line under the ice, unless first attaching a new one in its place. In the most extreme entanglement situation, namely the possibility that the diver's primary tether line cannot be untangled, the back-up diver will attach the contingency strap to the primary diver's harness carabiner before disconnecting the entangled tether line.

Tethering Systems

There are two primary types of tethering for ice diving: (1) a solo diver down, or (2) a buddy-pair on a single, yoked line or on two separate lines. Public safety divers typically go solo because it is safer with the proper training and because the search is more accurate and covers more area in the same amount of time than when two divers are placed in at one time.

Recreational standards typically forbid solo diving. Even when permitted, most recreational divers prefer to share the experience with a friend. There are more skills for both tenders and divers to learn if two divers are put in the water simultaneously, because it involves more task loading. However, solo divers must also have confident and competent basic skills, and it is recommended that they have previous experience with tethered diving and the use of pony bottles, prior to engaging in solo ice diving. Recreational divers who are not familiar with blacked-out environments, solo-tethered/tender-directed diving, and the use of pony bottles, should make their first ice dives during training with an ice diving divemaster or instructor.

There are many questions for recreational divers to review before deciding whether to tend using a single tether line with a yoke for two divers or two separate lines. Keep in mind that every line that goes into the ice hole compounds the possibilities of entanglement. Tethered diving is not that common in recreational diving and therefore needs to be kept simple.

Two lines

As divers move through the water, they have to be aware of their tether lines and their buddy's location. It takes training to prevent tethered divers from crossing over each other's lines. Two lines entwined become one line that does not work well for either the divers or their tenders. This simple crossing of tether lines can have a dramatic effect on retrieving divers since one diver or tender will move more quickly than the other. Hence, each diver is pulling against the other. Another potential problem is that divers on separate lines may separate and not be close enough to communicate or help each other. Intertwined tether lines can also hinder communication between diver and tender.

With a two-line system, a tender is needed for each of the primary divers, and another tender is needed for the backup diver. This not only places more people on the ice, but also requires additional trained personnel. Even in the professional world, one tender cannot easily handle two tether lines and should not be asked to do so. Generally, tenders for recreational divers have very little time to train and become proficient in tending skills. The coordination of two tenders sending separate signals to two separate divers and then coordinating their signals with each other requires more training than one tender using a single yoke line. It involves keeping both divers at approximately the same distance from the hole so that they can help one another if they have to. The two-line system makes it more difficult to keep the divers in the same area if signals are misunderstood or if one diver has an ascent/descent problem and is unable to signal the other diver.

Buddy-lines are commonly used to prevent divers from separating, and to allow each diver to signal the other. The difficulty with this technique is that it is a yoke-type situation, but with each diver having restricted movement of one hand. This system does not enable a normal swimming profile and can restrict hand movement in an emergency. Furthermore, this technique does not prevent the divers from entangling their tether lines.

If the backup diver is needed, the situation requires a third tender, a third set of lines in the ice hole, and a third diver needing to be controlled and directed underwater. This entails a third person trying to communicate during an emergency. This scenario requires extensive training and coordination of surface crew and divers alike. If the divers become separated and both need assistance, how many backups will be available?

Tenders need to stay at least 5ft to 6ft (1.5m to 2m) apart when working two separate lines. Divers need to coordinate so that diver "A" will always remain on the right side of diver "B". These two procedures will reduce the chance of lines intertwining. One diver has to be the leader to direct the pair's movement and to give line signals to the tenders should the two tether lines entangle.

Yoke

The yoke line system is one line coming to a "Y" at the divers' end. If the line will allow splicing, splice the "Y" and affix it with a figure-eight knot. In lines where splicing is not possible, a pair of figure-eight knots approximately 3in to 4in (8cm to 10cm) apart connects the two lines. The physical "Y" should be approximately 6ft to 10ft (2m to 3m) long on each leg. This allows the two divers ample room for freedom of movement, yet maintains a reasonable buddy system and joint communication capability.

Designate one diver as the primary who returns the majority of signals. Either diver can send emergency signals to the surface or to the buddy. Both divers can receive all signals as long as they both keep their yoke lines taut.

The yoke system restricts the chances of the dive buddies being too far apart and greatly reduces the risk of the divers entangling their lines. At worst, their yoke lines could be entwined together.

Yoke diving reduces the number of surface crew needed and the number of lines entering the hole, and makes tending and line signal coordination much simpler. With this system, only two tenders need to communicate in case of an emergency. The overall tender and diver training process is much simpler.

The primary disadvantage of the yoke system occurs when one of the divers has a problem that prevents him/her from surfacing (*e.g.*, entanglement, entrapment, and unconsciousness) and the buddy cannot solve the problem alone. The buddy has to have a contingency strap and the training to be able to move to the main line, snap the contingency strap onto it, disconnect from the yoke line, and go back to the surface with the contingency strap moving up the main line. The diver can then get what is needed and return to the diver in need, or can have the backup diver make the assist. This technique takes training. When the two-tether line system is used, there is strong probability that the lines will be entangled into one, anyway, when a problem occurs.

When a backup diver goes down to assist a buddy, whether the buddy is yoked or attached to a two-tether system that is entwined, the backup diver faces more potential problems than if descending to assist a solo diver. Imagine two divers with their lines or yokes entangled who are now stuck on the bottom because one or both divers is entangled. They have kicked up two times more silt than a single diver would have kicked up in their efforts to help or get out of each other's way. The water is now black. The backup diver has to determine which diver is in trouble and how to work around that diver without becoming entangled with the second diver. What if both divers need help! Contingency divers need more training and skills to rescue one of two divers diving simultaneously than a backup needs to rescue a solo-tethered diver.

Lastly, with solo diving, it is very feasible to have two divers up for each diver down. This is a standard which is required by OSHA. To meet this standard would require a buddy system for divers topside.

Tethering Equipment

There are five basic components of hard tethering equipment:

1. Harness
2. Locking carabiner
3. Duct tape
4. Tether line
5. A tool to secure the dry end of the tether line to the ice or shore

Photo 6-1 Ice diving tethering requires a harness, locking carabiner, duct tape, tether line, and an ice screw with a strap and carabiner.

Harness

A proper diving harness is required for safe ice diving. The angle of the diver to the tether line is critical for keeping the line taut. Visualize a tender holding a line connected to a diver who is laying on the bottom searching with his arms out in front of him. In your mind's eye see the diver with his body perpendicular to the tether line. The question becomes how can he keep the line taut in this position? How can he move sideways, away from the tender to keep the line taut? Moving sideways enough to keep a line taut is not easy, and can certainly hinder search efficiency.

There is a significantly easier and more efficient method for divers to keep their lines taut. All a diver has to do is position perpendicularly to the tether line, and then rotate their body 45° in a direction away from the line, away from the tender. The diver's head and body angle will then be 135° away from the tether line. This angle allows the diver to simultaneous pull the line taut and progress forward with little effort. With good instruction and four or five training dives this position will become second nature and divers will have to do nothing but concentrate on searching.

Because it is such a mouthful to describe this position, we find it simpler to say, "divers are positioned 45° away from their tenders," which is how the position will be referred to in the rest of this book. We have found that most students understand this description better than the description of "divers are positioned 135° away from their tenders."

To understand why the harness is essential, consider the other options. Using hand loops or holding the line are not acceptable for ice diving for several reasons. Holding a line or loop reduces circulation in the hand and makes it become cold faster, thereby reducing the diver's ability to maintain a hold on the loop. When this happens, an accidental disconnect can easily occur. Even without the hand becoming cold, a diver can accidentally drop the line. This is most likely to happen when the diver is having a problem and has the greatest need to maintain direct line access to contingency divers. Divers who hold onto hand loops essentially eliminate the use of one hand. This cuts searching capability in half for public safety diving and, more importantly, leaves the diver with only one hand to manage problems. In an emergency situation, the tender needs to take gentle tension on the diver's tether line, which further locks the hand that is holding the loop. This decreases self-rescue capability and increases the risk of an accidental disconnect. By the standards used in this book, a diver who is merely holding a line or a loop in the end of the line is *not* considered to be tethered.

A procedure that is commonly taught involves tethering into a harness, adding a loop in the line one or more feet from the harness, and then holding that loop during the dive. Although the harness solves the accidental disconnect problem, holding a loop still leaves the divers with only one hand free. The loop becomes a security blanket, and divers are likely to maintain a grip on the loop even when they need both hands free to search or save themselves. The loop, when not held, becomes a needless entanglement risk. Divers follow this procedure because they do not know how to properly dive a taut line. They feel the need to hold a hand loop in order to receive and give line signals to their tenders and to know where the line is.

There are several other less dangerous, but nevertheless important, aspects to using hand-held lines or loops that make it not a good tethering technique, even if a harness is worn. To perform an accurate and thorough search while holding a line or loop, divers must maintain their hands in the same position in relation to their chests throughout each search sweep. When making a turn, however, they will have to switch the line to their other hand. If divers change their hand position by as little as 6in (15cm) on each sweep or turn, after six sweeps, the pattern can be off by as much as three feet.

Hand loops in the line, whether attached to a harness or not, typically result in a slacker line than a harness-tethered line. In order to maintain a line taut enough for both the diver and tender to feel and recognize a tether-line snag or entanglement within a few seconds, the diver must be positioned 45° away from the tender. To maintain an accurate pattern, hand-loop holding divers must maintain their arm either continually outstretched behind them or bent, with their hand holding the loop against their chest. For divers to keep their hands in either of these two positions continuously while keeping the line taut is difficult and exhausting. Once there is slack line between the diver's hand and chest, the diver is more likely to swim in on himself, thereby creating slack line. Slack lines lead to poor patterns, increased risk of entanglement, and loss of line signal communication.

We have worked with many hundreds of divers who were taught to hold hand loops while using harnesses and who believe they are keeping taut lines. They soon discover that, because their lines are only marked every 10ft (3m) or not at all, their tenders do not realize that they are automatically taking up the diver-created slack. Because profile maps of the diver's every movement are not made (*e.g.*, left sweep at 37.5ft, right sweep at 35ft etc.), no one realizes that the tender is taking up slack. Everyone assumes the divers are keeping their lines taut.

Some divers tie a line around their waist. This puts the tether attachment-point too low and forces the diver into something of a rappelling position that is more vertical than horizontal. Because the line is not held in place around the waist, it can shift, with the result that the tether point can end up behind the diver, caught under tanks, and out of reach. Keeping the line taut, which requires a continual swimming away from the line, can keep constant pressure on the kidneys, diaphragm, and other abdominal organs. A properly harnessed line will keep the pressure on the ribs and away from the abdominal organs. A waist-line may slip over the weightbelt and cause accidental belt loss or the inability to ditch the belt.

Another common tether point is a D-ring placed on the BCD or weightbelt. BCD attachment points are typically found on a shoulder strap, which is a poor tether point for several reasons. In order to keep the line taut, the BCD will continually be pulled and shifted around the diver's body as the diver swims away from the line and changes directions. BCD designers who put tether points on BCDs never intended those lines to be kept taut enough for accurate patterns or effective line signaling. To prevent accidental weightbelt loss or an inability to drop a belt that has shifted so the buckle is out of reach, nothing should be attached to the weightbelt. A weightbelt tether point is far too low. D-rings attached to the tank band are a poor choice as well.

A harness without a hand loop leaves both hands free for searching, handling a recovered body or evidence, working on the surface, managing problems, and self-rescuing. A harness provides direct line access to the hole, tender, and contingency divers even if the diver loses consciousness. Harnessed tether points produce the most accurate search patterns because the tether point is always on the same place on the diver's body throughout the dive and is on the same place on every diver on the team. Harnessed divers can be assisted into and out of ice holes, and up or down steep embankments. Harnessed divers will not experience fatigue, cold stress, and muscle soreness from attempting to maintain a taut line by hand-holding it; rather, they can use their entire bodies to gently swim away from the line at a 45° angle from the tender. A properly fitting harness and line will not place pressure on the abdominal organs, will not damage the BCD, and will not affect the weightbelt.

What to look for in a harness

- Full body harnesses are not required, do not provide necessary benefits, may restrict leg movement and circulation, and are more costly. Divers wearing properly fitted, well-designed diving chest harnesses under their BCDs will not come out of those harnesses. If divers need leg straps on the harness because of a very strong current, they most likely should not be performing the dive.
- The harness girth strap and tether point should sit across the solar plexus region, where the body is more muscular and the rib cage is strong. This prevents

diaphragm and other abdominal organ compression. Do not wear a harness over the lower rib cage or abdomen area, as this can restrict breathing. The harness should keep the diver in a comfortable and effective horizontal position regardless of the angle of tending. Rappelling harnesses are not recommended for water use because their tether point is too low.

- Harness webbing should be slightly stiffer than vehicle seatbelt webbing. Soft webbing has a tendency to twist and to dig between the ribs when there is strong tension on the line. The harness does not need to be as stiff as some commercial diving harnesses designed to carry bail-out bottles. Stiff commercial harnesses are expensive and may be less comfortable than a typical recreational or public safety dive harness.
- Harnesses should come in at least four different sizes, because proper fit is essential to keep the tether point stationary and in the correct location and to prevent restriction.
- The harness should have size adjustment capability on each shoulder and in the girth strap so the wearer can easily adjust for different exposure suits or different drysuit undergarments.
- The back of the harness should have an "X" design, not a "Y" design. The "Y" design hits the back of the neck and can put pressure on the cervical spine and cause discomfort.
- Because wearing the correct harness size is so important, we suggest for training programs and public safety dive teams that sizes be distinguished by colors that are easily seen, even from a distance.
- The closing mechanism on the harnesses should be more than just a strap through double D-rings. Velcro™ added to the end of the strap will provide more insurance that the harness will not open accidentally.
- Harnesses should have both front and back D-ring tether points, so that both divers and tenders can wear them.

Photo 6-2 An "X" design is more comfortable and helps keep the front tether point in one place as the diver keeps a taut line.

- The harness's construction should be such that the tether point does not move when the diver changes direction while keeping the line taut. Some harnesses allow the tether point to shift up under divers' armpits as they try to keep taut lines. Not only is this uncomfortable; it also hinders the diver's ability to keep the lines taut and decreases the accuracy of search patterns.

Sizing a harness

Have tenders don the harnesses on their fully suited-up divers without closing the girth strap. Each diver should then place a flat hand over their solar plexus region. The tender should then close the girth strap snugly over the diver's hand while asking the diver to take and hold the largest inhalation possible. This will ensure that no matter how hard the diver is breathing, the girth strap will not become restrictive. The girth strap should run across the solar plexus. If it is lower, the diver will be forced into more of a vertical position and may have pressure exerted on the diaphragm or other organs. If the strap sits higher, it may cut off circulation under the arms. Lengthen or shorten the shoulder straps to allow proper girth strap placement.

Attachment points

As stated above, the attachment point to the harness is in the solar plexus area. This location keeps the carabiner and line free from the diver's BCD cummerbund and weightbelt. When the diver is lying horizontally in the proper swim position to keep the line taut (45° angle away from the tender), the tether line will lay from the attachment point along the diver's side and down past the hip. This gives the diver the ability to reach back with an arm and reach the line easily when desired. If the attachment point is lower, less line will run along the diver's body, and the diver will have reduced ability to "feel" the line while searching. A lower position will make it more difficult for the diver to reach the line.

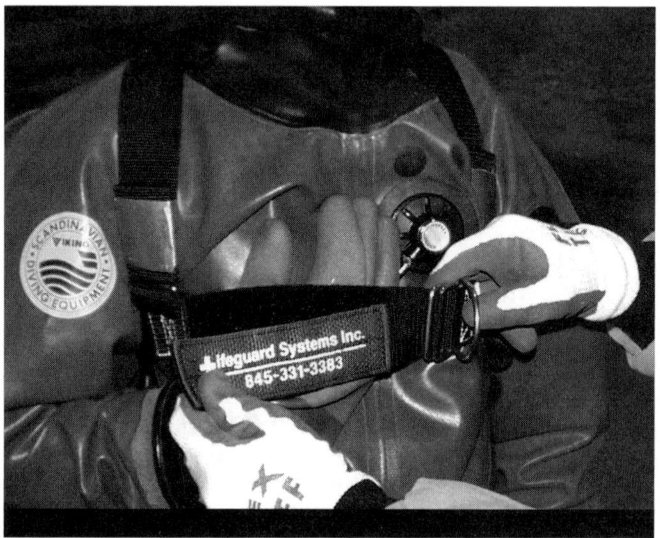

Photo 6–3 Tenders always close the harness. The diver places a hand over the solar plexus while holding in a large inhalation as the tender snugs the girth strap closed.

Use locking carabiners to connect the attachment points and the diver. After gently tightening the locking swivel closed, wrap the lock in the locking direction with a quick twist of duct tape, and finish taping with a slight back flip of the tape for ease of removal after the dive. This will prevent line abrasion against the rough texture of the lock, will prevent the line from rubbing the lock open and tripping the carabiner open, and will prevent the lock from icing up and freezing closed. If duct tape is not used, and the lock should freeze closed, have hand-sized re-usable heat packs available to thaw it open. Do not allow tenders to remove their gloves and use their hands to melt the ice on the lock.

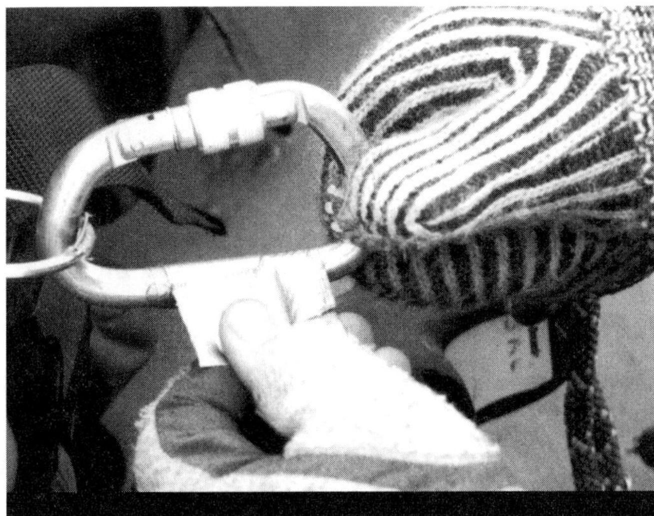

Photo 6–4 Keep a piece of duct tape with an easily-removable tail on each harness carabiner so tenders at the hole can properly secure their divers into their dive lines. Notice the glove liners this tender wears under his mittens.

Non-locking carabiners, speed swivel carabiners, plastic hooks, and other types of non-locking snaps and hooks are unacceptable. Black carabiners, on sunny days, are thought less likely to freeze than silver or gold ones, so they would be preferred for use on the ice or shore.

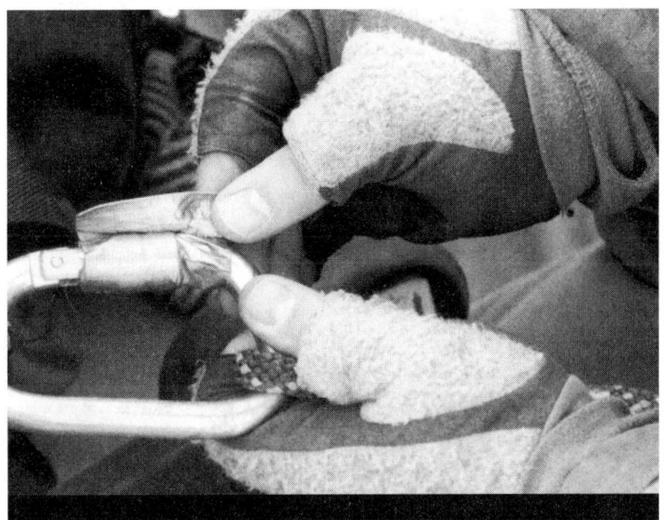

Photo 6–5 Wrap duct tape in the direction of the carabiner locking mechanism and leave a tape tail for easy removal.

The most common method for attaching the line to the carabiner is to make a figure-eight knot or a rescue-eight with approximately 4in (10cm) of tail secured with duct tape.

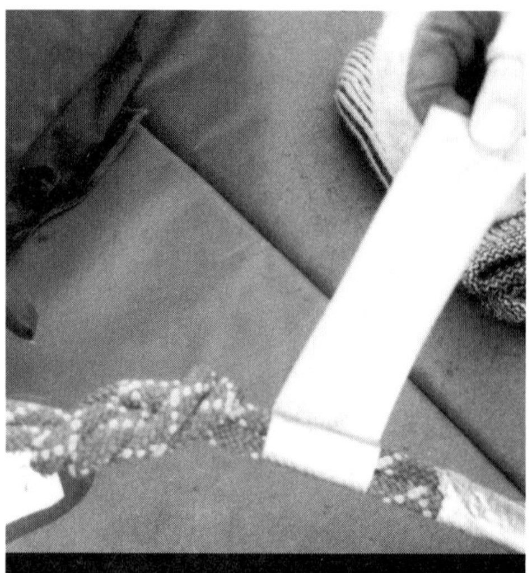

Photo 6–6 The hardwire communication line is secured to the carabiner with a figure-eight knot with the running end duct-taped back to the main line.

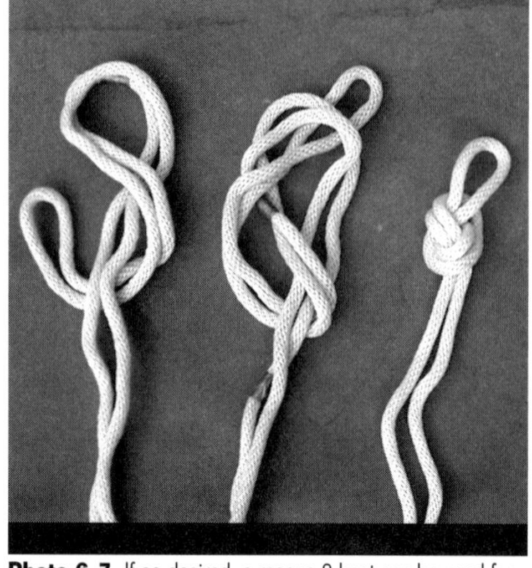

Photo 6–7 If so desired, a rescue-8 knot can be used for the diver's tether line to provide two loops to put the harness carabiner through.

Photo 6–8 If the hole is more than 400ft (122m) from shore, it may be more efficient to stage line tenders on the ice between the shore and the hole to assist with the transport device lines. Each of these tenders wears a flotation suit and is secured with webbing to an ice screw.

Photo 6–8a The divers and tenders will snap their tethers into the forward ice screw. These tethers are connected to shore by the ice board pulley system that is secured to the second screw in the mariner's hitch. All tenders and divers are thereby tethered back to shore.

All tether lines should have a properly made loop at the tender's end so that it can be secured to the ice with an ice screw.

Tether lines

Except for rigging, ice diving is not much different from most other tethered diving situations. Unless the divers are working in moving water or ice-flow situations, ice diving does not require line rope in diameters larger than 3/8in (10mm). Standard tether line, such as multi-filament polypropylene braid, will work well. Line of 3/8in (10mm) diameter is preferred. Thinner line is too thin to feel and work with gloves and is more difficult to manage during entanglements, and is less strong. Tether line, such as 1/2in line, has reduced line signal capabilities beyond 100ft (30m). The increased drag created by the added size and weight of the line makes keeping the line taut much more difficult and noticeably increases diver fatigue. Larger diameter lines create more drag and a greater workload that is similar to the effects of tying knots or loops in the line. Thicker diameters increase the chance of diver back-feed (returning on his own axis), because the line can bow from the drag. When the line bows, the search pattern becomes inaccurate, and the risk of entanglement increases. Heavy perlons are not usually used due to their negative weight and drag. Their use is really overkill, since there is no real static load in tethered diving. The greater concern in regard to lines is poor tender training and compromising direct line access between divers and tenders.

Shore-to-hole lines

All divers and tenders are ultimately tethered to shore. The ice screws that secure their tether lines or straps are connected to shore with lines. An example would be the line used in the ice transport device pulley system.

If the holes are more than 300ft (91m) from shore, line reels are recommended as long as they have a braking system to prevent the reel from out-spinning the line as it comes off the reel. Hollow-braid polypropylene line is preferred because it is less likely to freeze. Periodically, snap the line down hard on the ice to expel water from the line. Less water means less ice in the line. Ice severely decreases

Photo 6-9 The best way to store tether lines is in bags. Every diver goes out with a line bag. Used lines are taken back to shore to thaw if frozen. The longer transport device lines are stored on reels. These reels do not have brakes, so care must be taken to control their spin speed.

Photo 6–10 The rescue tether reel has a clutch cover to slow the reel down as line is pulled out. Made of plastic, it is lightweight and is less likely to freeze than metal reels. A handle on the other side allows rapid line retrieval.

line flexibility and increases line weight. Hollow-braid polypropylene line will remain the lightest line in ice operations. This becomes increasingly important as operations move 400ft (122m) to 600ft (183m) or more from shore.

Line material

There are a variety of materials available when selecting lines for use in ice diving. A little knowledge of rope will help divers and teams make the best choices. A rope becomes a "line" when it has been cut to a particular length, has one or more knots in it, or is put to a particular use.

There are two main types of rope structure: three-strand laid line and braided line. The three-strand line is made by a process called "laying up". Laying up consists of twisting together a number of fibers in opposite directions to form a yarn. The yarns are then twisted together to form a strand, and typically, three or more strands are twisted together to make a finished rope. Laying up produces a stiff rope suitable for heavy-duty work that will stretch under load. Laid lines are not a good choice for tending lines because they are too rigid and therefore difficult to store and work with, and they have too much stretch.

The preferred type of tending line is made by braiding, rather than twisting, the yarn. A braided rope is constructed of a braided sheath that covers a braided core. Braiding produces a more flexible rope that is then pre-stretched and will not stretch any further. Braided line fits easily into deployment bags, which are the most efficient carriers for diver tether lines for both ice and warmer weather diving.

If the fibers run the full length of the rope, the rope is described as a filament construction.[1] If, on the other hand, the rope is made of short pieces of fibers spun together, it is called spun rope. You can see the difference between spun and filament rope. Spun rope is softer and "fuzzier". Spun rope is more elastic than filament and therefore less recommended as an ice diving tether line.

Not only should diving line be flexible, it also should float. Polypropylene ("Polypro"), which is a polyolefin fiber, "is the only fiber that floats well."[2] Polypropylene has excellent abrasion resistance and a high resistance to sunlight and hydrocarbons, such as those found in fuels. It is also very cost effective.

Polyethylene fibers float, but they are relatively weak, stretch easily, do not hold knots well, and are difficult to work with, especially with gloves on. They make very inexpensive lines, but do not be tempted when a price for tether lines looks exceedingly good. Polyester lines are strong and resistant to mechanical and chemical forces, but they do not float. Nylon, which is made of polyamide fibers, is very strong and does not rot easily, but nylon also does not float. Fishing line is made of nylon, and every diver has seen or felt it lying on the bottom. Hemp lines are unacceptable because dry rot is inevitable and will run rampant throughout the line.

Photo 6-11 Author Hendrick demonstrates knot tying in flexible, 3/8in (10mm), multi-filament, polypropylene, duct-taped, distance-marked tether line. Fall is a good time to develop or reinforce rope skills. Once knots can be tied completely with closed eyes, add winter gloves.

Kernmantle rescue rope is the best choice when rope will be used as a lifeline, as in climbing and rappelling. It is constructed by taking a kern (a high-strength inner core) and covering it with a mantle (an outer braided sheath). Kernmantle rescue rope is often heavy and expensive and, therefore, less than optimal for diving tether lines.

To date, there is only one case history in which an ice diver disconnect was possibly due to a line breaking. The requirements for high- or even low-angle rescue lines should not be used for diving lines—the two jobs are completely different.

Ice diving does not normally involve dynamic tension load and therefore does not require high-load tensile strength ropes. Lines with a tensile strength of 1,200lb to 1,600lb (544kg to 726kg) are sufficient. When purchasing diving tether lines, look for: flexibility, flotation capability, resistance to water and hydrocarbon absorption, and a 3/8in (10mm) diameter with 1,200lb to 1,600lb (544g to 726g) of tensile strength. With proper maintenance, a 3/8in (10mm) multifilament, double-braided polypropylene line will work exceedingly well for years.

Line length

In non-overhead environment, tethered diving situations, it is possible to work with a maximum line length of 150ft (462m) for contingency divers and a recommended maximum of 125ft (38m) for experienced primary divers. The reasons for a 150ft (46m) non-overhead diving maximum distance are as follows:

- 150ft (46m) is a long way from help. Backup divers should be able to reach their primary divers in less than ninety seconds.
- A signal does not travel farther than 150ft (46m) through a 3/8in (10mm) line, even in the best of conditions.
- To move a signal greater than 125ft (38m), there must be more motion on the line, which could force the diver out of the search pattern.
- The additional length and weight of the longer line causes a bowed, slack line that increases the chances of diver backfeed on their own line. Attempting to keep a taut line over 125ft (38m) can be exhausting for the diver.

Ice tether lines should have shorter maximums than lines used in warmer weather dives for several reasons. A diver in non-overhead water who is 100ft (30m) from the tender in 30ft (9m) of water is only 30ft (9m) from the surface, but that same diver in the same location under the ice is 100ft (30m) from the surface. For ice divers, the length of their tether lines dictates how far they are from air and the surface. In addition, ice divers are far more likely to experience free-flowing regulators than warm-water divers; therefore, out-of-air emergencies are more likely to occur.

The maximum recommended length for experienced primary divers with at least fifteen logged ice dives is 100ft (30m), which makes the maximum length for contingency divers 125ft to150ft (38m to 46m). Note that this extra 25ft to 50ft is sufficient for divers with taut lines because they will feel a disconnect in seconds and thus will not travel far from their last known point. This also assumes no current.

Do not confuse maximum line length with maximum depth. Maximum line length does not mean that divers should go to 100ft (30m) of depth. Rather, 50ft (15m) plus a possible 10ft (3m) extension for experienced ice divers, is a good maximum limit.

Line marking

In tethered diving, leave nothing to chance. In executing a tethered dive, it is important to know exactly where the diver is as much of the time as possible. Using marked lines allows the exact distance and general location of the diver to be known. Diver location is especially important in public safety diving, because the smallest distance mistake could make the difference between recovering and not recovering the object of the diver's search. With a marked line, 65ft is 65ft (20m).

All tether lines should begin at the diver, with a rescue-eight and at least 4 inches (10cm) of tail taped back. Markings should be every 5ft (1.5m), with special recognition at 25ft (8m) intervals. The line in the bag is normally only 150ft (46m) long.

Lines can be bought with the numbers laced into them. If the line is not pre-marked, add markings with duct tape. Color-coding can be used in addition to distance marks on the lines. Use different colored tape for each 25ft (8m) section, creating dynamic visual markings, as well as numbers on the lines. For example, mark the first 25ft (8m) section in green, the second blue, the third black, the fourth yellow, and the last red. No matter what color system you apply, mark and color-code all lines exactly the same way.

A simple line-marking system that has been used for more than twenty years by thousands of divers consists of the following steps:

1. Tie a figure-eight knot with a 3in (8cm) loop in the bitter end (the end of the line).
2. Duct tape the excess bitter end back to the main running line.
3. From the top of the loop, measure 5ft (1.5m) back along the line.
4. At the 5ft (1.5m) point, place a 1/2in (1cm) strip of duct tape around the line. Wrap the duct tape around itself tightly three or four times.
5. Repeat this process for the next 5ft (1.5m) section and continue every 5ft (1.5m) to the 25ft (8m) point.
6. At the 25ft (8m) point, place a 1in to 1.5in (3cm) piece of duct tape past (farther from the tender) the 1/2in (1cm) piece to signify the first 25ft (8m) section.
7. Continue marking every 5ft (1.5m) with a 1/2in (1cm) piece of tape to the 50ft, 75ft, 100ft, and 125ft points (15m, 23m, 30m, 38m).
8. At the 50ft (15m) point, place two narrow pieces of tape on the bag (tender) side and one wide piece on the diver's side.
9. At the 75ft (23m) point, place three narrow pieces of tape on the bag side and one wide piece on the diver's side.
10. At the 100ft (30m) point, place two wide pieces of tape and no narrow pieces.
11. At the 125ft (38m) point, place two wide pieces of tape on the bag side and one narrow piece on the diver's side.

The idea behind the taping system is that when the tender looks at the line, he sees the shorter tape first and reads that the diver is on the short side of 100ft (30m). When the tender sees the longer tape first, he knows the diver is on the longer side of 100ft (30m).

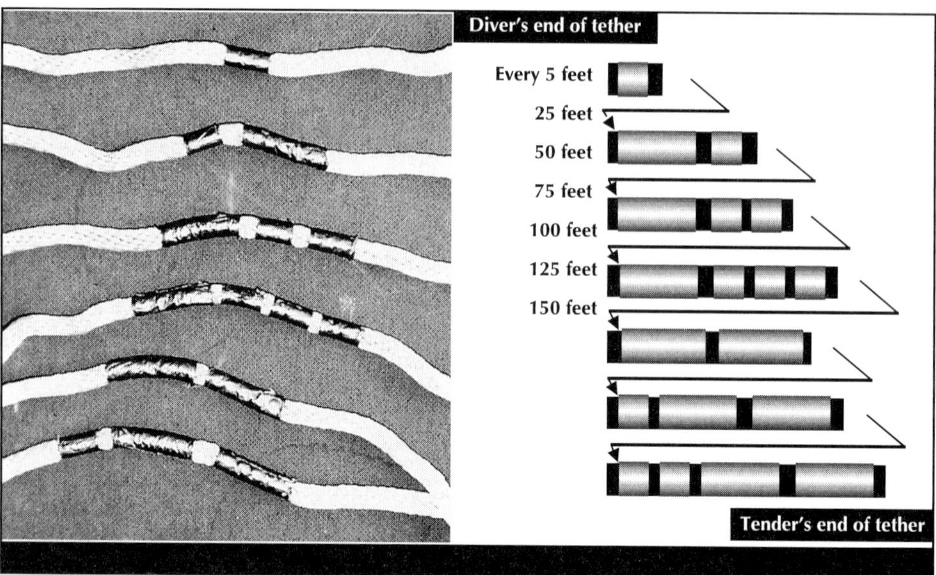

Photo 6–12; Figure 6–1 Use a narrow piece of tape every 5ft (1.5m), and these combinations of narrow and wide tape at 25ft, 50ft, 75ft, 100ft, 125ft, and 150ft increments (8m, 15m, 23m, 30m, 38m, and 46m).

Do not tie knots or loops in tether lines to mark distances. Knots weaken the overall strength of the line and have a tendency to become entangled in anything they encounter. Knots and loops increase the drag, creating a greater workload for the diver. The greater the drag, the less efficient the swim pattern. Increased drag tends to cause the diver to back-swing on his own axis; this distorts the swim pattern or control of the diver. Knots or loops typically result in less distance accuracy than is obtained with the taping technique.

Care and cleaning of ice lines

Checking the line before use. Checking for line fatigue requires both visual inspection and physical handling of the line. Start at the looped end of the line and begin gently feeling the line for weak, worn, or broken spots. The line should feel the same all over. Feel for sections that are thinner than other sections or line that is separating in the center. Bad sections of line can be cut off if near the ends; otherwise, use the line only for non-safety purposes. Tying a knot in the line will not strengthen the weak section.

In braided lines, look for feathers, which are sections in a line where the strands are sticking out. At any feathers you find, check the inner core for weakness. Feathers are typically found in sections where the manufacturer spliced the line, and a little excess was left in the external coat.

After using a line, clean, dry, and store it. This is particularly important after using lines on dirt, ice, and snow. Divers and tenders track dirt to the hole, and the dirt, ice, and snow can become impacted in the lines. One concern in line care is people standing on the lines. Grinding dirt into a line with boots and body weight is always a problem. On an ice diving site, tenders and crew will most likely be wearing ice creepers or spiked shoes, which compounds the problem. Keeping the lines neat, clean, and out from underfoot is the job of everyone on the site. In cold weather, ice diving lines can freeze, which means that ice has formed in the core or inner lining of the line. As ice freezes and expands, it can damage the inner portion of the line. This is especially true if frozen lines are worked and re-worked. Keep lines off the ice whenever possible. Do not force frozen lines into bags or onto reels. Soften frozen lines first by immersion or by placing them in a warm environment.

The best method for cleaning lines is to use a large bucket of warm water. Slowly dunk the lines, allowing the dirt to fall off. If a bucket is not available and a hose is used, hang the lines and then rinse slowly. Forced water only washes away surface dirt and possibly further impacts inner dirt. If there is a great deal of dirt on the line, the line can be washed in a line washer on gentle flow, prior to being washed in a bucket. A line washer is a T-shaped section of PVC piping that connects to a garden hose.

If water is not enough to clean the line, use a small amount of mild soap, such as Ivory Flakes™. Make sure all of the soap is rinsed out before drying. Water temperature can be as hot as your hands can handle.

The authors of *On Rope* also offer the suggestion of washing line in a big round commercial washing machine with a front opening. Make sure the door window is glass, not

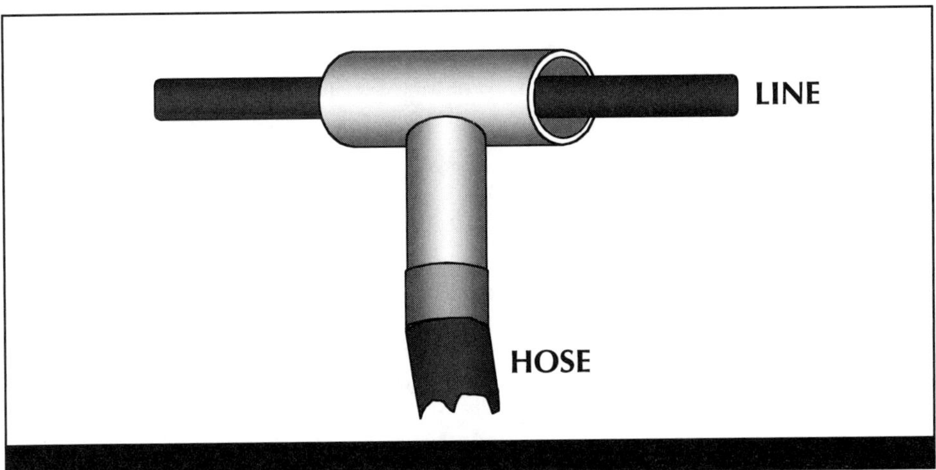

Figure 6–2 A line washer is a useful and inexpensive tool to clean lines

plastic, because plastic may damage the rope by abrasion. Put the rope in a large mesh bag and place it inside the machine. The authors add an important note of caution. They advise against using a top-loading washer that has a central rotating agitator, because the rope "tends to become very tightly snarled about the agitator and severe abrasion can result..."[3]

Hang lines to dry with air circulating around the entire surface area. The potential for dry rot increases when lines are dried without air circulating in this way. When lines are left lying on the ground to dry, they continue to pull cold and moisture out of the ground. Contact with any flat surface area will keep the lines wet and damp and they will not dry properly. The life of the line is dramatically shortened by this treatment. In warmer months, avoid direct sunlight when drying lines.

After the line is dry, it is a good idea to wash and dry it a second time. Dirt impacts under the topcoat during ice operations and may not be fully removed until the line is fully defrosted.

Once the line is clean and dry, pack it in dry bags or on a dry reel. The use of rope bags has become quite popular in the water rescue industry in recent years. Rope bags are easy to use and simple to pack. They offer quick line deployment, and keep the dive area neat and organized.

Reels for the most part have not been as useful. They are usually made of metal and freeze in the ice environment. Reels also have a tendency to overrun under inexperienced handling. Overrun occurs when the wheel rolls faster than the line can exit the reel; the line becomes caught within itself. There is, however, a reel on the market made from PVC with a patented braking system that works well. A reel is good for line lengths of 300ft (91m) or longer. With shorter lines, the cost and weight of a reel outweigh its benefits. Reels are excellent for surface lines from the shore to the dive hole.

Coiling is the least desirable choice for the ice diver. Coiling leaves the line exposed to the elements; and, in the hands of a novice line handler, all too often line becomes entangled. Coiling is not conducive to maintaining a neat dive area nor to keeping people from standing on the line.

Once the line is ready for storage, store it on a shelf in a dry place off the ground.

Bagging line

A properly packed line bag improves efficiency and ensures proper line deployment during each use. Try this procedure for packing a line bag:

1. Open the bag completely.
2. Hold the open top of the bag with one hand and allow the running end of the line (portion out of the bag) to lie between the thumb and forefinger.
3. Reach your free hand into the mouth of the bag and gently push the line from the top to the bottom of the bag.
4. Release the line and repeat the process.

Tether and rescue throw line bags function on the concept of spaghetti webbing. When correctly packed, the line releases freely 100% of the time.

Proper use of lines

Do not leave lines on the ice all day, especially when the temperature is below freezing. Detach lines and send them back to shore with each diver. This gives the team a chance to put them in a warm place to thaw out a bit so that they will be more flexible. Since divers should not be on the ice without a tether, this ensures a tether line for each diver as they access and depart the ice. To prevent being pulled into the ice hole, never wrap a line around your hand or arm.

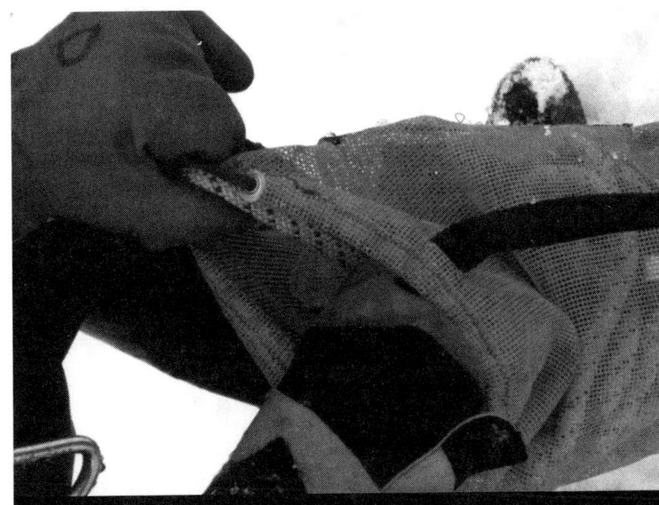

Photo 6-13 To stuff line into a bag, put the line through an "okay" signal with one gloved hand that also holds the bag (left), and pull the line in with the other hand (right).

Securing tether lines

Secure all tether lines to a fixed object such as an ice screw. There is always the possibility that a tender with cold hands could drop a line. A tender could also slip and fall and let go of the line.

Do not attach the tether line directly to the fixed object; instead, use a primary attachment, such as flat tubular webbing, for tie-offs to trees, ice screws, or vehicle bumpers. Next, use a carabiner to secure the tether line to the webbing. Do not attach tether lines to vehicles or other moveable objects unless there is no other safe option. If vehicles are the only

Photo 6-14 In-line mariner's hitch with two ice screws. The ice board pulley system is attached to the screw closest to shore. Tender and diver tether lines are attached to the carabiner on the other screw.

Photo 6-15 An ice screw can be used to pull an ice board with divers back and forth all day when properly placed in as little as 1.5in (4cm) of ice.

alternative, secure the keys with a tender and place a sign in the window: Do Not Move, Tether Lines Attached. Never assume that no other person will have a set of keys for that vehicle. Do not use sharp or abrasive attachment points that could chafe the webbing. If you have nothing better, place some type of strong material, such as a heavy canvas tarp, around the object and place the webbing over the material.

It may sometimes become necessary to rapidly change a tie-off location or to add an extra line bag. For that reason, non-locking carabiners are useful to secure the line to the webbing. Use two carabiners side-by-side and reverse the gates, one up, one down. Do not use knots, which can freeze and become difficult to untie. If adding additional line to the tether line, simply snap a new line bag in with another carabiner. If a locking carabiner is used, be sure to duct tape the lock to prevent it from freezing closed.

When working around the ice hole, it is important to ensure the safety and efficiency of the support crew. Tenders can find themselves in some ambiguous positions around the hole, and they have the greatest potential for falling, slipping, or being pulled in. Early in the set-up, establish a tie-off point for the tether lines. Use this same ice screw for tender tethers. The preferred technique is to place two separate ice screws approximately 12in to 18in (30cm to 46cm) apart and tether them together with webbing. Generally two ice screws, and sometimes in thin ice three ice screws, are used to help take up the load of personnel and equipment transported between the shore and the diver point-of-entry. You cannot have a pre-planned exact distance between ice screws because the distance is dependent on ice conditions. Therefore, you cannot have a pre-set section of webbing with an exact length to use between the ice screws. You need to have a section of webbing longer than the distance between the first and last screw, and you need a rigging system that allows you to "use up" the extra webbing length without creating any knots. This is easily accomplished with a mariner's hitch.

Using flat tubular webbing, attach the tenders' tethers to the rear D-rings on their harnesses and back to the main tether point or ice screws. This webbing should be long enough for the tender to work freely around two sides of the hole. If the tender slips, falls, or is pulled into the water, the webbing enables a rapid retrieval either by the tender or other hot-zone personnel. Accidental immersion of the tender is a major concern. In the worst-case scenario, the tender is accidentally immersed and then quickly pulled under the ice cap by the unknowing diver.

Maintaining direct line access without compromising the line

The compromise of the diver's tether line is a major concern at all levels of ice diving. Proper line handling requires that tether lines never be bent between the diver and tender and never rub against sharp, flat, or square edges without edge protection or guides.

Take extreme care during training to ensure that tenders understand how to properly handle lines to reduce, if not eliminate, line rubbing or friction. Extreme cold weather can make chafing an even greater concern. These guidelines will help avoid compromising the line:

- Tenders should always be observant.
- Tenders need to keep their hands low to the ice to reduce the angle of the diver's tether line as it moves under the ice.

Photo 6–16 To keep tenders both properly tethered and mobile around the hole, the tender's webbing can be secured on a runner system webbing between two pitons.

Photo 6–17 Tenders keep their hands low to the ice to reduce the chances of the line snagging on uneven chunks on the underside of the ice roof or the ice block. Tenders wearing drysuits can even reach arms-length into the water to work divers under the ice block.

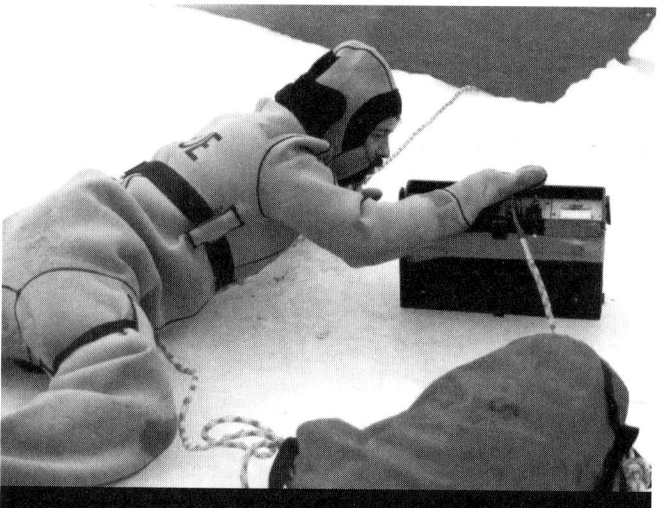

Photo 6–18 A tender adjusts the volume of an OTS communication system speaker box. Learn how to work with equipment without taking your eyes off the diver!

- Tenders need to be prepared to lie on the ice to keep the line at the lowest possible angle in order to reduce rubbing and chafing and to keep a direct (straight) line to the diver. This is especially true during training sessions with thick (10in [25cm] or more) ice or during extended operations at the same hole when ice erodes into a jagged edge.
- Tenders need to be mobile to keep the divers' tether lines from going under the ice or behind or up underneath the tenders and the ice. Such a situation instantly compromises the line and confuses what directional signals to give divers, since they are facing in a different direction than normal to the tender.
- Mobility should not compromise the tenders' safety. If they fall or are pulled in by the diver, their body should not go under the ice roof. Every public safety tender should have at least one training session in self-extraction wearing an exposure suit. Have them actually fall in and pull themselves out with their tether straps. Be as realistic as possible.

Summary Questions

1. Ice divers are ________ ________ divers with one way in and one way out.
2. Hand-held loops or ropes around waists are completely ________ and ________ for ice diving operations.
3. _______ disconnect a tether line under the ice unless a new one is attached first.
4. Public safety divers often dive _____ because it is actually safer in low/zero visibility and because the search will be more accurate and effective.
5. List at least two problems with a two line, two diver down system.
6. List the five basic components of a diver's tethering systems.
7. Explain why hand-held loops are not to be used under the ice and why they are a poor choice for all low/zero visibility search diving.
8. Where should the tether line be connected on the diver?
9. Describe where the harness girth strap should sit and explain why.
10. List at least four features to look for in a harness.
11. Describe how to attach the tether line to the diver's harness.
12. Describe the most effective tether line diameter and explain why.
13. What are the most effective ways to carry and store line less then 300ft and more than 300ft long?
14. What is laid line and why is it not a good choice for tether lines?
15. What is braided line and why is it preferred for water/ice operations?
16. What is filament construction and why is it preferred over spun rope?
17. What is the preferred line material for tether lines and why?
18. What is the recommended maximum primary diver line length and depth for divers with plenty of tethered diving experience and at least fifteen ice dives, and what is the maximum contingency diver line length?
19. Tether lines should be marked every ____ feet to know exactly where the divers are.
20. Why should loops or knots not be used to indicate line markings/diver distance out?
21. Describe how to care for lines after use.
22. Describe how to secure lines to a stationary object.
23. Explain how to tether tenders.
24. Describe techniques to maintain direct line access and an uncompromised line.

Notes

[1] Brion Tross, *Chapman's Knots* (New York: Hearst Marine Books, 1990), p. 12

[2] Mario Bigon and Guido Regazzoni, *The Morrow Guide to Knots* (New York: Quill, 1982), p. 12

[3] Allen Padgett and Bruce Smith, *On Rope* (National Speleological Society, 1987), p. 23

Chapter 7

Cold Stress, Immersion Hypothermia, Drowning, and Other Immersion Factors

There is only one job you have to do every time – Go Home!

Note: *This chapter will present readers with medical terminology and the corresponding definitions. Although an understanding of medical terminology is not very useful in the field when working on a patient, it is useful when reading literature that does not provide definitions.*

This understanding also helps bridge the wide communication gap that often exists among recreational divers; fire, law enforcement and EMS personnel; and emergency room physicians. Do not assume that your emergency department has effective or up-to-date protocols on immersion hypothermia, hypothermia, drowning, or gas bubble injuries (*i.e.,* barotrauma and decompression sickness). Every dive team, and even recreational dive instructors, should open the lines of communication with their local receiving emergency departments for mutual information sharing. This is important for the victims that public safety dive teams may be working to rescue, and for the direct and indirect safety of divers and tenders.

The ways in which doctors can cause safety issues for dive teams are not always obvious and therefore should be discussed. The following account illustrates how one emergency room physician decreased the safety of dive teams through lack of knowledge. A possibly intoxicated young man jumped off a 25ft (8m) high rock cliff into 60ºF to 70ºF (16°C to 21°C) water at the same spot where several other people had drowned in recent years. The victim was recovered just over three hours after the time he drowned. The on-call emergency department physician ordered ALS to begin treatment and continue treatment during transport, despite being told about the time submerged. Of course, local newspapers later printed this information. The local dive teams have in their SOPs a rescue-mode time of one hour, after which they can, if conditions warrant, shut down and later start the recovery operation. This emergency room physician set a new precedent of a rescue mode being more than three hours. When we spoke with the hospital's emergency room staff, we requested to see their protocol regarding work on a patient submerged for over three hours. The staff handed us a textbook that stated, "A patient is not dead until warm and dead." That was it. That was their justification. They did not know that the record for submersion time survival is an hour or less for a very young child in frigid water. The hospital did not realize that their actions were harmful to local dive teams. The moral to this story is to become educated and then arrange meetings with both pre-hospital and emergency department personnel to share information and to make sure effective patient management protocols are in place.

Cold-Water Immersion and Its Effects

Tenders and divers need a good understanding of how cold affects them on land and in the water in order to prevent cold stress and its resulting problems. This understanding will enable them to recognize the signs and symptoms of cold stress and to manage the condition in the field or en route to a hospital PSD. Recreational ice dive leadership personnel need a good foundation of cold-water drowning physiology to be able to provide in-field and perhaps pre-hospital patient care.

Anyone can drown. Even Olympic or long-distance swimmers can drown under certain circumstances. One of the most common causes of drowning among people who can swim is cold water. Sudden immersion in very cold water without appropriate personal protection equipment (PPE) can result in cold shock, which can be lethal because of the resulting gasping reflex and because of possible cardiovascular problems, such as ventricular fibrillation.

Anyone who has jumped in a cold pool or ocean can describe the gasp reflex—a sudden, involuntary inhalation that can result in aspiration (fluid inhalation) if the person's airway is underwater at the time. Such aspiration can lead to a laryngospasm, or a sealing of the trachea by the movement of the epiglottis and pharynx. This is a defense mechanism to protect the lungs from fluids or particulate matter. A laryngospasm can be quite frightening, because the person cannot inhale or exhale, and panic may rapidly ensue, thereby decreasing the person's ability to survive in the water. A laryngospasm can also cause death due to asphyxia (suffocation). If a person accidentally punctures the ice or falls in the water while wearing a personal flotation device (PFD), the chances of the person's airway being submerged is greatly reduced, hence the chances of survival are greatly increased.

This gasping reflex, hyperventilation, can last for one to two minutes with an uncontrollable 50Lpm to 60Lpm ventilation volume at a rate of sixty to seventy breaths per minute. This rapid breathing greatly decreases carbon dioxide levels, which can result in muscle cramps, decreased cerebral blood flow, tetanic convulsion, and decreased levels of consciousness. A study found a 525% increase in pulmonary ventilation in the first few minutes of immersion.[1] This severe hyperventilation may be a major factor in a good swimmer's sudden inability to swim when immersed in cold water, and it can cause a cold-induced asthmatic attack in surface personnel with such a predisposition. [2] This is one of the reasons why people with asthma should not serve as ice rescue technicians, public safety ice diving tenders working on thin ice, or ice divers.

Cold water can cause other physiological responses that further increase the risk of drowning, including: intense shivering, muscle tensing, and muscle weakness. All of these responses occur rapidly when a person is immersed in cold water without proper personal protective equipment (PPE). These symptoms greatly increase the potential of drowning.

Cold water can be very painful to a person not wearing appropriate PPE. Pain can cause disorientation, poor judgment, fear, diminished survival skills, and panic. To appreciate the level of this pain, fill a large bowl with ice and water, and immerse your hand and forearm. See how quickly you pull your arm out. We offer this voluntary experience in ice rescue classes to help students think twice before diving or tending without proper PPE. In public safety diving operations, tenders need to be prepared for accidental immersion.

Photo 7–1 Immersion in ice water without proper PPE can cause extreme pain that can render even an Olympic swimmer into a drowning victim. Anyone who would respond to an ice site should experience this with a bowl of ice water. Take blood pressures and pulse rates before and after to see if there is a rise.

If you are able to hold your arm in the bowl for more than a minute, attempt to perform tasks that require finger dexterity. Close your eyes and have someone test your hand's sensation and strength. Hand immersion in ice water can cause such tremendous vasoconstriction that the blood loss is comparable to hand amputation. Total unprotected-body immersion in icy water also results in massive peripheral vasoconstriction that can greatly inhibit survival skills such as basic treading ability.[3]

This peripheral vasoconstriction increases cardiac work, with the result that cardiac or vascular problems can be significantly exacerbated. Subjects resting head-out in 30°C (86°F) and 34°C (93°F) water had 50% greater cardiac output than in the same temperatures of air.[4,5,6] The study found that cardiac output increased in both air and water during mild exercise, but was significantly higher when subjects were immersed in water. Ice divers may be faced with more than mild exercise if something goes wrong.

Cold-water immersion can also increase the incidence of irregular heartbeats called arrhythmias. Although most of us occasionally have them without noticing or being harmed by their occurrence, the increase of arrhythmias in cold water could become a problem and, in some extreme cases, could result in unconsciousness or death.

Ice diving instructors should carefully screen the medical waivers of prospective students. Public safety dive teams should ensure that divers and tenders have up-to-date medical examinations and physician-signed approvals. It is important to ensure that the approving physicians understand what ice diving entails. Do not assume that they do. Divers should stay in good cardiovascular shape between medical examinations and should tell instructors or team captains if any medical problems develop.

The struggle to survive greatly increases oxygen consumption, as does the increased metabolism caused by thermoregulation to increase heat production. Shivering itself can increase oxygen consumption by 50%. At the same time, immersion can make it more difficult to breathe and to acquire the oxygen needed. Immersion can reduce vital capacity by as much as 15%. (Vital capacity is the total volume of gas from the top of inhalation to the bottom of expiration.)

A person who is in a head-out immersion, such as a person who has punctured the ice roof and is hanging in the water holding the ice roof for dear life, experiences negative pressure breathing. In a head-out position, the ambient pressure at the mouth is less than the ambient pressure at the chest because of the increased pressure of water. For example, at sea level the air pressure just above the water's surface is 14.7psi (1bar). Every foot of freshwater is .432psi (.029bar), therefore a person's chest submerged one foot experiences 15.132psi (1.029bar). The result is that we have to use a greater effort to inhale. "Upright immersion has been shown to adversely affect respiratory mechanics."[7]

Photo 7–2 Head-out, vertical immersion causes negative breathing and an ensuing chain of events. (Notice the hood vents at the top of the diver's head and the light attachment bracket on the top of the mask.)

Negative-pressure breathing also initiates a chain of events beginning with atrial engorgement (too much blood in the top two chambers of the heart) and ending with water being filtered out of the blood with resulting urine production.[8] This means a head-out immersed person will have a lower blood volume and "thicker" blood. Think about how this could affect divers, immersed tenders, or drowning victims that public safety dive teams are trying to save.

A study of human performance capacity and cold-water immersion demonstrated reduced abilities and decreased time to reach exhaustion in the immersed subjects. "In all probability, cardiac output was also impaired, which resulted in decreased oxygen delivery."[9] Cold-water immersion increases the need for oxygen, but decreases our ability to get it. [10]

Cold exposure in air also stresses the body. Cold exposure activates the immune system, increasing natural killer cell count and activity, circulating levels of interleukin-6 and other immune factors. Subjects given a pretreatment of exercise while immersed in 18°C (64°F) water showed the largest increase in leukocyte, granulocyte, and monocyte response (immune system), when compared to subjects with pretreatments of warm water immersion or passive heating.[11] The cold-water exercise pretreatment could be seen as analogous to divers who are not wearing appropriate PPE and who work hard in the water, and then stay in the cold after the dive instead of rehabilitating in a warm environment. The results of this study also show the need for tenders to wear proper PPE.

Let us now take a closer look at cold stress, immersion hypothermia, and drowning as they pertain to divers, tenders, and recovered victims. The mission of this chapter is to help divers and tenders make rational decisions, such as, "I will not step on the ice or enter the water without proper PPE", and to help rescuers provide optimal care to immersion hypothermic patients.

Most people define hypothermia as "the body's inability to maintain normal core temperature." The medical definition of hypothermia is a body core temperature of less than 95°F (35°C). The difference between these two definitions can present a problem for safety, in that a person using the medical definition might infer that anything less than a 3°F (1.5°C) loss is not a serious problem, as it does not constitute hypothermia. This is far from correct.

How a person feels can be just as important as the actual quantity of heat lost. Heat loss from extremities, without any core temperature loss, can cause life-threatening problems. Cold can cause diminished sensation, strength, and motor functions in unprotected parts of the body, which can increase the risk of injury and death. For example, divers may lose the ability to release their weightbelts to save themselves because of cold hands. Tenders without any core temperature loss might have such cold hands, because of improper gloves, that they would be unable to effectively hold a tether line, help divers with their equipment, or handle a patient without causing further injury. Think about the fatality described earlier, in which a tender may have left his two divers underwater by tying their tether lines to the bumper of their vehicle because the tender was perhaps cold and felt the need to get inside to warm up.

Personnel who simply feel cold, regardless of core temperature loss, are more likely to shortcut standards, use poor judgment, or rush. An inner voice continually tells them, *"I'm cold, I want to get out of here, I want to get back in the warm vehicle; why am I here? When is this going to end? Can't we do this any faster? Why are they taking so long?"* Instead of taking the time to properly put the harness on a diver, they don it improperly and the diver's breathing is restricted, which he does not realize until he is underwater and 50ft (15m) from the ice hole. Simply feeling cold increases stress and task loading. A drop in skin temperature can result in the release of the hormone norepinephrine (noradrenaline) which is part of the "fight or flight" mechanism. In one study, young athletes in a head-out immersion in 14°C (57°F) water for one hour had a four-fold increase in plasma noradrenaline concentration.[12] That release of norepinephrine could be added to a previous release caused by anxiety about making a first ice dive or caused by the ice rescue incident itself. The result can be rushing, shortcutting, and a lack of good judgment in addition to the physiological effects. The release of norepinephrine also causes a decrease in blood volume through urine production, as we will describe in the section below on cold and immersion diuresis.

Cold-water immersion can cause impaired mental function before the person is clinically hypothermic. This is in addition to the panic caused by pain. Mental functioning can become impaired as cold-induced hyperventilation reduces the concentration of carbon dioxide in the system and causes a reduction of cerebral blood-flow. Less cerebral blood-flow produces impaired cerebration. It has also been found that impaired memory during immersion continues after the person is out of the water.[13] Cold-water immersion victims have been known to hallucinate, especially during exercise, and to experience personality changes, ranging from becoming quiet to acting belligerently. These mental impairments have implications for the validity of witness statements during interviews after cold-water immersion. Law enforcement and other personnel responsible for interviewing should be made aware of this.

The effects of cold and feeling cold can result in physical, mental, and even emotional stress well before medically-defined hypothermia occurs. Therefore, this book will operationally define "cold stress" as the direct or indirect effects of heat loss rather than specifying a core temperature loss.

The Four Processes of Heat Loss and How They Relate to Ice Rescue

The first step in preventing and managing cold stress for both rescuers and victims is understanding how we lose heat.

Cold is the absence of heat. If you submerge your hand in cold water, as described in the optional exercise above, you will eventually feel as though the cold transfers up your forearm. In reality, heat is being transferred from your forearm to your hand and then to the water.

When it comes to water, immersion in anything colder than 92°F (33°C) can cause heat loss. We lose heat in 80ºF (27°C) water at the same rate as in 42ºF (6°C) air![14] If you don't picture yourself being comfortable working outdoors in nothing but your underwear in 42ºF (6°C) air, then you should not consider working in 80ºF (27°C) water without a proper exposure suit. "The unprotected diver will be affected by excessive heat loss and become chilled within a short time in water temperatures below 72°F (23°C)."[15]

Let us look at how water steals our heat so much faster and in such greater quantities than air.

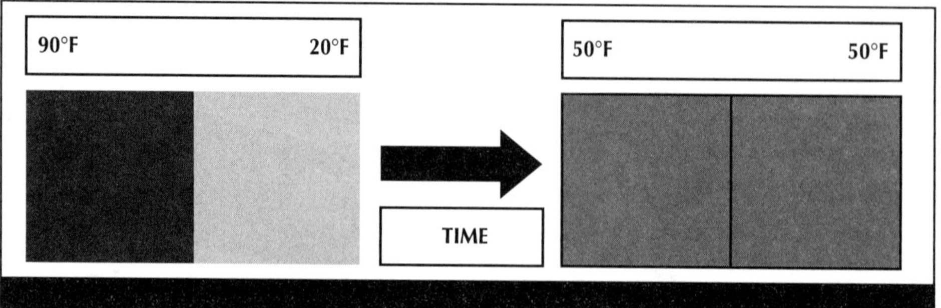

Figure 7-1 A warm object in contact with a cold object will lose heat to the cold object, until both objects are the same temperature.

Conduction

Imagine one warm and one cold object touching each other. Heat will travel from the warm object to the cold object until both are the same temperature. This is heat transfer by conduction. The molecules of the hot objects move around more than the molecules of the cold objects. When the two objects come in contact, heat is transferred by the warmer object's molecules hitting against the colder object's molecules, which are then set into greater motion. The wider the temperature range is between the two objects, the greater the force of heat being transferred from the hot to the cold object.

For example, hands lose heat to wet line and metal gear. Backup and 90%-ready divers lose heat to the ice if they are sitting or lying on it.

Try it now. Touch something that is cooler than your hand and keep your hand on the object until it feels the same temperature as your hand. The object conducted heat away from your hand and dropped your hand's temperature, while raising the object's temperature until their temperatures became the same. Touch the object and your hand with your other hand. The object should feel warmer than it originally did, and your hand should feel colder.

A good training aid for classes or drills is to take a metal teaspoon and heat it for ten seconds with a lighter. Hold it in room temperature air and time how long it takes the heated area to feel cool to the touch. It generally takes over two minutes. Then re-heat the spoon and put it in room temperature water for two seconds. The spoon will feel cool. This demonstrates how fast water will rip the heat out of our bodies if we are not wearing appropriate PPE. Water conducts heat away from an object more than twenty-five times faster than air. This means your body heat will be transferred out of you more than twenty-five times faster in water than in air.

Water has a tremendous heat capacity. It takes a thousand times more heat to raise the temperature of water one degree than it does to raise air temperature one degree. Therefore, water will continue to steal your heat far longer than will air. Water will drop your temperature to a much lower degree, because it needs so much more heat to raise its own temperature. Remember, conduction doesn't stop until both objects are the same temperature. Your body will not raise the temperature of the lake; rather the lake will bring your body down to its temperature.

How does conduction affect divers and tenders? Insulation afforded by 1/4in (6mm) neoprene ice rescue suits inhibits the ability of water to conduct heat away from a tender's body. Air and undergarments provide insulation for divers wearing drysuits made of any material other than full-thickness neoprene. A thick pair of wool or fleece socks under the ice rescue suit or drysuit is recommended, because the boots consist of un-insulated rubber. Heat loss from feet should not be ignored.

To stay prepared, always keep an extra pair of thick wool socks along with the extra gloves, hats, boots, and ice creepers in your car or rescue vehicle. Keep a pair of thick wool or fleece socks and glove liners in the bag of each exposure suit on the dive team's vehicle. Avoid wearing cotton during winter calls.

Warm gloves with insulation are vital to prevent rope, metal equipment, wet bodies, and other heat-stealing objects from conducting heat away from hands and thereby making hands relatively useless. Good boots for operation-level personnel are important for keeping the cold winter ground from stealing the heat out of their feet. We lose 15% of our heat from our hands, and 15% from our feet, so keep them insulated!

The torso is also of particular concern when it comes to PPE. One cause of the gasping reflex is believed to be a neurological response when cutaneous (skin) cold receptors are stimulated. Researchers found that the upper torso had "an increased receptor density and/or sensitivity"[16] over the extremities. The torso should be kept dry and properly insulated, whether the person is working in the water, on the shore, or on the ice.

How does conduction affect patients? A patient should not be laid directly on a cold backboard, which can conduct significant heat from the patient. Without a blanket, a backboard could conduct more heat away from the patient than the air above them. Transport devices stored in unheated compartments will be especially cold, as will devices constructed of metal.

Always remember: Insulation over, insulation under, and insulation all around. Secure a pad or non-cotton blanket on the board as insulation under the patient.

If a motor vehicle accident patient is placed on an uninsulated, cold backboard, the patient is likely to start or continue shivering, which increases the need for oxygen. If their ability to take in and transport oxygen has already diminished due to trauma, shock, near-drowning, or other conditions, shivering might be the last straw that results in cardiac arrest. Think about it.

A person who lacks the ability to move, either because of injury or because of immobilization, can become colder faster because they cannot use voluntary muscle movement to generate heat. To make matters worse, the patient experiencing hemorrhagic shock is also more likely to become hypothermic. An immobilized trauma patient lying on an uninsulated backboard can be in serious risk of severe shock and even death because of heat loss that often can be prevented.

To make matters worse, hypothermia increases bleeding time, which means that a hypothermic trauma patient is at greater risk of experiencing hemorrhagic shock.[17] Hemorrhagic shock will worsen the hypothermia. Rescuers must take all possible measures to prevent this vicious cycle. Avoid any sleds, boats, or transport devices that put patients or rescuers in contact with metal. Metal sleds, boats, and ladders steal the heat from rescuers and patients. In addition, as some children's tongues have painfully discovered, wet skin can freeze to metal. Metal transport devices are also more likely to freeze to the ice itself, resulting in greater exertion and frustration for rescuers.

Convection

Blow on a hot drink. Cool off in front of a fan. Your body is surrounded by a cushion of warm air or water heated by the process of conduction. When wind or movement carries away this warm air or water, your body loses heat as it warms the new surrounding medium. Every time the warm cushion is removed, the body must heat up the new surrounding medium. The transfer of heat by wind or fluid movement is the process of convection.

Try it now. Blow forcefully on your hand and feel the cooling effect. The windchill factor is caused by convection. The faster the wind blows, the greater the heat loss and the lower the equivalent windless air temperature. As long as the ambient water or air temperature is lower than our skin temperature, we will lose heat by convection. Internally, the exchange between tissues and blood is a form of convection.[18] The blood moves through the tissues (perfusion) and either gives or takes heat away depending on which is warmer, the tissue or the blood.

How does convection affect divers and tenders? In winter water-related calls, SOPs/SOGs should include setting up a wind-protected staging area with vehicles, tarps, and other windbreaks. The few minutes that these procedures take are well worth it and can actually save time in the end. Remember, cold rescuers take longer to do the job safely, and are at increased risk of causing injury to self and others. Wind speed and estimated operation duration should dictate the extent to which windbreaks are set up.

Photo 7-3 A wind-block protects a contingency diving lying on a gym mat, reducing convective and conductive heat loss.

Water does not have to be fast-moving for convection to occur. Actually, it can be quite still. As soon as a person moves in water, convection takes place. Movement in the water causes turbulence that convects heat away from the skin. The trickle of cold water entering a wetsuit quickly teaches that lesson. Smaller people in one-size-fits-all "dry" ice rescue suits are easily flooded and put at risk for significant heat loss.

Photo 7-4 Standard neoprene ice rescue suits were not designed for swimming or immersion in moving water, but with ankle weights will work as in-water tender suits.

Moving water will take more heat out of an immersed person than still water. Keep that in mind when deciding on effective PPE. Properly fitting immersion suits are important for tenders, especially in moving water. Standard neoprene ice rescue suits are not designed for moving water in any season, and should definitely not be the suit used in frigid moving water. These suits only serve as drysuits if the suit perfectly fits the rescuer's body, or if the rescuer is only immersed from the neck down. Sadly, most suits are designed for larger people, while the first choice for surface ice rescue technicians are lighter, smaller people. Bayleysuit™ offers small-sized ice rescue suits that enable rescue teams to take advantage of lighter surface ice rescue technicians. Working in moving water also requires greater maneuverability and swimming ability, which these suits do not afford. True drysuits with winter-weight undergarments are the preferred choice here.

We lose approximately 25% to 30% of our heat from our head. Heat escapes from our ears, nose, lips, scalp, carotid artery neck area, and face to the surrounding cold air. Fleece or wool balaclavas (hood socks) are recommended for surface personnel because they cover the wearer's ears, nose, neck and mouth.

How does convection affect patients? Extra care must be taken to reduce heat loss from convection in the following scenarios: the patient must be transported over a long distance; the wind is blowing; the transport device is an airboat, an unprotected fast boat, or a helicopter. The thermal recovery stabilizer provides effective protection from air movement.

Photo 7-5 This tender's PPE includes a warm hat over a combination head sock and facemask, a warm one-piece suit, waterproof gloves with liners, thick-soled boots, and a class-V PFD.

In the case of accidental immersion, don't kick around furiously while waiting for rescuers. Instead, conserve energy by bobbing and huddling. Swimming or working hard in cold water without proper personal protective equipment accelerates death. If you can, float with as little movement as possible. If you are negatively buoyant and therefore cannot float, obtain training in bobbing techniques. If a person is buoyant enough, huddling can further slow down heat loss.

By the process of convection, a victim struggling in cold water will lose heat more quickly than a still victim, even though the struggling victim may feel warmer. The latter is true because struggling increases metabolism and causes warm blood to be sent to the cold peripheral tissues where we have temperature sensors. In the water, an increased feeling of warmth can coincide with greater heat loss. Divers or tenders exerting themselves in cold water need to keep this in mind.

Evaporation. Evaporation is the process by which a liquid changes to a vapor. This change requires significant heat if the liquid is water, because a large quantity of heat is required to raise the temperature of water. Perspiring uses the process of evaporation to keep us cool in warm environments or when our core temperature is higher than normal. We can lose heat through evaporation even if our skin temperature is greater than the ambient air temperature. Evaporation from the skin and respiratory tract is normally 20% of total heat loss. If the skin is wet from an external source, such as rain or immersion, that percentage increases.

If we sweat while wearing something that does not permit the fluid to evaporate into a vapor off our skin, we do not lose much heat. We lose heat from evaporation mainly from the process of the phase change (fluid to vapor). Cotton, which readily absorbs fluids, is great to wear in the summer because it allows us to lose much evaporative heat. For the same reason, cotton is a poor choice for cold environments. It takes a long time to dry, which means a longer time for evaporative heat loss to occur. Materials such as wool and fleece do not allow as much evaporative heat loss from sweat or wetness due to previous immersion.

When we are immersed or submerged, the fluid against our skin prevents evaporative heat loss.

How does evaporation affect divers and tenders? When the ambient air temperature is cold, more body heat is required for skin to dry. That is why it is important to dry off if you exert yourself and perspire in the winter. For example, many of us have noticed that during a winter call we might actually feel warm when we exert ourselves, but once the work is over and the vehicles are cleaned and repacked, we suddenly feel wet and cold. Before zipping up that jacket, dry off your skin, especially around the torso and underarms.

It is very important to remove wet exposure suits and wet gloves as soon as possible. Wet clothing or neoprene next to skin functions like a big sweating machine, stealing body heat to dry itself. As long as the body is wet, it will lose heat. If a suit must be worn more than once, be sure to put on dry wool or fleece undergarments each time the suit is donned. Wet feet and hands lose temperature more quickly than dry ones, making them more prone to cold injuries such as frostnip and frostbite. Waterproof foot protection is a must. Boots should be high enough to prevent snow from entering them.

Tenders should wear gloves that are either waterproof or made from pure wool or polar fleece. This will enable them to handle wet lines, technicians, and equipment without a great deal of evaporative heat loss to their hands. Shredded newspaper can help if hands and feet become wet. Stuff shredded newspaper in gloves and boots for added insulation, and replace it with a dry batch of newspaper when it becomes wet from perspiration or water.

A sports chamois is an excellent tool to keep in a jump kit all year round because it is good for repeated uses. In one long call, you might need to dry yourself off several times to keep warm. One chamois can do the job of many towels and can also be thrown in the washer and dryer.

If your inner shirt becomes wet, change it if at all possible. Teams performing strenuous, long-distance wilderness rescues should have extra shirts to change into after removing damp undergarments.

Rescuers cannot provide optimal patient care when they are cold; how does evaporation affect patients? EMS personnel can probably picture a diabetic or cardiac patient in their ambulance, soaking wet from perspiration. If the wet undergarments are not removed, that patient can stay cold through a long transport drive, even in a very warm ambulance and even after being wrapped in blankets. Get the wet stuff off.

Patients pulled from the water are already hypothermic from conduction and convection in the water. After they leave the water, heat is additionally lost through evaporation. Gently remove wet clothing, and wrap patients as soon as possible in non-cotton blankets, or put them in a thermal recovery stabilizer. Space blankets should only be used as wind protection on the outside of wool blankets. Space blankets work by reflecting heat; if the patient is already hypothermic, there is little heat to reflect. Space blankets have negligible insulation and no fluid-absorbing capabilities.

Breathing

We lose heat directly from our core by convection and evaporation every time we breathe in cold environments. When we breathe, inhaled gas is moistened and brought up to the body's core temperature by the time it reaches the trachea. Then, every six to ten seconds we exhale that warmed gas and replace it with new cold, dry gas that robs us of more heat and moisture. Rapidly breathing ambient air has been shown to cause such a warm fluid loss from our core that "the rate at which water is returned to the airways may not be adequate to keep the periciliary [area around the miniscule hairs found in the airway] fluid isotonic. These findings support the proposal that the intrathoracic airways could become significantly dehydrated during hyperpnoea [rapidity of breathing]."[19]

Divers breathe cold dry gas because moisture is removed from the breathing gas before it is pumped into the cylinders. As depth increases, so does gas density and the amount of body heat necessary to warm it. As we learned earlier, increased heat production requires increased oxygen consumption, and increased oxygen consumption requires an increased respiratory minute volume. For example, divers at rest in cold water may increase their respirations from 9Lpm to 36Lpm of air. More air consumed means more heat loss by convection and evaporation. One of the reasons why the maximum dive time is shortened in ice diving is to help prevent out-of-air emergencies and cold stress.

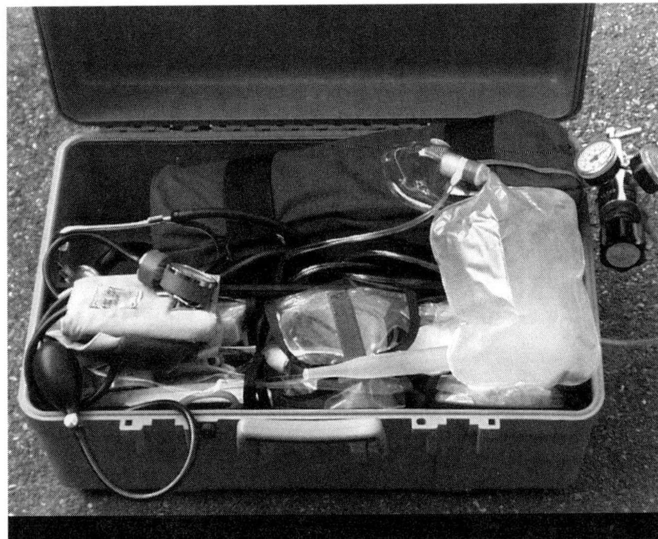

Photo 7-6 All dive sites should have an oxygen system with trained personnel. If possible, have a humidifier with distilled water and heat packs to be able to add some warmth and moisture so that cold, dry gas is not administered directly to a hypothermic patient's core.

If local protocol allows, moisten and warm oxygen before administering it to cold patients. The last place you want to cause direct heat loss is in the core. Oxygen is normally stored in unheated ambulance compartments. Even if it is used from a warm jump kit, the decrease in pressure from 2,015psi (137bar) to ambient 14.7psi (1bar) can cause a drop in temperature in the oxygen regulator first stage to 22°F (-6°C). Moisture is removed from the oxygen as it is compressed into cylinders, so that it is not only cold when we receive it, but dry as well.

Warming and moistening the oxygen prior to administration can be as simple as taping re-usable heat packs to the outside of an inexpensive oxygen humidifier, or, in the field, by running oxygen through a thermos of distilled water heated with a few heat packs.

Radiation

Radiation is the emission of infrared energy, whereas heat is transferred by electromagnetic waves. Unlike the heat lost in conduction, convection, or evaporation, radiation can only move in straight lines. Another unique feature of radiation is that heat loss from radiation does not depend on matter. It can cross a vacuum. An example of radiation is the sun's heat, which travels through space and then through the atmosphere before it reaches us. We will lose heat directly from our skin by radiation as long as ambient temperature is less than skin temperature.

Proper PPE will help reduce heat loss due to radiation. Be sure to protect the major heat-loss areas, such as the ears, nose, mouth, carotid neck region, and any place that has hair, such as underarms, top of the head, and groin.

Patient handling and heat loss on land

It is time to apply what we have just learned to the real world. For public safety personnel, picture in your mind's eye a winter motor vehicle accident. The patient in the vehicle is not wearing a coat, hat, gloves, or boots. A cold plastic cervical collar, stored in the ambulance outside compartment, is put on the patient over the carotid region. Cold, dry oxygen is administered directly to the patient's core. While fire department personnel cut the car apart, the new EMT stands outside, excited and ready with the backboard and KED. These devices are as cold as they can be. Next, clothing is cut away to expose the injuries. The cold KED is applied by EMTs who have stood outside for ten minutes with nothing on their hands except thin latex gloves. Then the patient is laid on the very cold backboard, which may also be wet if it has been snowing or raining. Inside the ambulance, the medic administers IV fluid that was stored in a jump kit in a cold ambulance with the result that cold fluid is sent directly to the core. The patient crashes and personnel wonder why. Now imagine that the patient was pulled out of an ice hole or a car that was partially submerged in freezing water.

What can be done differently? Prior to a call, insulate the backboard with a taped-on blanket or a commercial board pad. On the MVA scene, keep the KED and insulated backboard inside the warm ambulance until they are needed. Before starting extrication, wrap the patient in blankets, from head to toe, getting as much blanket as possible under the body and extremities without moving the patient. A hat and gloves can be put on the patient if trauma is not present in head or hands. Wrap the oxygen tubing around a re-usable heat pack a few times, and then wrap that with a towel and tape it. If the patient is shivering, place wrapped, hand-sized heat packs under the armpits. Do not lay the cold, metal oxygen bottle against the patient. Duty crew medical personnel not directly involved in the Jaws of Life operation can stay warm in the ambulance. In addition to warm exposure wear, other medical personnel should wear warm gloves over their latex gloves while waiting. Get a warm oxygen humidifier ready in the ambulance, and make sure the IV fluids are warmed to at least room temperature. The backboard should only be brought out of the ambulance to the vehicle when it is time to move the patient. If the patient's clothes are wet, remove them entirely once inside the ambulance. If you have a thermal recovery capsule, use it.

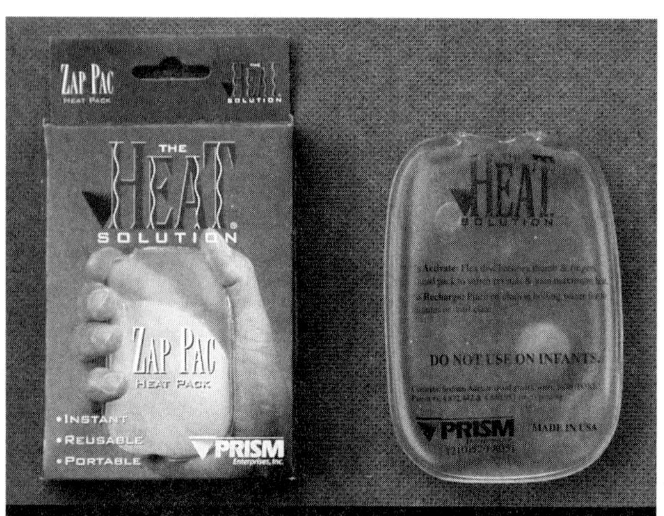

Photo 7–7 Heat packs are very useful for pre-warming oxygen, warming cold hands that need to perform pulse checks or fine-motor tasks, re-warming the arm pits or other key areas of a cold-stressed person, and helping manage equipment freeze-up problems. Make sure never to place heat packs directly on hypothermic skin.

Immersion hypothermia patients are just as fragile as motor vehicle accident patients. Immersion hypothermia patient management requires extra care, pre-planning, and training to ensure that patients are not dropped, yanked, or banged. Review the water-related incidents you have been part of or have watched. The average patient pulled from the water is dropped, yanked, and banged at least three times before being placed in the ambulance. This is completely unacceptable in any other type of incident that does not pose an immediate life threat to rescuers. Once a patient is pulled from the water, there is no building about to collapse or fire about to engulf rescuers between the water and the ambulance. So why is uncontrolled rushing and lack of patient handling skills the norm for water incident scenes—particularly ice-related ones? The irony is that rushing increases the risk of exposure to the threat—ice water. We believe the causes of water-scene poor patient handling include:

- Cold-stressed rescuers.
- Lack of ice and water training.
- Poor understanding of cold stress, immersion, and hypothermia physiology.
- Lack of standard operating procedures and guidelines.
- Lack of inter-agency pre-planning and drilling, which often results in ineffective incident command systems, or no command system, and chaos.

Let us examine cold stress, immersion/submersion, and hypothermia a little more closely.

As we have just learned, we lose heat by four different processes: conduction, convection, evaporation, and radiation. We also lose heat differently from the various parts of our bodies, and these areas lose heat at different rates depending on exercise level, ambient temperature, time of day, and other factors.

Photo 7-8 The point of diver-to-surface personnel pass-off is where victims are typically handled too roughly. Training sessions with horizontal, slow, gentle pass-offs supporting the head are imperative.

For example, compare the head with the rest of the body. Although some of us have bigger heads than others do, most heads account for 7% to 10% of our total surface area. Yet approximately 25% to 30% of heat loss occurs from our heads while we are resting. One of the reasons for this large percentage is that the blood vessels in the skin tissue covering our heads do not protectively vasoconstrict when exposed to cold, as do blood vessels in other parts of our bodies. Heat loss increases as ambient temperature decreases. For example, at 32°F (0°C), we can lose 35% of our heat from our heads.

Our core temperature is on average 98.6°F (37°C), but it can drop one to two degrees in the early morning sleeping hours and can rise as much as six degrees during exercise. Skin temperature is usually lower than core temperature and typically runs around 92°F (33°C) when indoors. If we lower our skin's temperature, the temperature gradient between skin and environment decreases. This reduces heat loss from the skin and, subsequently, the core. If the skin ("shell") was the same temperature as our core, it could not serve as a protective barrier against the cold ambient temperature.

There are several processes by which the body lowers skin temperature. To help us understand these processes, Rennie presents a useful thermoregulatory model of the body.[20] He considers the body a thermodynamic engine with a liquid core surrounded by a musculo-skeletal framework, bound with a layer of fat. Fat can be considered to have a constant thermal conductance—it provides a specific layer of insulated protection. Muscle tissue, on the other hand, varies in thermal conductance, depending on activity. So muscle tissue activity can determine rate of heat loss from the body.

When at rest, muscle tissue acts as insulation, thereby increasing the thickness of the core's protective shell. When muscles are at work, however, two processes occur. First, metabolism increases, and therefore heat production increases. That heat is quickly transferred by convection to the blood, which transfers it to the core and peripheral tissues. When at work, muscles require fuel, which is carried in the blood. Second, blood flow increases to the muscles and also to the extremities and the body's shell. If the body is covered with sufficient PPE, the body may have a net heat gain, which stabilizes its core temperature. In contrast, if the body is not sufficiently protected, there will be a greater heat loss from this peripheral vasodilation and muscle action.

If, peripheral circulation is restricted by peripheral vasoconstriction, the body's insulative shell layer is thicker. The peripheral tissues, which now have very little blood flow serve as an insulation barrier protecting the core. Remember, earlier we learned that the greater the heat gradient between two objects, the greater the heat transfer from the warmer to the colder object. With less blood in peripheral tissues, the temperature of these tissues decreases, and so the temperature gradient between the body and the environment decreases. Hence, the force of heat transfer slows, and heat loss slows.

If a peripherally vasoconstricted person begins to exercise or moves into a warmer environment, peripheral blood vessels will dilate, moving blood from the core to cold peripheral tissues. If a hypothermic person simply moved to a much warmer environment and did not increase internal heat conduction by exercising, core temperature could actually lower if the peripheral tissues are cold enough. Warm core blood that loses heat to cold peripheral tissues may return to the core at temperatures lower than core temperature. If this occurs, the core will lose heat to the returning blood, and core temperature will drop. As we will learn later, this process of peripheral vasodilation can also cause a drop in blood pressure, called "after-drop shock" or "rewarming shock". This result is one reason why internal rewarming may be more effective than external rewarming, and why hypothermic patients should not be actively externally rewarmed in the field if a hospital can be reached quickly.

The body has four main defenses against heat loss in cold environments. The first is to increase heat production by increasing metabolism through shivering (thermogenesis), exercise, and chemical reactions (non-shivering thermogenesis). The hypothalamus receives information from the skin and from core thermal receptors. With the right stimuli, muscle tensing and shivering occur. Shivering can be affected by the temperature of the spinal cord, cooled blood returning to the spinal cord, inspired-air temperature, mental activity, and stimulation of other receptors in the body.

Thin people, having less fat insulation, begin to shiver sooner than people with more fat and may begin shivering at a higher body temperature. People with higher body fat composition may have the same rectal temperature during immersion in moderately cold water as thinner people, but thin people have higher oxygen consumption. One study showed that "despite the thermal strain of cold-water immersion, the low-fat subjects were able to maintain a similar rectal temperature compared to the high-fat subjects due to a significantly greater shivering thermogenesis."[21]

The higher oxygen consumption needed to support the shivering, however, may result in less chance of survival in an immersion situation if rescue is delayed.

It is important to note that a fraction of the population does not shiver at all.[22] Just because someone is not shivering does not mean that the person is not cold. Alcohol can minimize shivering, which is one of the many reasons it should not be consumed when exposure to cold is a possibility.[23]

When the body can no longer stabilize its temperature, it must shiver or exercise to increase heat production. "Exercise produces metabolic heat at a rate of about three watts per watt of mechanical power."[24] Sadly, though, without a proper exposure suit, heat loss almost always exceeds heat production during immersion in water. Experiments in water-filled tanks showed that physical activity, including shivering, leads to a faster rate of heat loss than does remaining still.[25, 26]

The body's second defense against heat loss is a layer of fat in the body's protective, outer shell. A third is the process of thickening the shell layer by vasoconstricting peripheral blood vessels. Alcohol inhibits that defense by promoting peripheral vasodilation. The drinker may feel warmer, because skin thermal receptors will send back "warm" sense perceptions to the hypothalamus, but the person is likely to be losing heat faster. [27]

A fourth defense involves conscious behavior. People respond to feeling cold in several ways. One way is to exercise, which, as we have shown, could result in net heat gain or loss, depending on the degree of PPE worn, the thickness of the body's fat layer, and ambient temperature. Behaviors that almost always help include donning appropriate PPE and removing wet clothing. Drinking warm, noncaffeinated, nonalcoholic beverages provides heat directly to the core and is also a good method of hydration. Simply moving to a warmer environment is a defensive behavior. The use of heat packs can provide additional heat to hands and other body parts. Note, though, that external heat should never be placed directly on hypothermic skin as severe burns can result.

Defensive measures can also be taken prior to cold exposure. Eating sufficient nutritious meals before cold exposure can make a difference. Studies have shown that subjects eating highly protein-deficient diets were more likely to suffer from cold stress than subjects who ate more balanced meals. Getting enough rest and sleep prior to cold exposure and being properly hydrated are other helpful behaviors. Setting up shelters, when possible, obviously lessens the effects of cold. The greater an individual's awareness and knowledge, the more effective their behavioral responses will be.

There actually may be a fifth defense against heat loss. This is called counter-current heat exchange. The defense is well developed in some arctic animals and may apply to humans as well. Imagine warm arterial blood (oxygenated blood leaving the heart) perfusing cold peripheral tissues, which in turn transfer the heat to the surrounding environment. In addition, imagine the now-cooled returning blood (venous blood) stealing heat from the warmer core. Now, instead of this heat-loss scenario, picture that, in the peripheral tissues, a vein filled with cooled blood runs alongside a warm blood-filled artery. Instead of losing heat only to the cold peripheral tissues, the arterial blood now loses a part of its heat to the cold venous blood that is

Photo 7-9 A transport device should be used to prevent unnecessary exertion to reach the diver point-of-entry, as demonstrated here.

returning to the core. The result is that the body conserves more heat.

Cold water can quickly exhaust even highly fit individuals. The US Navy tells us, "In cold water, the ability to concentrate and work efficiently will decrease rapidly. Even in water of moderate temperature (60°F to 70°F [16°C to 21°C]), body heat loss to the water can quickly bring on diver exhaustion."[28]

Photo 7-10 If there is no other choice but to have divers walk back to shore, make sure they do not wear or carry their weightbelts and that a tender stays with them until they reach the rehab station.

Unnecessary exertion should be prevented before, during, and after the rescue operation. "Prior physical exercise may predispose a person to greater heat loss and to experience a larger decline in core temperature when subsequently exposed to cold air."[29]

The use of the ice board for transporting equipment and personnel between shore and victim will greatly reduce exertion in both surface and subsurface operations. It is important to plan how equipment and personnel will be transported from vehicles to the shore area. Rescue teams should look at all the potential sites in their district and practice accessing them as easily and efficiently as possible. They should factor in the possibility of deep snow during this planning.

Time of day is another factor that may affect cold stress, as most people have a lower core temperature in the early morning hours than later in the day. A study of immersed victims at 0700 and 1500 showed that vasoconstriction and shivering were not affected by time of day. "These observations raise the possibility that cold-water immersion may increase the risk of hypothermia in the early morning, because of a lower initial core temperature."[30] Therefore, extra care should be taken during these hours.

Individual differences

Everyone is different, so do not compare yourself to anyone else. The US Navy acknowledges inter-individual and intra-individual factors. A note on their Water Temperature Protection chart explains "This chart can be used as a guide for planning. However the individual diver's physical condition, body fat, and recent experience in cold water will influence how long each diver can withstand exposure to different and extreme temperatures."[31]

As stated earlier, percentage of body fat is an important inter-individual variable. Plasma beta-endorphin levels in the blood have been used as an indication of physiological stress. One study immersed people up to their chests in 18°C to 27°C (64°F to 81°F) water for two hours.[32] People with high percentages of body fat had lower levels of these endorphins than did people with low levels of body fat. The latter group was more physiologically stressed by the cold-water immersion than the group with a higher body fat percentage.[33]

Body size is a factor separate from percentage of body fat. Animals (including humans) that maintain a relatively constant core temperature despite greatly varying environmental temperatures are called warm-blooded. In order to do this, warm-blooded animals have to generate enough heat to warm their bodies. The larger the body, the more heat that has to be generated and vice versa. In other words, heat production is generally proportional to body mass. A large person can create and store more heat.

Another factor is the surface-area-to-body-mass (S-M) ratio. Consider two people with the same body mass and same body fat percentage. One person has very long, thin extremities with a short torso, while the other person has a very large torso with short, thick extremities. The former person will have a greater S-M ratio, and, all other things being equal, will lose heat more rapidly than the stockier person. In simpler terms, consider a meatball that is too hot to eat. Round shapes have low S-M ratios, so they lose heat slowly. If you want to eat the meatball right away, what can you do to speed up the cooling process? Smash it flat into a pancake shape to increase the surface area without increasing the mass. Immersed infants will cool more quickly than adults, because infants have a higher S-M ratio and because they produce less heat than adults.

Our hands have a higher S-M ratio. Mittens keep fingers together, so they are warmer than when they are separated in gloves. It is important to note that while mittens keep hands warmer, they may inhibit function. If the mitten wearer cannot perform tasks such as closing a diver's carabiner or putting the pony regulator mouthpiece in the diver's neck ring, there is a good chance he/she will remove the mittens. As a result, bare hands will be used to do the job, and may become functionless from the cold exposure. It is better to wear hand protection that will be kept on continuously. Divers should not wear mittens unless they have shown that they can perform all standard and contingency skills with them.

Some people's hands become colder much more quickly than other people's, especially if they have had a previous cold injury. Well-insulated dry gloves are important for such people both above and below the water.

Ears are another body part with a large S-M ratio, which is why they are considered heat radiators on the head and should be protected.

Amount of muscle can also be a factor in varying the degree of cold stress. The results of a study of forearm tissue conductivity in regard to heat loss during cold-water immersion (15°C to 36°C, 59°F to 97°F) showed that muscle tissue accounted for 92% of the total forearm insulation.[34] There is a great inter-individual variance in the proportion of muscle tissue in the body.

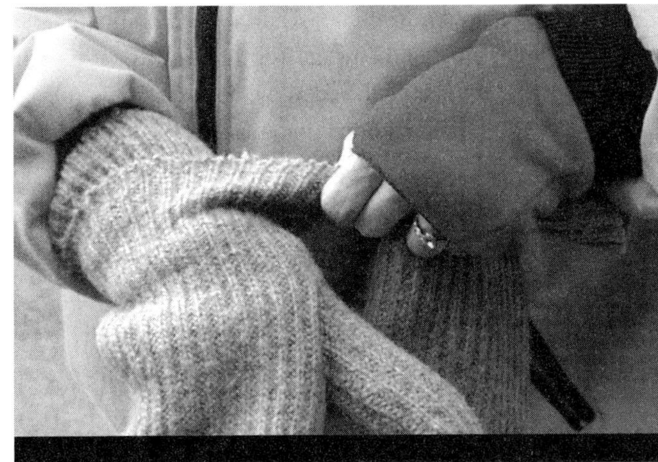

Photo 7-11 Double-layer wool mittens keep hands warm even when soaking wet and ice-covered. Wear some type of half-glove or liner underneath if it should be necessary to remove the mittens to perform tasks requiring dexterity.

Ice diving instructors and departments with dive teams must pay close attention to individual cold tolerance and must ensure that individuals dress appropriately for their degree of tolerance or intolerance. It is important to create a social atmosphere that praises, rather than degrades, an individual for admitting to feeling cold.

Facial Immersion

Technicians, especially divers, should perform facial immersion procedures immediately prior to water/ice operations. We have observed that facial immersion for at least fifteen seconds decreases the chances of a gasp reflex occurring when a diver's mask is dislodged underwater. Planned breath-holding facial immersions in cold water can cause bradycardia (lowered heart rate). The degree of bradycardia is inversely proportional

Photo 7-12 Kiss the water hello. The diver breathes on the pony mouthpiece attached to the neck strap for fifteen seconds without a mask for facial acclimation. The primary second stage is kept dry. Notice the diver remaining horizontal to decrease vertical position problem.

to water temperature.[35] Pre-operation facial immersion, what we call "kiss the water", may increase the ability of rescuers to survive if something goes wrong.

Facial immersion can lower the heart rate because facial receptors, especially those found around the nose and mouth, send direct signals to lower heart rate.[36] Heart rate is also lowered indirectly by the peripheral vasoconstriction, which sends greater volumes of blood to the core and therefore back to the heart. This is called increased venous return. A greater venous return, meaning more blood flowing into the heart, results in a greater volume of blood pumped out of the heart during each ventricular contraction (heart beat). This pumped-out volume is called stroke volume.

The heart tries to pump out a constant volume of blood each minute when a person is at rest. Therefore, if stroke volume increases, heart rate will reflexively decrease. If the heart rate does not reflexively decrease, the greater stroke volume causes the heart to take in and pump out a greater total volume of blood per minute. Pumping more blood is what the body should do during exercise, not during rest.

These circulatory functions are altered in unprotected victims of accidental immersion/ submersion. The direction and degree of the alterations may be related to how victims are immersed or submerged and what kind of PPE they have. Cold-water research that simulates conditions to produce these circulatory effects in immersed subjects is typically conducted in a controlled, unstressed manner, indoors, with physically fit individuals in water temperatures above 10°C (50°F). Most of these studies involve head-out immersion, not full submersion. The findings of these studies demonstrate blood pressures ranging from slightly higher than normal to below normal, but the fear, panic, pain, and exertion of actual accidental immersion may override the cold water-induced heart rate drop reflex, and can significantly raise heart rate and blood pressure.

One study that involved very cold (41°F, 5°C) moving water also supported this theory that less-protected individuals will have fewer beneficial cardiac alterations. Healthy men were submerged three times, in three different PPEs: a cotton overall, a wetsuit, and a drysuit. Significantly, more subjects developed the bradycardia when wearing the drysuit. The researchers concluded "the increasing cold stress experienced by individuals during cold-water submersion reduces the incidence of diving bradycardia".[37] One hypothesis is that the subjects, when wearing proper PPE/drysuits, were better able to lower their heart rates to correct higher stroke volumes than when not wearing appropriate cold-water PPE.

Divers who do not take the time to "kiss the water" may begin a dive with a higher than normal heart rate, and even breathing rate. The act of kissing the water helps divers start the dive in a better physiological and mental state, and subsequently, if something does go wrong, they are better able to manage the problem.

Immersion and Cold Diuresis

Two procedures that should be included in surface and subsurface ice operations are (1) pre-operation and post-operation blood pressure checks for divers, and (2) pre-operation and post-operation hydration for both divers and tenders. If tenders could possibly become immersed, such as in thin-ice public safety diving operations, their blood pressures should be checked as well.

One of the reasons why hydration is necessary has to do with why we feel the need to urinate when we become cold or are immersed in water. Cold, as described earlier, causes peripheral vasoconstriction, which pools blood into the core. More blood in a smaller area results in greater pressure. Furthermore, water exerts a little less than .5psi (.034bar) per foot of pressure on the body, which causes further pooling of blood in the body's core. This is because, when standing on land, gravity pulls blood to our legs. Because veins can expand much more than arteries, they can hold more of the blood being pulled into them by gravity. Thus blood pools in the leg veins. A standing person will have greater blood pressure in the legs than in the arms, because ambient air pressure is not strong enough to push the pooled leg blood back up to the core. Water pressure, on the other hand, is strong enough to counteract the increased blood volume and pressure in the legs. Therefore, when immersed, more blood remains in the core than when standing on land.

All of us have experienced the result of this increased blood volume in our cores from cold or immersion, which is the urge to urinate. The process is called cold or immersion diuresis (urine production from cold/immersion). When blood pools in the core, the upper two chambers of the heart swell from the increased volume of blood entering them. The atria respond by secreting atrial natriuretic factor (ANF), which causes diuresis (urine production).[38] Simultaneously, this blood pooling causes the hypothalamus to suppress the anti-diuretic hormone (ADH). The suppression of ADH allows the kidneys to filter more water from the blood, thus creating urine.

Photo 7-13 All tenders and divers on a dive team should be capable of performing pre- and post-dive blood pressure checks.

Once water is filtered out of the blood and urine is created, the volume of blood is diminished, thereby reducing blood pressure. The colder the water, the higher the rise in

Photo 7–14 Do not overheat shelters. Keep them at normal room temperature to prevent after-drop shock. This diver was cold-stressed and possibly dehydrated, and upon entering the shelter, felt lightheaded and weak. Notice the yellow propane heating vent tube.

blood pressure, the more water filtered from the blood, the more urine created, and the greater the reduction in blood volume. If you do not feel the need to urinate after ten or so minutes of immersion in cold water, there is a good chance that you are dehydrated.

A diver's blood pressure should be checked prior to suiting up to ensure that it is not too high. It should be checked again post-dive after the diver has been suited down by a tender to make sure that it is not too low. Low pressure is the concern at this point because, when divers are removed from the water and placed in a warm vehicle or shelter, peripheral vasodilation shunts blood from the core to the peripheral tissues. Now the lowered volume of blood is unable to fill the larger space. A noticeable drop in blood pressure can occur if a diver is not properly hydrated. This is one of the reasons that external rewarming techniques of hypothermic patients can cause "after-drop" or "rewarming" shock. This is simply shock caused by a drop in blood pressure from the movement of a low blood volume to a larger space, namely the peripheral tissues and the core.

If you have ever felt light-headed after stepping into a warm place from a very cold one, you are experiencing this type of shock. Older individuals will experience these diuretic affects more quickly than young adults.

It is recommended that divers drink 8oz (237mL) or more of water prior to and after diving. Hydrate with warm fluids that do not have caffeine or high sugar content, if possible. Caffeine is a diuretic that has no physiological benefit during exercise in cold water.[39] It is important that tenders ensure their divers drink fluids even if they are not thirsty, because the release of ANF reduces thirst and immersion itself has been known to decrease thirst.[40]

Research has been conducted to determine whether drinking a glycerol solution prior to diving helps inhibit the plasma fluid loss from immersion diuresis. The glycerol solution will produce this effect on land, but it has not been shown to be effective for immersed subjects.[41] Immersion diuresis overrides the effects of the glycerol.

The release of epinephrine and norepinephrine compounds the immersion/cold diuresis effect. Therefore, anything that causes more of these two hormones to be released, such as stress, fear, or major exertion, should be monitored and if possible prevented. To compensate for these effects, drink enough fluids to make urine clear and copious. Water-soluble vitamins may give urine a bright yellow color, but it should not be dark yellow and especially not orange-hued.

Immersion and cold diuresis are also often accompanied by losses of sodium and potassium, which means that leg cramps are more likely to occur. [42,43] A leg cramp under an ice roof is worth preventing. Proper hydration, pre-dive stretching, and consumption of plums or a banana at least a few hours pre-dive is especially important to divers who know they are susceptible to muscle cramps.

One study suggested that physically fit subjects may experience milder immersion diuresis than less fit individuals.[44] This is yet another reason for divers to maintain good cardiovascular fitness.

There are several ways for plasma volumes (blood volumes) to decrease and hematocrit (the concentration of cells such as red blood cells in the blood) levels to increase. One is the cold/immersion diuresis process described above. However, fluid from the blood is not only filtered out by the kidneys, it is also shifted from the blood to the tissues.[45] A study of subjects bicycling with head-out immersion indicated that the exercise caused decreased plasma volumes and increased hemoconcentration (hematocrit/hemoglobin), and that the latter was completely reversed by thirty minutes of resting.[46] Once urine is created, the fluid in the urine cannot be put back into the circulatory system. The plasma volume loss in the bicyclers occurred between the blood and tissues and was reversible. The same study also found that the degree of hemoconcentration, or the thickness of the blood, was proportionate to exercise intensity.

What we learn from the study is that diving procedures should involve minimal exercise, proper hydration, and resting rehabilitation for at least thirty minutes after diving. Without this rehabilitation, divers will not be at their peak fitness to perform their next in-water session.

To manage the effects of cold-water immersion, divers and in-water tenders should do the following:

1. Remain well hydrated—before and after each dive.
2. Maintain good cardiovascular fitness.
3. Use procedures that prevent unnecessary exertion and hanging at the surface.
4. Perform pre-dive and post-dive blood pressure checks.

5. Ensure that divers are mentally, physically, and emotionally ready to dive before suiting up.
6. Enter the water slowly. Try sitting on the ice roof with legs in the water for a few seconds, then making a slow, gentle roof turn with the tender's assistance as described in chapter 9.
7. Kiss the water.
8. Wear appropriate thermal PPE.
9. Obtain the best training possible to decrease mental, emotional, and physical stress.
10. Keep a well-trained chief tender with the diver from the time of suiting up to the rehabilitation period, which should be at least thirty minutes.

Divers and their chief tenders should remain together until the divers have been checked out by EMS, rehydrated with warm fluids, and changed into dry clothing. [47] If there are not enough personnel and the chief tender is needed elsewhere, the chief tender can turn the technician over to another qualified individual who can assume those responsibilities.

An understanding of the causes of cold stress and immersion hypothermia is crucial for keeping divers and tenders as safe as possible. This knowledge enables teams to develop standard operating procedures and guidelines to help prevent problems and to provide the foundation for the risk/benefit analysis and on-scene plan of action. An understanding of hypothermia teaches us that contingency and self-rescue procedures must be reflexive, because mental capabilities may be well below normal. "Mental impairment is an early effect of general body cooling."[48] We cannot just assume "we will be okay." Hypothermia can kill. Wear sufficient PPE, and never shortcut with training, equipment, or personnel. This is true for all cold-weather operations, be they highway motor vehicle accidents or ice diving operations.

Mammalian Dive Reflex

Now that we have a better understanding of immersion hypothermia, we can look at some of the misconceptions surrounding the mammalian dive reflex. Almost everyone has heard of miraculous saves of children and even adults who have been submerged in cold water for extended periods up to an hour.[49] Credit for these saves is due in large part to the father of cold-water drowning saves, Dr. Martin Nemiroff. When asked how these rescues happen, the majority of people, including trained public safety personnel, will say, "the mammalian dive reflex." When an incident commander makes a risk/benefit analysis on an ice-drowning scene, knowledge of such golden-hour miracles often comes into play.

Photo 7–15 The mammalian dive reflex can be considered a misnomer when applied to humans. Unlike humans, marine mammals can slow their oxygen consumption and have extended breath holds during cold-water submersions. When the reflex is applied to humans it refers to those in cardiac arrest, while it is applied to living marine mammals.

Aquatic mammals and humans each have reflexes that are initiated by cold-water immersion, but the reflexes are very different. The reflexes of the two groups should not be combined under the single term "mammalian dive reflex." A major distinction is that the reflex, in aquatic mammals, involves the ability to perform impressive breath holds with durations of as long as two hours. For certain whales and seals, this behavior is a normal part of everyday existence. It is not just done in extreme situations.

In humans, the mammalian dive reflex is often incorrectly thought of as keeping drowned (dead), submerged humans in a physiological state that may allow them to be successfully resuscitated and brought back to normal or near normal lives.

So let us first compare live humans to live aquatic mammals. The reflex for the latter involves an ability to conserve oxygen for extended periods while breath holding underwater. This is done by decreasing heart rate and metabolism. Metabolism is the process that enables living beings to combine oxygen with fuel to create water, energy, and waste products. In humans, for example, glucose ($C_6H_{12}O_6$) is combined with oxygen (O) to produce water (H_2O), carbon dioxide (CO_2), and energy. By slowing their metabolisms, aquatic mammals reduce their need for oxygen and subsequently, protect their cells from dying from hypoxia (too little oxygen) during extended breath holds.

A lowered heart rate means that the heart is working less and therefore needs less oxygen to keep its cells alive. Some marine mammals can drop their heart rates by 90%, unlike humans, who can drop theirs by 50% but may not do so at all. Humans immersed in cold water may have reduced heart rates in controlled, calm conditions if the water is not cold enough to become painful, but they certainly do not conserve oxygen. Instead, a human's need for oxygen in cold-water immersion grows as the thermoregulatory center raises the metabolism to increase heat production in an attempt to maintain core temperature. Shivering further increases the need for oxygen, as does an increased heart rate in response to fear, pain, or exertion.

Aquatic mammals significantly decrease blood flow to their flippers by vasoconstriction. Human vasoconstriction is dependent on, and inversely correlated with, temperature. Human breath-holding ability, and the ability to survive underwater, is lessened by immersion in cold water, while aquatic mammal breath holding is immensely lengthened.

Weddell seals show significant increases in hemoglobin concentration and hematocrit when performing breath-hold dives. In 1990, Korean Ama (female repetitive breath-hold divers) and a control group were studied to determine whether humans had that same dive reflex response. The control group was unaffected by diving, but the Korean diving women showed increases in hemoglobin and hematocrit levels.[50]

In drowned humans, the pooling of blood in the core and the greatly reduced metabolism that occurs in cold water after death is believed to be part of the reason long-term drowning victims (ten or more minutes of submergence) can sometimes be successfully resuscitated and kept alive. This is not an example of the mammalian dive reflex.

Therefore, if a victim is simply hanging on the edge of the ice roof and breathing as their body cools prior to sinking, the likelihood of survival could be increased. If, however, the victim is struggling so that the body is using up oxygen and cannot slow its metabolism or reduce oxygen requirements, the odds of survival decline.

Nitrogen Narcosis and Hypothermia

It is useful to examine the effect on the body's thermoregulatory process of breathing higher partial pressures of nitrogen. We learn that hypothermia prevention includes a maximum ice dive depth standard shallower than 100ft (30m).

In normal air conditions, the human body can have a core temperature change of plus or minus 0.7°C (1°F), without the initiation of sweating or shivering, respectively.[51] This range of 1.4°C (2°F) is called the "null zone." This null zone can be widened by breathing 30% nitrous oxide (30% N_2O). This indicates that an increased partial pressure of nitrogen causes a loss of fine thermoregulation.[52] When the null zone expands, it shifts downward, so that the threshold for shivering decreases, but the threshold for sweating stays the same. Thus the body's defense mechanism of increasing heat production through shivering does not occur until a person's core reaches a lower than normal temperature. Nitrogen narcosis has been shown to decrease heat production in 28°C (82°F) and 15°C (59°F) immersions.[53,54] Breathing 30% N_2O has proven to be a good substitute for hyperbaric nitrogen and has similar effects to breathing air at 210fsw (64msw) on specific performance tasks.[55]

When shivering (heat production) is not initiated, the only other defenses available to the body are vasoconstricting peripheral blood vessels and initiating behavioral responses such as exiting the water. The latter are more effective. A study of healthy male subjects who refrained from exhaustive exercise or alcohol consumption for twenty-four hours was conducted to see if breathing 30% N_2O during head-out immersion in 28°C (82.4°F) and 15°C (59°F) water affected their perception of cold.[56] If the men were desensitized, protective behavioral responses were less likely. All of the subjects reported that the immersions were more comfortable when breathing the 30% N_2O. They all gave higher thermal comfort votes at all normal to below normal core temperatures, and all reported feelings of euphoria.

Perception of thermal comfort is determined by core and skin temperatures. The study found that the source of the perception of thermal comfort affected by 30% N_2O was core temperature. The hypothalamus, the thermostat of the body, appears to be affected by the anesthetic effects of 30% N_2O. In essence, nitrogen narcosis increases the risk of hypothermia by inhibiting thermogenesis (shivering) and by inhibiting the protective behavioral responses because cold perception is dulled.

Divers do not have to go below 200fsw on air to feel the effects of nitrogen narcosis. Factors such as exhaustion, alcohol, cold stress, anxiety, and increased carbon dioxide levels may increase susceptibility to nitrogen narcosis at much shallower depths.

Hypercapnia and Hypothermia

Another factor that can lower the threshold for shivering is increased carbon dioxide levels, or hypercapnia. Let us look first at how hypercapnia can occur in divers and then at what general effects it can have.

Although divers in all environments need to be able to monitor their own physical and mental status during dives, self-monitoring becomes even more critical during ice dives. A very useful indicator is breathing rate and quality, which can easily be monitored by tenders during non-overhead environment dives. The only accurate method of breath monitoring by tenders in ice diving is through a hardwire or wireless communication system, which, although recommended for overhead diving, is not always available. A line-signal system can be used to monitor breathing rates, but it is not as accu-

Photo 7-16 The diving supervisor is called over to listen to a high diver breathing rate that the backup tender records.

rate, because divers are aware that their breathing rate is being monitored, and because the quality of breathing cannot be assessed. Self-monitoring is also exceedingly critical in ice diving because divers cannot simply ascend to the surface should they feel the need to. This is because ice divers are often solo divers, and because cold can increase the severity of other problems.

Increased levels of carbon dioxide in the blood indicate that something is wrong; they can also initiate a chain of events that can lead to serious injury or even death. Hypercapnia impairs the thermoregulatory response and can hinder problem-solving and self-rescue skills.

Hypercapnia can result from external or internal causes. Externally high levels of carbon dioxide can occur from excessive dead-air space in an underwater breathing apparatus, failing carbon dioxide scrubbers in closed-circuit re-breathers, or poor helmet ventilation in surface-supplied diving. Endogenous, or internal, levels of carbon dioxide can rise during overexertion or when regulators impede respiration.

Divers in general appear to have a higher degree of carbon dioxide retention than non-divers.[57] It is important that divers be able to recognize the symptoms of increased levels of carbon dioxide. One study found the most common symptoms reported by divers were breathlessness, increased depth of breathing (above normal for the workload), and increased rate of breathing (above normal for the workload). Other symptoms included: headache, dizziness, sweating, flushed sensations, nausea, irritability, lack of concentration, narcotic sensations, visual distortions, numbness, and mental disorientation. The study also found "Despite no overall significant difference in post-test symptom ratings between cold and warm immersions, there was a tendency for more pronounced dizziness and visual distortions during cold carbon dioxide exposures than during warm carbon dioxide exposures."[58]

The same study found that hypercapnia caused significant restriction in blood flow to divers' forearms already inhibited by the cold-water immersion. Increased carbon dioxide levels can cause a greater decline in temperature. Hypercapnia was found to lower the shivering threshold and increase core cooling rates in people mildly stressed by cycling underwater.[59]

Long-Term Effects of Cold-Water Immersion

The ears may be an area of special concern for divers who spend a great deal of time in cold water over a period of years and who do not dive with dry hoods. Long-distance swimmers, surfers, and working divers can develop "surfer's ear", or exostosis, which is a broad-based growth that emerges from the bony portion of the external auditory canal. Ordinarily, these growths have little or no effect, but with prolonged exposure to cold water, the growths can become large enough to obstruct the external ear canal. A study of 194 Japanese military divers showed that the incidence of exostosis increased with the length of the diving career, and that the group of divers who worked in cold water had a greater incidence than the group that worked in warmer water.[60] Prevention of extosis is another good reason to wear a dry hood when ice diving.

Repeated cold immersions have been found to induce changes in thermal homeostasis (the ability to maintain a normal core temperature). Young sportsmen were subjected to one-hour, head-out immersions in 14°C (57°F) water three times a week for four to six weeks. Results showed a "cold acclimation," in which subjects showed lowered central and peripheral temperatures at rest and during cold immersion. "The metabolic response to cold was delayed and subjective shivering was attenuated."[61] Subjects also showed a lowered sensitivity to cold. The cold-acclimated subjects saved 20% of their total heat production during cold-water immersion. All these changes were noticed after only four cold-water immersions and lasted for at least two weeks after the immersions. The study also noted a tendency toward increase in body fat for the cold-water immersion subjects, and a tendency toward increased vasoconstriction of peripheral tissues.

What does this mean for ice divers? Divers who dive often (several dives per week), and who do so in wetsuits may develop some cold-acclimation physiological responses. These divers need to understand that other divers may feel colder more quickly and should respect those feelings. An example of this is when the captain of a dive team or a sport diving instructor becomes cold-acclimated and then loses the ability to understand the needs of team members or students who are not cold-acclimated.

The next step is to learn more about sign/symptom recognition of cold stress and hypothermia and what can be done in the field to begin managing these conditions.

Other Physiological Concerns

Carotid sinus reflex

The word carotid comes from the Greek "Karotides", which comes from the word "karoo," to put to sleep. Compression of the carotid arteries can result in unconsciousness and, in very rare cases, death. Heart rate is regulated and affected by a variety of factors, one of which is the blood pressure in carotid arteries. Carotid sinus receptors measure this blood pressure and signal the cardio-inhibitory center in the brain when carotid artery blood pressure is high. When stimulated, the cardio-inhibitory center slows the heart rate through nervous impulses. When blood pressure drops in the carotid arteries, the carotid sinus receptor stops signaling the cardio-inhibitory center, and heart rate increases.

Compression of the carotid arteries by a tight hood, drysuit neck seal, or wetsuit provides the carotid artery receptors with a false indication of high blood pressure. The receptors signal the cardio-inhibitory center in the brain, which in turn alerts the heart to slow down. As the heart slows down, the brain receives less blood. As the suit or hood continues to compress the arteries, the receptors keep sending false signals to the brain and the heart continues to slow down. The diver will typically begin to feel lightheaded, nauseous, and weak as blood flow is reduced to the brain. If the compression is not alleviated, the carotid sinus reflex can result in unconsciousness or death.

Photo 7-17 Do not let divers dive with over-tightened neck seals to prevent the diver from experiencing carotid sinus reflex.

Although the condition is potentially dangerous, prevention is simple and easy. Never wear anything that will put any amount of pressure on the carotid arteries. If divers are learning to dive in a drysuit, they should be monitored to ensure that no lightheaded feeling or discomfort is felt at any time.

Summary Questions

The numerous questions in this chapter are provided so that ice instructors can use this book and quiz section as a patient-care final exam. During both surface and diving ice rescue courses, the chapter can assist EMS personnel in receiving EMS CEU credit in those states that offer it. Students can answer these questions during the class and, after grading by the instructor, can use the answer section and book to debrief.

1. An understanding of the effects of cold and immersion on our physiology gives us the ability to ______ these problems from happening, ________ signs of these problems, and provide __________.
2. Sudden immersion in cold water can cause a _______ reflex that can result in aspiration if PPE does not keep the airway out of the water.
3. Aspiration can cause the epiglottis and pharynx to seal the trachea, this situation is called a ___________.
4. List four possible consequences of the decreased carbon dioxide levels caused by hyperventilation that can occur during sudden, unprotected cold-water immersion.
5. Unprotected cold-water immersion can feel very _______.
6. Cold-water immersion can cause intense shivering, muscle ______, and muscle weakness.
7. List at least four possible consequences that pain can cause.
8. Cold ambient temperatures cause peripheral vaso_________.
9. Head-out immersion can increase cardiac output by ______ percent more than the same temperature air.
10. Cold-water immersion can increase the incidence of _________ heartbeats.
11. Shivering can increase ________ ________ by 50%.
12. Immersion can ________ (reduce, increase) vital capacity, and head-out immersion can cause ______ pressure breathing.
13. Cold water ______ the need for oxygen, but ______ our ability to deliver oxygen to our tissues.
14. The stress of cold exposure, especially in conjunction with exercise, activates the _______ _______ with increases in natural killer cell count and other factors.
15. State the medical definition of hypothermia.

16. Simply feeling cold can cause task loading and stress, which can be worsened when a drop in skin temperature causes the release of the hormone ___________.

17. Explain why and how mental function can be impaired by cold stress.

18. Cold is the _______ of heat.

19. We lose heat in _____ °F water as fast as we do in 42°F air.

20. Define conduction.

21. Provide at least two examples of tender and diver heat loss by conduction.

22. Water conducts heat away from an object _______ times faster than air of the same temperature.

23. Water requires a great amount of heat to raise its temperature because it has such a tremendous _____ ______.

24. To reduce conductive heat loss from feet, wear thick _______ socks under ice rescue suit boots.

25. Skin cold receptors on an unprotected torso are believed to be one cause of the _______ reflex when they are stimulated.

26. In regards to patient handling, remember _______ over, ______ under, ______ all around.

27. Blood loss can ________ (increase, decrease) the chances of becoming hypothermic, and hypothermia _______ (increases, decreases) bleeding time.

28. Avoid transport devices that are made of ________ to decrease the amount of conduction.

29. Define convection.

30. Explain how convective heat loss can occur in still water.

31. List four important heat radiator regions from our necks up that need proper insulation from wind and cold.

32. An unprotected victim immersed in cold water should kick and exercise to create heat to increases chances of survival. True or False?

33. Define evaporation.

34. We normally lose ____ percent of our total heat loss by evaporation, and this percentage _______ if we are wet.

35. Cotton is a _____ choice to wear in the winter.

36. It is important to ______ wet clothing from patients and our own bodies when working in air to decrease evaporative heat loss.

37. A sport _____ is useful to dry ourselves, patients, and equipment during an operation.

38. We lose heat from our core by evaporation and convection when we ______ in cold environments.

39. If local protocol allows it, _____ and ______ oxygen before administering it to cold stressed patients.

40. _______ is the emission of infra-red energy.

41. List at least five ways public safety personnel can allow a motor vehicle accident patient to become cold stressed and even hypothermic.

42. List five reasons why victims pulled from the water are typically repeatedly banged, yanked, and dropped.

43. Our core temperature can drop ___ to ___ degrees during early morning sleeping hours.

44. We lose approximately ___ to ___ percent of our heat from our head at rest.

45. Our skin temperature is normally ______ than our core temperature.

46. At rest, muscle tissue ______ (increases, decreases) the thickness of a body's protective insulation, while during exertion, muscle tissue can _____ (increase, decrease) heat loss to a body not wearing appropriate PPE.

47. Describe an effect of peripheral vasoconstriction in an unprotected body in a cold environment.

48. Peripheral vasodilation from external rewarming can result in a ______ core temperature in a person with cold peripheral tissues.

49. List three ways the body can increase internal heat production (metabolism).

50. Thin people, as compared to people with more fat, typically shiver _____ (earlier, later), shiver at a ____ (higher, lower) core temperature, and may have a ______ (higher, lower) oxygen consumption during cold-water immersion.

51. Everyone shivers when cold. True or False?

52. List two physiological effects of alcohol consumption on defenses against cold.

53. List the four main defenses against heat loss in cold environments.

54. List seven defensive behaviors to help prevent hypothermia.

55. State a safety rule in regards to heat pack use.
56. Physical exercise prior to cold exposure may predispose a person to heat loss. True or False?
57. Shivering and vasoconstriction do not seem to be affected by time of day. How does this relate to the risk of cold stress?
58. People with _____ fat percentages are more physiologically stressed in cold water than are people with ____ fat percentages.
59. A ______ (large, small) person can create and store more heat.
60. Describe effective hand protection.
61. An atmosphere must be created that _____ individuals for admitting when they feel cold stressed.
62. Planned breath-holding facial immersions in cold water can cause _____ heart rates.
63. Technicians should perform facial immersions prior to diving or performing surface ice rescues. True or False?
64. Cold-water immersion/submersion without sufficient PPE can lead to _____ (increased, decreased) blood pressure and heart rate.
65. The urge to urinate that occurs during cold-water immersion is caused by _______ and ________ _______.
66. Explain two processes that cause cold- and immersion-induced core blood pooling.
67. Explain the function of technician pre- and post-operation blood pressure checks.
68. Explain after-drop/rewarming shock.
69. Explain why it is important to hydrate in the winter even when not thirsty.
70. Immersion and cold diuresis can decrease sodium and potassium levels, which can increase the risk of muscle _______ if diet is not used to correct the loss.
71. Exercise while immersed can also cause fluid from the blood to shift to the _______, which can be reversed by thirty minutes of rest.
72. List ten steps to prepare in-water technicians for the effects of cold-water immersion.
73. Give differences between the effects of breath holding in cold-water immersion in humans and marine mammals.
74. The null zone of temperature change where humans do not shiver or sweat is a change of plus or minus ____.
75. Increased levels of nitrogen can _______ (widen, narrow) this null zone, which means that increased levels of nitrogen can ______ (increase, decrease) the chances of cold stress.

76. Breathing increased levels of nitrogen can give feelings of _____ (increased, decreased) thermal comfort and euphoria, which may decrease defensive behaviors.

77. Define hypercapnia.

78. What can hypercapnia do to blood flow in divers' forearms and to the body temperature that initiates shivering?

79. Explain what "surfer's ear" is.

80. Repeated cold-water immersion can result in a degree of cold-acclimation. True or False?

81. Explain carotid sinus reflex.

82. What can cause a carotid sinus reflex during ice operations?

83. List symptoms of a carotid sinus reflex.

Notes

[1] K.E. Cooper, S. Martin and P. Simper, *Factors Causing Hyperventilation in Man During Cold Water Immersion: Human Performance in the Cold* (Undersea Medical Society, eds. G. Laursen, R. Pozos and F. Hempel, 1982): 52-57.

[2] W.R. Keating, M.B. McIlroy and A. Goldfien, *Cardiovascular Responses to Ice-Cold Showers, J Appl Physiol.* (1964, 19): 1145-1150.

[3] L. Jansky, P. Sramek, ? Savlikova, B. Ulicny, H. Janakva and K. Horky, "Change in Sympathetic Activity , Cardiovascular Functions and Plasma Hormone Concentrations Due to Cold Water Immersion in Men," *Eur J. Appl. Physiol* (1996, 74):148-152.

[4] K.S. Park, J.K. Choi and Y.S. Park, "Cardiovascular Regulation During Water Immersion," *Appl Human Sci* (1999, 18): 233-241.

[5] This increased cardiac output was mainly due to increased stroke volume (the volume of blood expelled by the left ventricle), which was indicated by a significantly greater left ventricular end-diastolic volume.

[6] A. Gabrielsen, L.B. Johansen and P. Norsk, "Central Cardiovascular Pressures During Graded Water Immersion in Humans," *J Appl Physiol* (1993: 75:2): 581-585.

[7] N.A. Taylor and J.B. Morrison, "Lung Centroid Pressure in Immersed Man," *Undersea Biomedical Research* (1989, 16:1): 3.

[8] R.A. Hebden, B.J. Fruend, J.R. Calybaugh, W.M. Ichimura and G.M. Hashiro, "Effect of inspiratory-phase negative pressure breathing on urine flow in man," *Undersea Biomed Res* (1992, 19:1): 21-29.

[9] S.M. Horvath and F.B. Julian, "Overview: Exercise in a Cold Environment Human Performance in the Cold," Undersea Medical Society, eds. G. Laursen, R.Pozos and F.Hempel(1982): 17.

[10] I.B. Mekjavic, C.J. Sundberg and D. Linnarsson, "Core Temperature 'Null Zone'," *J Appl Physiol* (1991, 71:4): 1289-1295.

[11] I.K. Brenner, J.W. Castellani, C. Gabaree, *et al*, "Immune Changes in Humans During Cold Exposure Effects of Prior Heating and Exercise," *J Appl Physiol* (1999, 87): 699-710.

[12] L. Jansky, P. Sramek, J. Savlikova, B. Ulicny, H. Janakova and K. Horky, "Change in Sympathetic Activity, Cardiovascular Functions and Plasma Hormone Concentrations Due to Cold Water Immersion in Men," *Eur J Appl Physiol* (1996; 74): 148-152.

[13] S. Martin, *Cold Water Immersion* (Ph.D. diss., University of Calgary) cited in Cooper K.E., Martin S., Simper P. *Factors Causing Hyperventilation in Man During Cold Water Immersion: Human Performance in the Cold* (1983).

[14] Y. Mebane, *Hypothermia* (Diving Medicine, eds A. Bove, J. Davis (1990, W.B. Saunders): 96.

[15] Naval Sea Systems Command, *US Navy Diving Manual,* 0910-LP-708, Revision 4, (January 1999): 3-12.2.

16 W.E.A. Burke and I.B. Mekajavic, "Estimation of Regional Cutaneous Cold Sensitivity by Analysis of Gasping Response," *J Appl Physiol* (1991,71:5): 1933-1940.

17 R. Valeri, R. Dennis, A. Melaragno and M. Altschule, "Resuscitation of Hypothermic-Hypovolemic Hypotensive Baboons," *Human Performance in the Cold,* eds. R. Laursen, R. Pozos and F. Hempel, Undersea Medical Society, Inc., 1982): 105-128.

18 M.B. Ducharme, W.P. VanHelder and M.W. Radmoski, "Tissue temperature profile in the human forearm during thermal stress at thermal stability," *J Appl Physiol* (1991, 7:1): 1973-1978.

19 E. Daviskae, I. Gonda and S.D. Anderson, "Local Airway Heat and Water Vapour Loss," *Respir Physiol.* (1991, 84): 115-132.

20 D.W. Rennie, "Human Thermal Balance at Rest and Exercise in Water: A Review," (9th Symposium on Underwater and Hyperbaric Physiology, UHMS, Bethesda, MD, 1987): 95-107.

21 E.L. Glickman-Weiss, F.L. Goss, R.J. Robertson, K.F. Metz and D.A. Cassinelli, "Physiological and Thermal Responses of Males with Varying Body Compositions During Immersion in Moderately Cold Water," *Aviat Space Environ Med* (1991, 62): 1063-1067.

22 R.S. Pozox, "Cold Stress and Its Effects on Neural Function," (*Human Performance in the Cold,* eds. Laurset, Pozos, Hempel, 1983): 25.

23 Ibid.

24 W. Van Dorn, "Thermodynamic Model for Coldwater Survival," AAUS Polar Diving Workshop, San Diego, CA (eds. Land and Stewart, May 1991): 10-17

25 J.S. Hayward, J. Eckerson and M.L. Collis, "Thermal Balance and Survival Time Predicition of Man in Cold Water," (*Canadian J. Physiol.,* Pharmacol. 53): 21-32.

26 W.R. Keatinge, *Survival in Cold Water: The Physiology and Treatment of Immersion Hypothermia and Drowning* (Blackwell Scientific, Oxford).

27 Ibid.

28 *U.S. Navy Diving Manual, Vol. I* "Air Diving" (Best Publishing Company, Rev. 3, 0994-LP-001-9110, 1993): 4-14.

29 J.W. Castellani, A.I. Young, J.E. Kain, A. Rouse and M.N. Sawka, "Thermoregulation During Cold Exposure; Effects of Prior Exercise," *J Appl Physiol* (1999, 87): 247-252.

30 JW Castellani, AI Young, JE Kain, A Rouse, MN Sawka. *Thermogulatory responses to cold water at different of day* (J Appl Phyisol. 1999, 87) pp. 243-246

31 *U.S. Navy Diving Manual, Vol. I* "Air Diving" (Best Publishing Company, Rev. 3, 1993, 0994-LP-001-9110): 4-15.

[32] E.L. Glickman-Weiss, A.G. Nelson, C.M. Hearon, F.L. Goss and R.J. Robertson, Are Beta-Endorphins and Thermogreulation During Cold-Water Immersion Related?" *Undersea and Hyperbaric Medicine* (1993, 20: 3): 205-213.

[33] E.L. Glickman-Weiss, F.L. Goss, R.J. Robertson, K.F. Metz and D.A. Cassinnelli, "Physiologic and Thermal Responses of Makes with Varying Body Compositions During Immersion in Moderately Cold Water," *J. Aviat. Space Environ. Med* (1991, 62): 1063-1067.

[34] M.B. Ducharme and P. Tikusis, "In Vivo Thermal Conductivity of the Human Forearm Tissues," *J Appl Physiol* (1991, 70:2): 2682-2690.

[35] E. Schagatay, B. Holm, "Effects of Water and Ambient Air Temperatures on Human Diving Bradycardia," *Eur J Appl Physiol* (1996, 73): 1-6.

[36] J. Bookspan, *Diving Physiology in Plain English* (Undersea and Hyperbaric Medical Society, Inc., 1995): 38.

[37] M. Tipton, "The Effect of Clothing on 'Diving Bradycardia' in Man During Submersion in Cold Water," *Eur J Appl Physiol Occup Physiol* (1989, 59): 360-364.

[38] N.C. Trippodo, A.A. MacPhee and F.E. Cole, *Partially Purified Human and Rate Atrial Natriuretic Factor* (Hypertension 1983, supplement 1): 81-88.

[39] T.J. Doubt and S.S. Hsieh, "Additive Effects of Caffeine and Cold Water During Submaximal Leg Exercise," *Med Sci Sports Exercise* (1991, 23): 435-444.

[40] S. Sagawa, K. Miki, F. Tajima, H. Tanaka, J.K. Choi, L.C. Keil, K. Shiraki and J.E. Greenleaf, "Effect of Dehydration on Thirst and Drinking During Immersion in Men," *J Appl Physiol* (1992, 72:1): 128-134.

[41] D.A. Arnall and J.R. Goforth, "Failure to Reduce Body Water Loss in Cold-Water Immersion by Glycerol Ingestion," *Undersea and Hyperbaric Med* (1993, 20:4): 309-320.

[42] M. Epstein D. Pins, R. Arrington, A. Denunzio and R. Engstrom, "Comparison of Water Immersion and Saline Infusion as a Means of Inducing Volume Expansion in Man," *J Appl Physiol* (1975, 39): 66-70.

[43] P. Norsk, F. Bonde-Peterson and N.J. Christensen, "Catecholamines, Circulation, and the Kidney During Water Immersion in Humans," *J Apply Physiol* (1990, 69:2): 479-484.

[44] J.R. Claybaugh, D.R. Pendergast, J.E. Davis, C. Akiba, M. Pazik and S. Hong, "Fluid Conservation in Athletes: Responses to Water Intake, Supine Posture, and Immersion," *J Appl Physiol* (1986, 61): 7-15.

[45] J.L. Sondeen, S.K. Hong, J.R. Claybaugh and J.A. Krasney, "Effect of Hydration State on Renal Responses to Head-Out Water Immersin in Conscious Dogs," *Undersea Biomed Res* (1990, 17:5): 395-411.

[46] A.C. Ertl, E.M. Bernauer, C.A. Hom, "Plasma Volume Shifts with Immersion at Rest and Two Exercise Intensities," *Med Sci Sports Exercise* (1991, 23): 450-457.

[47] A diver may have five or six line tenders to help pull the diver back and forth to shore on an ice board, but each diver has only one chief tender. The chief tender's sole responsibility is to his/her diver from the time of dressing right through rehab. The chief tender does not, therefore, assist with recovered victims in public safety diving operations.

[48] F.E. Abyholm, "The Human Organism in the Cold," *Human Performance in the Cold,* eds. G. Laursen, R. Pozos and F. Hempel , Undersea Medical Society, 1982): 85.

[49] A case of a very young girl recovered after being submerged for approximately an hour in frigid water off Alaska with consequent survival after expert resuscitation efforts was reported by Martin Nemiroff, MD. Personal communication.

[50] W.E. Hurford, S.K. Hong, Y.S. Park, D.W. Ahn, K. Shiraki, M. Mohri and W.M. Zapol, "Splenic Contraction During Breath-Hold Diving in Korean Ama," *J Appl Physiol* (1990, 69:1): 376-379.

[51] I.B. Mekjavic, C.J. Sundberg and D. Linnarsson," Core Temperature 'Null Zone'," *J Appl Phyisol* (1991, 71:4): 1289-1295.

[52] I.B. Mekjavic and C.J. Sundberg, "Human Temperature Regulation During Mild Narcosis Induced by Inhalation of Nitrous Oxide," *J Appl Phyisol* (1992, 73): 2246-2254.

[53] Ibid.

[54] T.C. Passias, I.B. Mekjavie and O. Eiken, "The Effect of 30% Nitrous Oxide on Thermoregulatory Responses in Humans," *Anesthesiology* (1992, 76): 550-559.

[55] F.J. Biersner, "Selective Performance Effects of Nitrous Oxide," *Human Factors* (1972, 14):187-194.

[56] I.B. Mekjavic, T. Passias, C.J. Sundberg and O. Eiken, "Perception of Thermal Comfort During Narcosis," *Undersea and Hyperbaric Med* (1994, 21:4): 9-19.

[57] D. Kerem, Y. Melamed, A. Moran and P. Alveolar, "Carbon Dioxide During Rest and Exercise in Divers and Non-Divers Breathing 0_2 at 1 ATA," *Undersea Biomed Res.* (1980, 7): 17-26.

[58] D.M. Fothergill, W.F. Taylor and D.E. Hude, "Physiologic and Perceptual Responses to Hypercarbia During Warm- and Cold-Water Immersion," *Undersea and Hyperbaric Medicine* (1998, 25): 1-12.

[59] C.E. Johnston, D.A. Elias, A.E. Ready and G.G. Giesbrecht, "Hypercapnia Lowers the Shivering Threshold and Increases Core Cooling Rate," *Aviat Space Environ Med* (1996, 67): 438-444.

[59] M. Ito and M. Ikeda. (*Undersea and Hyperbaric Medicine* 1998, 25): 59-62.

[60] L. Jansky, H. Janakova, B. Ulicny, *et al.*, "Changes in Thermal Homeostasis in Humans Due to Repeated Cold Water Immersions," *Pflugers Arch* (1996, 432): 368-372.

Chapter 8

Cold Stress, Immersion Hypothermia, Drowning Recognition, and First Aid

Cold Stress and Hypothermia Recognition

"Hypothermia can occur in any dive situation."

—Lt. Cdr. G.H. "Hall" Koch[1]

"...[R]ecognizing hypothermia in its earliest stages is a serious problem in diving."

—National Oceanic and Atmospheric Administration[2]

Continued loss of body heat can result in diminished survival capabilities and fatal errors by both divers and tenders. Early cold-stress sign/symptom recognition is critical to prevent the occurrence of hypothermia. Recognizing the signs or symptoms of late hypothermia is critical for determining safe and optimal hypothermic patient management procedures, as improper handling of hypothermic patients can result in permanent injury or death.

We will begin by examining early signs of cold stress and hypothermia. Even before core temperature noticeably drops, cold stimulation to the hands, feet, and head can activate heat-generating heat-saving responses, such as shivering and peripheral vasoconstriction. Shivering, which can increase basal body heat production as much as five to seven fold, is easy to recognize and indicates that a dive should be aborted and that the shivering diver or tender should be rehabilitated.[3]

Most of us readily recognize when hands and feet are cold, because the progressive symptoms of discomfort, pain, numbness, and loss of function are salient and familiar, as is skin redness.[4] Heat loss from the body and core, on the other hand, can be very difficult to recognize without the presence of shivering. This presents problems, because dives should be terminated before shivering begins. But some people never shiver, and shivering stops with the onset of deep hypothermia.[5] When and how shivering occurs can be determined by the rate of body heat loss, rather than by the body's temperature. Rapid heat loss can cause a rapid onset of severe shivering; while more gradual heat loss over an extended period of time can delay the onset of shivering past the point that hypothermia is reached. This latter occurrence is sometimes referred to as silent hypothermia.

Silent hypothermia most often occurs in warm-water divers who do not wear sufficient PPE and who perform repetitive dives over several days. It can also occur in drysuit divers who do not wear sufficient undergarments during repetitive dives. Silent hypothermia can occur in tenders who do not take the necessary precautions to prevent heat loss and who work outdoors for several days at a time.

One of the symptoms of silent hypothermia is feeling chilled sooner with each cold exposure. For example, a diver may be able to ice dive for twenty minutes before feeling slightly chilled during the first dive, but may feel chilled in thirteen minutes on a second dive and six minutes on a third dive. The water did not become colder; rather, the diver did not have sufficient rewarming time during surface intervals, with the result that the diver started each dive with a progressively slightly lower body temperature. Hence the time to reach the state of feeling chilled progressively shortened.

Long slow cooling without discomfort has been shown to cause a slowing of choice reaction time, increased errors in decision-making, and other impairments in cognitive performance.[6] Victims of silent hypothermia may not notice that they have a reduced temperature or that they are impaired cognitively, which means that sign recognition by other personnel is very important.

Irritability and fatigue may be present prior to shivering, especially in cases of silent hypothermia. Surface personnel should watch closely for any out-of-character behavior in both surface personnel and divers, such as:

- Reluctance or lack of enthusiasm to leave a shelter or to dress to go outside.
- Greater enthusiasm to perform in-shelter duties.
- Failure to remove exposure-wear when inside a warm shelter.
- Frequently holding and drinking cups of warm fluids.

- States of being withdrawn, short-tempered, or impatient.
- Poor concentration and short-term memory loss, which may include forgetting procedural steps.
- Greater incidence of mistakes and shortcuts. Cold-stressed individuals focus on their feelings of being chilly, with the result that their minds are not fully focused on the performance of their duties.
- Difficulty with fine-motor or complex tasks.
- Closed postures (*e.g.*, hugging oneself) or voluntary warming movements (*e.g.*, foot stamping, hand rubbing).
- Higher than normal SAC or breathing rates.
- Personality changes.[7]

To recognize out-of-character behavior, it is necessary to know what in-character behavior is. The persons responsible for monitoring personnel, such as safety officers and tenders, need to be continually observant, from the arrival on scene through the debriefing and equipment cleanup. A safety officer who is new to a particular shift of personnel on scene can ask others what is in character. For example, a safety officer notices that a diver, who was rehabilitated and properly checked out by EMS two hours ago, is still in the shelter instead of outside serving as surface support. Is this behavior occurring because this person is cold stressed, or is it in character for him to avoid working unless forced to do so? The safety officer can ask the diver's tender if the diver normally stays out of sight unless assigned a job, or if the diver normally is self-motivated enough to find out what needs to be done.

Photo 8–1 Tenders must be able to recognize signs of cold stress, apprehension, and mental stress. A tender checks his diver to make sure he is doing well.

Another sign of cold stress and hypothermia is a change in the ability to estimate time. Scuba divers were asked to count up to sixty at a one-second rate at various times during a week of cold-water diving.[8] The rate of counting correlated significantly and positively with body temperature. The lower the body temperature, the slower the rate of counting. A few studies have found that cognitive efficiency and vigilance are not hindered by core temperature drops as much as 1.3°F (0.72°C), but these studies are in the minority. [9,10]

Cold-water diving was found to cause mental impairments for tasks that involved intense concentration and considerable short-term memory.[11] Researchers believe these problems were caused by the distraction of being cold. When a tender is deciding whether or not a diver performed a thorough search for a small object, diver cold stress should be considered. Missing a search object can be a sign of cold stress.

Other studies that found significantly impaired memory, cognitive function, and speed of reasoning continued, even after subjects were rewarmed enough to feel good. After core temperature losses from cold-water immersion, subjects were rewarmed in 41°C (106°F) bathwater until their skin was warm. Although they felt comfortable, their core temperatures still remained low. In this state, the subjects were found to have impaired memory that worsened when core temperature was below 36.7°C (98°F). At core temperatures of 34°C to 35°C (93°F to 95°F), for example, 70% of the data normally retained was lost. Speed of performing two-digit calculations was initially reduced by 50%, but given enough time, the subjects performed the calculations accurately.

Other studies support these findings. Diminished abilities in performing logical reasoning, word recall, word recognition, and simple arithmetic have all been demonstrated.[12,13] Be watchful of personnel who take an unusually long time for them to perform SAC rate calculations and other cognitive tasks and who have difficulty following simple commands. A study of Navy-qualified divers suggested a "significant distraction effect of cold water exposure on performance of higher-order tasks."[14] To determine cold stress when shivering is not present, safety officers should have tenders perform SAC rate calculations and estimate running-feet search-distance calculations as soon as possible after a dive. This exercise provides the officer with a valuable diagnostic tool.

Importantly, cold-stressed or hypothermic patients, even when shivering violently, may feel that they are mentally alert, even when they have significant mental impairments such as confusion. Their dilemma may not become evident to them until they attempt to perform a cognitive task and fail.[15] Therefore, it cannot solely be left to tenders or divers to decide for themselves whether they are cold stressed.

Loss of muscle strength and motor impairment are also caused by body temperature loss. [16,17,18] Therefore, it is important to watch how personnel perform tasks involving strength during an ice diving operation.

The *US Navy Diving Manual* divides hypothermia into two categories–mild and severe. Most texts divide hypothermia into three categories: mild, moderate, and severe, with corresponding temperatures and signs/symptoms. We feel that because signs and symptoms do not necessarily correlate with temperature, and because there are significant inter-individual differences, it is not helpful to have three categories. We recommend using two categories, and, when in doubt, assume "severe".

Mild hypothermia signs include controlled shivering, slurred speech, imbalance, and poor judgment. Severe signs include the cessation of shivering, diminished awareness, cardiac irregularities, and depressed vital signs.[19] Severe hypothermia is a medical emergency and requires immediate activation of the EMS system for transport to a medical facility. The person in charge of the ice diving operation must ensure that the necessary equipment and trained personnel are available onsite. Divers with mild hypothermia can be safely rewarmed in the field. EMS should be activated for immediate transport to a medical facility should more severe hypothermia occur.

Hypothermia can also cause peripheral tissue injury. If skin temperatures fall below 10°C (50°F), there will be a complete loss of sensation. If that condition is allowed to continue, a freezing injury can occur. Look for cold injuries during the secondary survey of a cold-stressed patient. As we will learn, areas of the body with frostbite must be handled as gently as a serious fracture—as gently and carefully as you would handle an extremely fragile million-dollar Ming vase.

Chilblains

Chilblains are lesions found on skin that has been repeatedly and directly exposed to dry, cool environments between 32°F and 60°F (0°C and 16°C). People who work outdoors in the winter on a regular basis, and those who leave their hands, ears, face, and other parts exposed may experience chilblains. It is a chronic problem, not something that happens once. The most obvious sign is the appearance of swollen, red lesions that feel hot to the touch. Other symptoms in affected areas include tenderness with itching and sometimes burning sensations.

If you are not sure whether what you are experiencing is chilblains or some other problem, see your physician. If it is chilblains, do not scratch or further irritate the condition. Protect the area and wear proper exposure wear to prevent reoccurrence or increased severity. If a patient appears to have chilblains, simply protect the area and make note of it to the receiving hospital.

Frostnip

Frostnip is the first stage of frostbite. If left untreated, frostnip will affect deeper layers of tissue, becoming frostbite and eventually freezing. Frostnip is not a serious problem if treated properly, and if reoccurrence is prevented. Typical locations are the ears, nose, cheeks, fingers and other exposed skin areas, because these areas are typically not sufficiently protected by exposure wear.

Frostnip is characterized initially by redness that later turns white. The affected area lacks sensation and will feel numb to the patient. If you touch an area of skin with frostnip, both the skin and underlying tissue will feel soft.

Treat frostnip by blowing warm air on the affected area or by laying a warm hand over it. If possible, perform this treatment indoors to prevent exposing your warm hand to cold conditions, or insulate the parts of the hand not directly on the frostnip area. Sometimes, the patient may feel burning sensations and tingling during the rewarming process. Explain that this is normal. Blood vessels are dilating back to normal and blood is returning to the area. Nerves are "waking up".

If rewarming does not quickly cause a change, assume the problem is frostbite. *Stop* the rewarming process, and immediately seek medical attention. When in doubt about whether the problem is frostnip or frostbite, assume the worst, which is frostbite. Do *not* rewarm in the field unless local protocol dictates doing so. Pre-hospital rewarming of frostbite can lead to severe, permanent tissue damage.

Frostbite

In 1902, 210 Japanese soldiers died during a march. Of the 210, 193 died because of frostbite.[20] Frostbite is a serious medical condition that can result in amputation or even death if not treated properly.

Let us look at how frostbite occurs. An important cold stress defense mechanism is peripheral vasoconstriction. Exposing hands to severe cold, for example, can result in a blood shift similar to physiological amputation. Frostbite generally occurs in unprotected areas that are first affected by peripheral vasoconstriction, such as ears, nose, face, fingers, hands, toes, and feet. Risk factors include alcohol consumption, use of some drugs, poor PPE, wetness, leanness, physical exhaustion, and trauma.[21]

As also discussed earlier, skin temperature drops as a result of this blood shift. Without the warmth provided by blood, the fluid in the interstitial space (space between the cells) and within cells can literally freeze and form ice crystals. Blood vessels are damaged, further impairing circulation. The skin and underlying subcutaneous tissues are frozen, with little or no oxygen supply.

It is extremely unlikely that a frostbite injury would occur during an ice diving operation, unless the temperature was very cold and personnel did not wear sufficient PPE. Public safety teams who are called out to perform a rescue on the ice may receive a patient with frostbite, but this would more likely be a person recovered on the surface than one who was submerged on the bottom.

When frostbite occurs, the frozen skin will appear white and waxy and will feel cold and hard to the touch. The underlying tissue may not be frozen and may still feel soft, with a resilience, or bounce. Frostbitten areas may lose sensation. Be very careful not to poke, rub, squeeze, or manipulate frostbitten areas, as such mishandling can cause ice crystals to rupture and result in further tissue damage. Handle the area as if it had a serious fracture. Do *not* attempt to rewarm these areas unless directed by local protocols or a physician at the receiving hospital.

Immediately transport the patient to a medical facility. Delays can result in amputations. Cover the affected area with a clean, dry dressing, taking great care not to rub, poke, scratch, shake, apply pressure to, or drop frostbitten areas. Do not allow the patient to smoke, as this can cause further vasoconstriction and decrease oxygen to the area. Do not allow the patient to drink alcohol. Be sure to notify the receiving hospital that the patient has frostbite and describe its extent so that they can be prepared.

Freezing: deep frostbite

Freezing is frostbite that has progressed to deeper tissues, resulting in the freezing of deeper blood vessels, muscle, bone, organs, or other tissues. Like frostbite, freezing involves ice crystal formation, capillary wall damage, plasma protein molecular changes, cell dehydration, and cell death. Because both superficial and underlying tissues are frozen, the area will feel hard to the touch, with no resilience or bounce. The skin color may appear mottled and blotchy, with white, blue-gray, and yellow-gray areas.

Transport patients to a medical facility as quickly as possible. Handle them as gently as possible. Gently cover affected areas with clean material.

Difficulties of rewarming frostbitten or frozen tissues

Cold tissues have lower metabolisms than warm tissues, so they need less oxygen. Frozen tissues require even less oxygen. Once rewarming occurs, the need for oxygen increases. Once frostbitten areas are rewarmed, their need for oxygen is at its greatest level. The ability to deliver oxygen, however, is severely compromised. Lack of oxygen can result in cell death, infections, and gangrene. For this reason, aggressive oxygenation must occur during the rewarming process. This is very difficult in the field. It is especially problematic if the lungs are compromised from aspiration and pulmonary edema. In the hospital, advanced ventilation procedures such as PEEP (positive end expiratory pressure) and CPAP (constant positive airway pressure) may be used, but even these may be inadequate.[22]

Hyperbaric oxygen, or oxygen at pressures greater than ambient 14.7psi, are considered helpful during and after the rewarming process.[23] A good example is a case of an eleven-year-old boy in good health who experienced deep frostbite in six fingers while hunting in -32°C for four hours.[24] He was given hyperbaric treatments at 2bar and 4bar for ninety minutes for fourteen days, at which time there was total recovery of the severe frostbite.

A typical treatment is 2.82atm (60ft, 2.82bar). Hyperbaric oxygen improves circulation by increasing the concentration, or partial pressure, of oxygen in the blood. This process in turn improves the flexibility of red blood cells and enables them to pass through narrower capillaries. Hyperbaric oxygen also enhances capillary growth and reduces the probability of infection. It can be used to treat the gangrene problems that can result from frostbite. Discuss this treatment method ahead of time with your hospitals and with the closest hyperbaric facility. For more literature and information on this topic, contact the Undersea Hyperbaric Medical Society.[25]

When freezing injuries are thawed, blisters and inflammation in the skin are common. Figure 8-2 illustrates the effects that can occur from thawing.

Another possible treatment to explore is the use of semi-occlusive dressings. This treatment was found to prevent the need for amputation in a case of deep frostbite in fingers.[26]

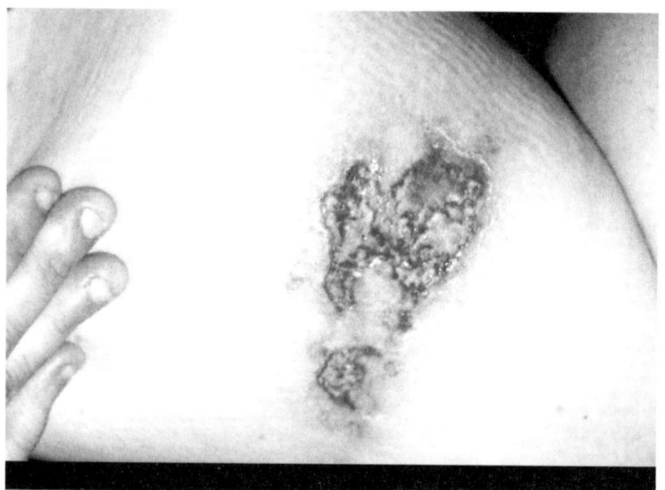

Photo 8–2 This freezing injury on the inner thigh occurred from using a prescribed ice pack wrap for too many hours over a twenty-four hour period after knee surgery. It initially presented as a large contusion and then progressed to an oozing wound. Treat any black and blue markings on hypothermic patients with great care and make sure to point them out to the receiving hospital personnel.

Other considerations

Snow blindness from exposure to ultraviolet light can occur if proper eye protection is not worn. Sometimes questions arise about the use of contact lenses versus eyeglasses during ice operations. Eyeglasses may fog if not treated with a solution. A study of military personnel wearing contact lenses during cold outdoor activities found no serious problems with them. The most common complaint from hard lens wearers was eye redness, while soft lens wearers most often complained about impaired vision.[27]

Some people have what is called a "cold allergy". When exposed to cold, these people will have reactions that mimic allergy attacks.[28] They may experience skin vasodilation, hives, nasal congestion, wheezing, coughing, and, in severe cases, shock. The allergy is caused by histamine release from skin cells and is believed to be hereditary. Standard allergy medications do not help cold allergies, but acclimation over a period of time to cold may help. In the few cases we have personally heard about, symptoms did not appear until middle age.

If the depth and dive time standards of this book are followed, the probability that a diver will develop decompression sickness is remote. Divers who exceed these standards are at greater risk because of the risk of hypothermia. Decompression sickness and hypothermia share many of the same signs and symptoms. These include: tingling, numbness, weakness, pain, loss of concentration, diminished motor and cognitive performance, and out-of-character behavior. When in doubt on an ice dive, assume the diver has both maladies and treat accordingly. Because hypothermia may increase the risk of decompression sickness, calculate your table profile at one greater time interval. Add an additional time interval if the dive was also arduous.

The incidence of arterial gas embolism is also problematic, not only because it is usually more serious than decompression sickness or hypothermia, but also because it too shares the same signs and symptoms.[29] When in doubt, assume that the diver has an arterial gas embolism and hypothermia, and treat accordingly.

The following "Could It Be Something Else?" chart is useful in helping determine whether a problem is related to decompression illness (*i.e.*, arterial gas embolism or decompression sickness) or hypothermia. The objective is to eliminate the presenting numbness, tingling, weakness, or pain by understanding its source. If cessation occurs within a few minutes, there is far less likelihood that the problem was related to decompression (DCI). If, in addition to the symptom cessation, no known risk factors occurred, such as an uncontrolled rapid ascent, breath holding during ascent, or extended range dive profiles, the likelihood of DCI is even more remote.

If drowning or near drowning is involved and the water is believed to have contaminants, it is important to send a water sample with the patient so that the receiving hospital can better help the patient. This is because aspirated water enters the circulatory system during the drowning process and tends to end up in the brains and lungs, causing cerebral and pulmonary edema respectively. Contaminants such as sewage or fertilizers from the water will be present in these important organs and can cause serious complications in the patient's recovery.

Dehydration

Dehydration increases the risk of cold stress and hypothermia.[30] Dehydration signs and symptoms include:

- No urge to urinate when immersed or submerged in water.
- Dry mouth and thirst (late symptoms).
- Headache and unusual fatigue.

COULD IT BE SOMETHING ELSE?

Possible Cause	Exclusion Test	Symptom Alleviated?
Loss of circulation: Too tight strap? Tight wetsuit? Tight drysuit seals? Watch or bracelet under suit? Body position? Weight on limb?	Remove restriction Lower extremity to below heart Gently massage affected area Move digits of affected limb	YES_____ NO_____
Vasoconstriction from cold?	Same as above, plus: Gently dry off. Drink warm fluids, non-alcoholic non-caffeinated Apply heat pack wrapped in wet towel for 1-2 minutes	YES_____ NO_____
Improper breathing (high/low Carbon dioxide): Panic or exertion? Unusually high/low air consumption?	Breathe normally 12-16 respirations per minute Gentle, relaxed breathing	YES_____ NO_____
Strained muscle: Heavy kicking? Hanging off downline in strong current? Carrying heavy object(s)? Pre-dive exertion	Flex muscle(s) in affected area and gently massage. Does that feel good? _____ Does that feel sore? _____ If yes to either, it is less likely a gas bubble injury.	
Injury: Banged by dive ladder? Banged by rock or coral? Kicked by other diver?	Examine for bruise or swelling _____ Point tenderness? _____	
History: Old injury in affected area? Explain _____ Ever experience this problem before? Explain _____ Medications? _____ When was problem first noticed? _____ If numbness and tingling present, which came first, second? _____		

Figure 8–1 Could it be something else? Because many signs and symptoms are similar for hypothermia and gas bubble injury problems, use this chart to help in the decision-making process.

- Irritability.
- Dark urine and pain in the kidney region (late signs and symptoms).
- Inability to sweat (late sign).
- Muscle cramping.
- If you pinch your skin and it does not immediately go back to normal, you are very dehydrated.

To maintain good hydration, ensure that your urine is clear, copious, and frequent.

Alcohol consumption is a problem because it masks signs and symptoms of cold stress and hypothermia and increases the chances of their occurrence. Alcohol can cause problems because it can have the following effects: [31]

- It is a diuretic that can cause dehydration.
- It limits judgment and increases risk-taking.
- It dilates peripheral blood vessels. The drinker feels warmer because core blood warms the skin's temperature sensors, but in reality, more heat may be lost.
- It interferes with the enzymes that enable hemoglobin to carry oxygen in the blood.
- It increases the probability of failure, injury, and death by reducing the ability to respond to a threatening situation correctly.
- It reduces dexterity.
- It interferes with the laryngeal reflex, thereby increasing the likelihood of drowning.
- It increases the probability of vomiting.
- It hides the signs and symptoms of cold stress and hypothermia and thereby increases the chances of hypothermic progression.
- It elevates hand heat loss.[32]
- It increases the risk of decompression sickness.
- It causes hypothalamic dysfunction.
- It decreases awareness of environmental conditions.

Many water-related rescues happen around 1600 to 1800 hours. Kids come home from school, change their clothes, have a snack, and go out to play. Adults hit happy hour after work; they drink and sometimes do very foolish, dangerous things. Approximately 50% of drowning incidents are alcohol-related. The potential for a water-related emergency is increased during early evening hours, as is the potential for volunteer rescuers themselves to have been drinking. (See table 8-1) The dive team's SOP/SOG should dictate that responders report any alcohol consumption prior to arrival on the scene to enable the officer assigning duties to make safe decisions. Recreational divers should not bring alcohol to dive sites. Have a cup of hot soup instead of a hot toddy.

Core Temperature		Signs and Symptoms[i]
°F	°C	
98.6	37	Sensation of cold. Peripheral vasoconstriction, pale skin, goose bumps. Increased oxygen consumption begins. Shivering begins. Normal oral temperature.
97	36	Sporadic shivering to bouts of full-body shivering to uncontrolled shivering. Increased metabolism. Feelings of confusion, with stumbling and disorientation.
95	35	Impairment of rational thought, confusion. Drowning possible. Maximum shivering.
93	34	Temporary amnesia. Sensory and motor impairment. Cardiac arrhythmias. Reduced sensation and motor performance. Speech impairment.
91	33	Delusions, hallucinations, partial loss of consciousness. 50% of immersion victims die.[ii]
90	32	Cyanosis, dilated pupils, respiratory alkalosis. Gross motor impairment. Beginning of cardiac irregularities.
88	31	Shivering stops.
86	30	Muscles "gel," and become rigid. Hypoventilation. Loss of consciousness. No response to pain. Severe reduction in pulse and respiratory rate.
84	29	Loss of consciousness. Ventricular fibrillation if heart is irritated or victim is handled roughly.
80	27	Appear clinically dead. Muscle rigidity gone. Pupil dilation. Ventricular Fibrillation. Loss of deep tendon, skin, and capillary reflexes.
79	26	Death likely.

[i] Compiled from various tables presented in NOAA, NAVY, EMS manuals, and A.D. Weinberg, "Hypothermia," Ann Emerg Med (Feb 1993, 22, Pt 2): 370-377.

[ii] NOAA

Table 8–1 Hypothermia Signs and Symptoms

Cardiac related problems

In mild hypothermia (34°C to 35°C, 93°F to 95°F), there can be an increase in pulse rate, blood pressure, central venous pressure, and cardiac output. Moderate hypothermia (30°C to 34°C, 86°F to 93°F) can cause bradycardia (a heart rate slower than 40bpm) as well as atrial or ventricular arrhythmias, with atrial fibrillation being the most common. Low blood pressure and a reduced cardiac output can also occur. In patients with severe hypothermia (less than 30°C, 86°F), the chance for ventricular fibrillation that can lead to asystole (no heart electrical activity) increases greatly.

Researchers have noted that the J wave (Osborn wave), best seen in lead V3 or V4, is observed in 80% of hypothermic patients and increases in size as core temperature decreases.[33,34] The level of blood oxygenation and the acid-base balance can be changed by hypothermia. For example, carbon dioxide levels can be lowered during the hyperventilation that can occur with cold-water immersion. As hypothermia progresses, respirations decline and hypoventilation can occur, leading to hypercapnia.[35] This, in conjunction with reduced hepatic metabolism of acid (caused by decreased blood-flow to the liver), increased lactic acid production (from shivering), and reduced blood-flow to skeletal muscles, can cause acidosis.

The decision whether to correct arterial blood gases in hypothermic patients is controversial. Metabolic acidosis in hypothermic patients may not respond to bicarbonate treatment.[36] Rewarming usually corrects metabolic imbalances once normal circulation returns.[37,38] In severe hypothermia, the heart is often unresponsive to drugs, pacemaker stimulation, and defibrillation.[39]

Furthermore, many drugs require a minimum temperature to work effectively. When repeated boluses of common intervention medications such as epinephrine or lidocaine are administered, the medications can accumulate and then reach toxic levels when the patient is rewarmed and the medications begin to function. It may be necessary to administer lower or less frequent doses of resuscitation medication to hypothermic patients.

"Medications, in general, should be avoided in the hypothermic patient in cardiac arrest until the core temperature is above 30°C (86°F)."[40] There are two reasons for this. One is related to fluid volume and the other is due to decreased medication effectiveness at low temperatures. Severely hypothermic patients are often hypovolemic, but administering IV fluids to correct the condition should be done cautiously. As we learned in the previous chapter, circulatory fluid is partially lost to third spacing (fluid from the circulation enters the interstitial tissue spaces), which means that it can be difficult to assess fluid requirements. In addition, establishing an IV line in the field can be very difficult due to vasoconstriction.

Pre-hospital-care drugs should not be allowed to freeze, as their ability to work may be impaired after thawing. Standard saline and dextrose IV fluids should be fine after thawing as long as no precipitation is evident. Temperature-controlled drug boxes may be necessary for ambulances and other emergency vehicles that are left in unheated environments for extended periods of time. A good box will keep drugs in the recommended storage temperature range of 59°F to 86°F (15°C to 30°C) in ambient temperatures that range from 20°F to 120°F (-8°C to 48.9°C). The box will also have a data logger to track and record the storage temperature of each medication, and inventory software will identify any medication that exceeds its recommended storage temperature range.[41]

A good rule of thumb is to avoid nonessential, advanced life-support procedures in the field and during transport to help prevent ventricular fibrillation. Have a defibrillator ready for use. Be aware that using a defibrillator for more than one standard set of three shocks on a cold heart is discouraged by more than one leading hypothermia expert.

CPR and intubating a patient are considered essential procedures when necessary. Prior to endotracheal intubation, ventilate the patient with oxygen (prewarmed and moistened if possible) to help prevent ventricular fibrillation. Be aware of the possibility that equipment such as oxygen tubing may be stiff and frozen. Heat packs are useful to manage these problems. During a long outdoor transport, tape the tubing of the intubation cuff-port to the patient's skin; if frozen, it could break off.

If a patient is severely hypothermic, perform ECG monitoring during transport and during resuscitation efforts. Be prepared in case monitor-lead adhesive pads do not stick to cold skin and conduction of electrical signals across the skin is impaired. "In cold environments in which continuous monitoring is desired, tincture of benzoin may be needed to maintain contact of the monitor leads. The QRS amplitude should be maximally amplified if no complexes are seen initially."[42] Also be aware of how extremely low temperatures can affect defibrillator and monitor devices. Battery power may become an issue. Most devices require a temperature above 15.5°C (60°F) to function properly.

Hypoglycemia

There is another set of signs and symptoms that tenders and divers should be aware of on ice dive sites. If a diver presents weakness, unusual hunger, numbness, chills, confusion, trembling, dizziness, lack of coordination, and limited consciousness after a dive, hypothermia and/or hypoglycemia could be the culprit. Hypoglycemia is a low blood-sugar (glucose) level. Because hypothermia and hypoglycemia share some of the same symptoms, it is possible one could be mistaken for the other. Some people are predisposed to becoming hypoglycemic, but anyone can experience some of the symptoms if a long, cold, arduous dive is performed on an empty stomach. If the diver has these symptoms frequently after dives that are not arduous or long or when obvious hypothermia is not present, the diver should seek a medical examination for a possible predisposition to hypoglycemia. Tenders who exert themselves could also be susceptible if they have not eaten properly. Exertion could be in the form of cutting holes in thirty-five inches of ice without a chainsaw. Prevention of hypoglycemia is simple: eat plenty of complex carbohydrates prior to the ice operation.

Cold stress and hypothermia first aid and treatment

"They are not dead until they are warm and dead" is a phrase many of us know. We are taught to provide pre-hospital care until the patient reaches the hospital, where the patient continues to receive care until the core temperature is raised. The incident recounted at the beginning of chapter 7, however, demonstrates that we cannot apply this rule to all of the victims we recover. "In hypothermia the most important differential diagnosis is death. Patients who are cold and could be resuscitated must be differentiated from patients who are cold because they are dead. Experience from abroad has shown that extreme hyperkalaemia may be a useful diagnostic tool."[43] In the field, we cannot use diagnostic tools such as checking for hyperkalemia, nor do we even need to know what hyperkalemia is. What we need is a system that informs us when to initiate care and when not to for cardiac-arrest, hypothermic/immersion patients. Make sure that you know what that system is; if one does not exist, it is time to get one started.

As ice diving operation personnel, we need to know how to provide first aid to cold-stressed and hypothermic patients in the field. EMS personnel need to know how to perform pre-hospital care during transport to a receiving emergency department. It is also important to have some understanding of hospital treatments to ensure that you or a teammate receive the best possible care if the need ever arises. If you are a member of a public safety dive team or are an ice diving instructor, meet with the hospital staff that would receive patients from an ice diving operation to ensure that they have up-to-date hypothermia and drowning protocols and equipment. If there is more than one hospital, it would be beneficial to find out which one is most capable of treating such patients. To do this, you need a basic understanding of the treatment modalities that are available, which is one purpose of this chapter. Do not assume that the local hospital can offer the most effective treatments. "It may be advantageous to transport the more severely hypothermic patient to a more advanced care facility even though transport time may be greater."[44] Choice of facility needs to be pre-set with the local EMS and hospital systems.

Be aware that victims have been known to regain heart function after prolonged immersions. Montgomery County Fire Department in Maryland, for example, in autumn on a mutual aid call, recovered a five-year-old girl after over two hours of submersion. Resuscitation efforts resulted in a spontaneous pulse that lasted for more than twenty hours. The heart is a fairly hearty organ. Although today's medical abilities cannot restore the nervous system after such durations, there may be an indirect benefit of continuing in rescue mode in colder months—if a heartbeat is regained, there is a greater chance that other lives can be saved with organ donations.

Rewarming procedures

There are three primary types of rewarming procedures: passive, active surface, and active core.

Passive rewarming. This involves using the person's own metabolism to raise the core temperature by helping the body reduce heat loss. Examples of passive rewarming procedures include: removing patient from the water, removing wet clothing, drying patient, placing him in a comfortable warm environment, relatively gentle exercising (on land), applying

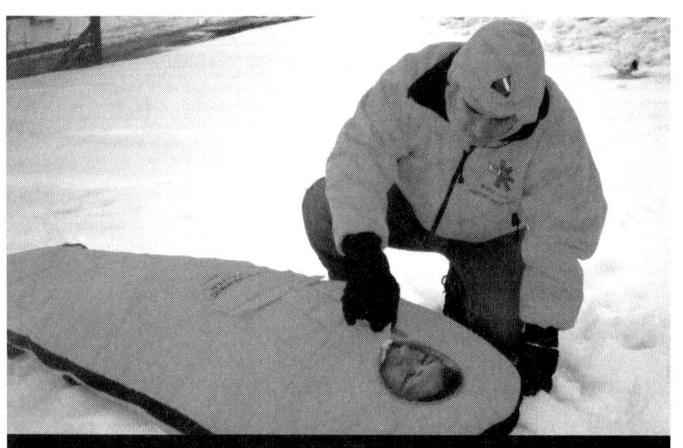

Photo 8–3 The hypothermic stabilizer bag is one of the most effective passive rewarming tools and works exceedingly well when transporting patients across rugged terrain, on airboats, long distances across ice, in rescue vehicles, or in helicopters.

additional insulating exposure wear, and placing in a thermal recovery stabilizer.

Intermittent leg exercises performed by cold, resting divers may initially cause a small after-drop of core temperature, but these after-drops can be overcome with continued exercise.[45] Waiting contingency divers may find that performing intermittent leg exercises prior to becoming cold can help prevent cold stress and cold feet.

Passive rewarming techniques are the most common ones used in the field. They require the least amount of equipment and training and do not involve the risk of after-drop (rewarming) shock problems, internal or external burns, or cardiac problems. The downside is that they may not be as effective as active rewarming techniques.

Active surface rewarming. These procedures consist of applying heat to peripheral tissues and hydrating the patient. Examples include: placing wrapped heat packs under the armpits, groin, and carotid artery areas; wrapping the patient in a heating blanket; applying a heating pad; and immersing in a warm bath or shower. There is strong disagreement among hypothermia experts about the advisability of active rewarming strategies, and various books present different procedures regarding what to do and what not to do. In fact, studies have produced contradictory research findings. Some books advise avoiding any form of external rewarming due to the possible complications mentioned above, while some experts claim that warm baths are imperative. By reviewing some of the findings, we can make better educated decisions.

Using battery-operated heating systems or heat packs to keep hands warm while working is useful. Auxiliary heating of the hands has been found to prevent or alleviate impaired performance of tasks involving manual dexterity.[46]

A short-wave therapy apparatus was successfully used to rewarm three severely hypothermic patients with core temperatures below 86°F (30°C) without after-drop complications, ventricular fibrillation, or burns.[47] The apparatus operated at a frequency of 27.12MHz and used eddy-current electrodes.

Warm airbeds have also been used with success. A patient with an initial 25.9°C core temperature was rewarmed on such a bed to 36.1°C in seven hours. There were no complications despite the patient having arrived with atrial fibrillation.[48]

In the field, it is not recommended to perform active surface rewarming on moderate or severely hypothermic patients who have had prolonged cold exposure. Doing so will produce vasodilation, which will allow the blood from peripheral tissues to return to the core. This blood will be cold and high in lactic acid, and this can cause after-drop and a decrease in blood pH. These resulting conditions can increase the risk of ventricular fibrillation and death. In the hospital setting, only truncal active surface rewarming is recommended.[49,50,51]

A sudden drop in blood pressure is also a major concern during the rewarming process. The exact cause of sudden blood pressure drop, known as rewarming shock, is not fully understood. Possible causes include: cardiac problems, peripheral vasodilation, cellular calcium overload, changes in myocardial myofilament responsiveness to intracellular calcium, and impaired high-energy phosphate homeostasis. In addition, altered capillary function, increased capillary leakage of plasma protein, changes in internal and external vascular volumes, and changes in autonomic vascular control may also be involved.[52] Each of these can contribute to low cardiac output during and after the rewarming process, which can result in death.

Active core rewarming. This involves heating the patient internally. It is the primary treatment for patients who are in a state of unconsciousness or cardiac arrest due to hypothermia.

If patients are fully alert and without nausea, warm, nonalcoholic, noncaffeinated beverages can be given to hydrate and provide a degree of warming. This is the most common method of active core rewarming for mild hypothermia patients, as it can easily be done in the field and works well in conjunction with passive rewarming techniques for cold-stressed or mildly hypothermic people.

Other methods of active core rewarming must be performed by properly trained medical personnel. Examples of these techniques include administering warm intravenous fluids, warm and humidified breathing gases, and warm fluids through a stomach tube. If the IV fluids were stored outdoors or in an unheated ambulance, rewarming the fluids prior to administration is strongly recommended. Trauma patients who need fluid replacement are particularly susceptible to losing core temperature when cold IV fluid is administered. One study found that "infusion heating devices are mandatory in patients with high fluid requirements."[53] Ideally, IV fluids should be warmed to 43°C prior to administration.[54] Intravenous fluids can easily be warmed by wrapping the tubing around a heat pack and placing the IV bag on a heat pack.

A very effective way to safely warm IV fluids is with a portable IV warmer that stores the IV bag. Res-Q™ makes an excellent bag that runs on a 12V portable or vehicle battery, has a thermostat to prevent overheating, and dispenses the IV fluid directly from the bag. A sleeve that covers the entire length of the IV tubing enables a steady and sure flow rate during extreme temperatures and wind chill conditions. This sleeve has an insulated end flap to cover the patient needle-entry area.

Heat packs taped to a simple, disposable oxygen humidifier will bring oxygen up to at least room temperature, which is better than cold, dry oxygen. A more effective and controlled method is using an actual oxygen rewarmer. Res-Q-Air™ is a portable and simple-to-use system that can be applied to a variety of resuscitative equipment, from bag-valve masks and mechanical ventilators to incubators. The system also complies with current hypothermia protocols and treatment guidelines.

At the very least, prewarming oxygen and IV fluids will prevent the heat loss that normally results when administering cold IV fluids and oxygen. As we will learn later, these methods actually offer a valid rewarming procedure.

There are other rewarming techniques. In the hospital setting, one successfully used technique is pleural lavage, which instills warm saline through a chest tube.[55] Similar procedures are thoracic and peritoneal lavage.[56] Lavage is a technique for circulating solutions that are externally heated into body cavities. Diathermy uses ultrasound and microwaves to rewarm the core. Extracorporeal circulation removes blood, warms it, and then replaces it, and has been found to be a reliable treatment for patients with severe hypothermia of less than 28°C.[57,58] Hemodialysis is a rapid and efficient rewarming technique for moderate and severe accidental hypothermia patients.[59] One study of twenty-three accidental hypothermia patients presented the following conclusion: "Due to lack of safe prognostic predictors, all accidental hypothermic victims with circulatory failure should be rewarmed by cardiopulmonary bypass before further therapeutic decisions are made."[60]

It has been argued that aggressive active rewarming should not be initiated until effective CPR has begun. As long as tissues are very cold, metabolism is lowered, and the need for oxygen has declined. This puts the tissues in a kind of suspended, animation in which they do not readily die even though they are not getting proper oxygenation. As tissues are warmed, they "wake up" and their need for oxygen returns. If they do not receive oxygen, they can die, and the probability of gangrene is increased.

Active rewarming has been demonstrated to be useful for patients with mild hypothermia. Research by Kuehn found that application of heat to the body's surface is generally preferred to methods involving core or internal heating.[61] He found that the optimal rewarming bath temperature is 37°C to 38°C. When temperatures are 3° or higher, patients may find the water uncomfortable and consequently leave the bath before reaching normothermia (normal temperatures).[62] Higher temperatures can also cause vasodilation, which can lead to hypotension or cardiac irregularities. Kuehn also noted problems with the effectiveness of showering, because rapid rewarming of the blood supply can cause shivering to abate. The cessation of shivering and feelings of warmth may cause patients to leave the shower before their internal temperature has improved to a significant degree.

It is very important to understand that warm external tissues do not at all guarantee that the core temperature is normothermic. A study of the use of hot water suits for deep saturation diving found that divers with skin temperatures as high as 38°C were hypothermic. Active surface warming kept the peripheral tissues warm, but the inhalation of cold, compressed, oxyhelium breathing gas caused significant drops in core temperature.[63]

Kuehn's research also found that respiratory warming using warmed, humidified air did not rewarm patients as effectively as placing them in a recumbent position and covering them with a warming blanket. In fact, if the breathing gas was made too warm, there was a risk of burning the hypothermic respiratory tract. Kuehn concluded that in a rescue or field environment, the safest and "least provocative" rewarming technique is to use rewarming blankets.

It is unlikely that rewarming blankets will be available in the field. Therefore, make sure that the blankets used are wool or fleece. A thermal recovery stabilizer capsule is even better. Metallized blankets, such as space blankets, are not a good option. A study of cooled male subjects found that metallized plastic sheeting incorporated into a casualty bag did "not provide significant additional thermal insulation."[64]

Marcus studied the efficacy of active core and surface rewarming techniques, including warmed breathing gas, a hot bath, a piped suit, and spontaneous rewarming.[65] All the techniques produced after-drop effects with a variety of temperatures and durations. Marcus found that auditory canal temperature was a useful method of determining the patient's temperature. The hot bath was the most effective technique for raising deep body temperature, with the piped suit being second-best. Inhalation rewarming did not provide any better results than spontaneous rewarming.

Another study used hot air (45°C) to rewarm subjects cooled to a 2°C temperature loss by immersion. This study found that "each additional 10Lpm of ventilation of hot, saturated air increased the rate of core rewarming from hypothermia by approximately 0.3°C [per hour]."[66]

Several studies found that inhalation of warmed, water-saturated gas was the preferred and most practical method of infield rewarming since it did not cause peripheral vasodilation and the resulting complications and only had a minimal after-drop effect. Inhalation rewarming was compared to placement of heating pads over areas of high heat transfer, immersion in a hot whirlpool, shivering, and a combination of methods.[67] The second study also found that inhalation rewarming was more effective than the passive shivering method.[68]

Another study measured the patient's cardiac temperature during rewarming and found no after-drop effect with inhalation (43°C to 45°C) or passive rewarming procedures, but did find an after-drop following fifteen minutes in a bath when skin temperature was greater than 30°C.[69] In addition to the after-drop, patients experienced sudden decreases in arterial pressure, vasodilation, increased heart rate; and increased cardiac output. Cardiac temperature increased twice as fast when inhalation rewarming was administered as when only passive rewarming was used.

One study of seventy-two rewarmings by a variety of methods, including heating pads, warm water suits, and body-to-body heat exchange, found that trunk warm-water immersion produced the smallest after-drop, the shortest recovery period, and the fastest rewarming. Inhalation therapy also produced small after-drops. Body-to-body heat exchange produced larger after-drops than spontaneous rewarming. The researchers concluded that "this leaves inhalation therapy alone as the recommended treatment for profound hypothermia in the field."[70]

Inhalation rewarming therapy (42°C gas) in a recompression treatment appears to be a valid rewarming technique. When body core temperature is between 28°C and 33°C, this technique may prevent further heat loss when hot-water immersion is not possible.

One reason why inhalation rewarming techniques may be so effective is that inhaling warm, humidified gas may rewarm the hypothalamus (the body's thermostat) and the lower brain stem (which controls heart and breathing functions). Rewarming these critical tissues may allow them to maintain vital cardiovascular and respiratory functions.

Contradictory research results regarding which rewarming techniques are most effective may be due to differences in subjects. One study found that seawater-cooled patient rewarming rates using inhalation of saturated 44°C air could be predicted by a combination of initial core and skin temperatures and the patient's body composition (height/weight ratio). This study also found that "comparisons of rewarming data obtained in different investigations and with other treatment methods are likely to be misleading unless experimental protocols and subject groups are carefully matched."[71]

A useful study was conducted using twenty US Navy combat swimmers.[72] Combat swimmers are more likely to have similar body compositions than random subjects. The swimmers were cooled in 4.5°C water for four hour and six hour durations. Core temperatures, measured by temperature-sensitive endoradiosonde pills, showed a loss of 0.3°C after four hours and 1.4°C after six hours of immersion. Two rewarming methods were used, immersion in warm air and in warm water. All subjects were removed from the water, dried, and dressed in a cotton sweatsuit and wool blanket. The air rewarming involved placing subjects in 38°C air, and the water surface rewarming was conducted with a 40°C warm bath. Both methods resulted in after-drop temperatures, with the greatest drop occurring after fifteen minutes of rewarming. The return of the subjects to normothermia took 1.67 times longer in air than in water. Very important to note is that the subjects reported feeling warm and comfortable shortly after the largest after-drop, which was long before normothermia was reached.

Divers and tenders who have reached states of noticeable cold stress or hypothermia should not be allowed to leave the warm rehabilitation staging area for at least two hours. They should not perform any outdoor duties or important indoor duties until they have been thoroughly rewarmed. Procedures and PPE should be evaluated during the debriefing session to discover why one or more persons reached the state of shivering and how such an occurrence can be better prevented during the next operation.

In regard to what type of rewarming procedure is best, "It remains to be determined whether aggressive rather than passive rewarming procedures are better."[73] For infield and pre-hospital care, passive warming, with the administration of warmed, saturated oxygen and, if necessary, warm IV fluids, seems to be the method of choice as far as current research stands. This may change in the next few years, so keep up to date with the literature and local protocols.

A very useful set of protocols to review is the *State of Alaska Cold Injuries and Cold Water Near Drowning Guidelines (Rev 01/96).* These guidelines present protocols for the management of hypothermia, freezing injuries, and drowning for every level of caregiver, from the public through paramedic and hospital personnel. The guidelines can be accessed free of charge on the internet at: http://www.hypothermia.org/protocol.htm. Martin Nemiroff, MD, deserves a great deal of gratitude for his contribution in making these guidelines what they are today.

In June of 2002 experts in all areas of water-related emergencies gathered at the first World Drowning Congress in Amsterdam. Author Zaferes attended the congress. The following summary comes from information presented during task force meetings and presentations.

- When a drowning victim is removed from the water, cardiopulmonary resuscitation should be started immediately, and hypothermia should be considered an additional, rather than the main problem.
- In head-out immersion cases, hypothermia is generally the main problem.
- Patients with moderate to severe hypothermia should not be actively rewarmed during pre-hospital care; rather, an effort should be made to prevent further cooling.

Photo 8–4 In rescue modes, divers can do their best to seal the victim's airway from the bottom-to-the-surface to decrease the amount of water that can enter the victim's lungs and stomach.

- Drowning patients should not be put at a head-down position in the field because of the increased risk of regurgitation.
- The hospital method that offered the fastest and best rewarming results was a femoro-femoral cardiopulmonary bypass.

The field of pre-hospital and hospital care of drowning and hypothermia patients is ever producing new information, techniques, and procedures. The Congress website is a good place to get up-to-date information: www.drowning.nl.

To summarize how to manage hypothermic patients in the field:

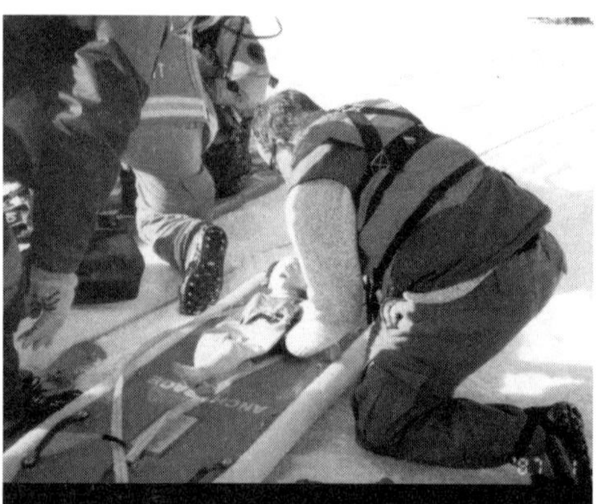

Photo 8–5 Gentle, gentle, gentle patient handling is imperative! Some type of ice transport device is needed to get the victim to shore.

- Handling must be extremely gentle to prevent ventricular fibrillation and possible resulting death.
- Keep the patient horizontal to keep blood in the core and help prevent cardiac problems. Do not let the patient walk.
- Gently move the patient to a warm, dry environment.
- Remove wet clothing, dry the patient off, and wrap the patient in proper insulation from head to toe. Place the patient on material that contains air pockets and has low heat conductivity, such as wool or fleece.
- If oxygen is administered, warm and humidify it first if possible.
- If IV's are administered, warm the fluid first if possible.
- If the patient is only mildly chilled, apply insulated heat packs to the armpits, groin, and carotid neck area.
- If there is no core temperature loss, mild leg and arm exercises can be performed. Exercise in a colder patient could cause cardiac problems.

Photo 8–6 Gently removing victims from the water horizontally takes pre-planning and practice.

- If the patient is fully alert and not nauseous, provide warm, nonalcoholic beverages that do not contain caffeine.
- Avoid aggressive, active surface rewarming procedures on moderate to severely hypothermic patients.
- Carefully monitor vital signs.
- Initiate CPR as soon as possible when called for.
- Be gentle!
- Seek appropriate medical attention.

Photo 8–7 If a victim is removed horizontally from the water, it is possible that blood in the core will drop to the lower extremities. This could make resuscitative efforts less effective.

Summary Questions

1. Cold stimulation to the hands, feet, and head can cause what defensive mechanisms?
2. If shivering is not present, it can be very _____ to recognize heat loss from the body.
3. Silent hypothermia can ______ the onset of shivering past the point when hypothermia is reached.
4. List at least five out-of-character behaviors to watch for.
5. Cold stress can result in impaired memory, cognitive functioning, concentration, and speed of reasoning even though the victim feels fully alert. True or False?
6. Shivering can cause what problems in addition to increased oxygen consumption?
7. Severe hypothermia is a ______ ______ requiring immediate transport to a medical facility.
8. What are chilblains and what causes them?
9. What are the signs, symptoms, and treatment of chilblains?
10. Describe frostnip signs/symptoms and treatment.
11. Describe the physiology and signs/symptoms of frostbite.
12. Describe pre-hospital care for frostbite.
13. What is a freezing (deep frostbite) injury?
14. Explain why frostbite and freezing injuries should not be rewarmed.
15. A person experiences vasodilation, hives, nasal congestion, and wheezing when exposed to cold. What might the person have?
16. A diver experienced a rapid ascent after his drysuit valve free-flowed. After the dive his right arm tingled and felt weak and numb. He found it difficult to concentrate when filling out his log book. He felt unusually fatigued and withdrawn. What might the diver's injury be?
17. If the hypothermic patient also aspirated contaminated water, what should be sent with the patient to the hospital?
18. Dehydration _____ the risk of hypothermia.
19. Give at least five signs of dehydration.
20. Give at least nine reasons to avoid alcohol consumption before, during, or soon after an ice diving operation.

21. Mild hypothermia can cause a ____ in pulse rate, blood pressure, and cardiac output, while moderate hypothermia can cause a _____ in them.

22. Explain some problems regarding the use of intervention medications during pre-hospital care.

23. What is a rule of thumb regarding advanced-life-support pre-hospital care?

24. List some considerations when using a defibrillator and monitor.

25. After cold and arduous dives, a diver often feels weak, unusually hungry, chilled, confused, a little dizzy, and appears to have a lack of coordination. Besides hypothermia, what else might the diver be suffering from?

26. What should the diver in question 26 do about the problem?

27. Define passive rewarming and give four examples.

28. Define active surface rewarming and give three examples.

29. Define active core rewarming and give five examples.

30. Explain why aggressive active rewarming procedures should not be done on a cardiac arrest patient until effective CPR has been started.

31. List some benefits and problems with bath and shower rewarming methods.

32. List some benefits and problems with inhalation rewarming.

33. Subjects often report feeling warm and comfortable well ____ (before, after) normal core temperature is reached.

34. Cold-stressed personnel should not be allowed to leave the warm rehabilitation environment to perform outdoor duties for at least ____ hours, and should also be able to produce clear and copious ___ and be able to ____ when feeling very warm.

35. Give the steps to managing hypothermic patients in the field and during transport

Notes

1 Lt. Cdr. G.H. "Hall" Koch, *Emergency Handling of Diving Casualties,* (DCIEM Operational Report No. 75-OR-1076, January 1975): 7.

2 James Miller, ed., *NOAA Diving Manual, 2nd Ed.,* (National Oceanic and Atmospheric Administration, U.S. Department of Commerce, December 1979): sec. 2.2.2

3 Ibid., sec. 2.4.

4 W.F. Fox, "Human Performance in the Cold," *Hum Fac* (1967, 9:3): 203-220.

5 E.H. Wissler and S.A. Nunneley, "Human Thermal Response to Accidental Immersion in Cold Water: A Theoretical Study," Scientific program preprints from the Aerospace Medical Association Annual Scientific Meeting (Washington DC: Aerospace Medical Association, 1983): 198-199.

6 P. Webb, "Impaired Performance from Prolonged Mild Body Cooling," *Underwater Physiology VIII,* eds. A.J. Bachrach and M.M. Matzen (Bethesda MD, UHMS, 1983): 25.

7 M. McMillan and R.S. Pozos, "The Warming Signs of Hypothermia," *Offshore* (1981, 41:14): 150.

8 A.D. Baddely, "Time-Estimation at Reduced Body-Temperature," *Am J Psychol* (1966, 79): 475-479.

9 A.D. Baddeley, W.J. Cuccaro, G.H. Egstrom, G.Weltman and M.A. Willis, "Cognitive Efficiency of Divers Working in Cold Water," *Hum Fac* (1975, 17:5): 446-454.

10 R.J. Biersner, "Motor and cognitive effects of cold water immersion under hyperbaric conditions," *Hum Fac* (1976, 18:3): 299-304.

11 H.M. Bowen, "Diver Performance and the Effects of Cold," *Hum Fac* (1968, 10:5): 445-464.

12 F.M. Davis, A.D. Baddeley and T.R. Hancock, "Diver Performance: The Effect of Cold," *Undersea Biomed Res* (1975, 2): 195-213.

13 S. Martin, K.E. Cooper and R. Sainsbury, "Psychological Effects of Cold Water Immersion," *Proc Ann Meeting Can Fed Biol Sci* (1976, 19): 1.

14 W.S. Vaughan, Jr., "Distraction Effect of Cold Water on Performances of Higher-Order Tasks," *Undersea Biomed Res* (1977, 4): 103-116.

15 S.J.C. Hamilton, "Hypothermia and Unawareness of Mental Impairment" (letter), *BR Med J* (1980, 280:6213): 565.

16 F.M. Davis, A.D. Baddeley and T.R. Hancock, "Diver Performance: The Effect of Cold," *Undersea Biomed Res* (1975, 2): 195-213.

17 U. Bergh and B. Ekblom, "Influence of Muscle Temperature on Maximal Muscle Strength and Power Output in Human Skeletal Muscle," *Acta Physiol Scand* (1979, 107): 33-37.

18 G. Bolstad, A. Brubakk, B. Holand and A. Pasche, "Effect of Cooling on Maximal Isometric Force in Human Skeletal Muscle During Saturation Diving," *Undersea Biomed Res* (1981, 8 [Suppl 1]): A32.

19 *U.S. Navy Diving Manual,* 0910-LP-708, Rev. 4 (published by direction of Commander, Naval Sea Systems Command, 20 January 1999): sections 19-7.2.1 to 19-7.2.2.

20 A. Matsuki, *Medical Aspect of the Winter March of the Fifth Regiment of the Eighth Military Division in the Winter of 1902* (Nippon: Ishigaku Zasshi, 1993 Sep, 39:3): 291-313.

21 J. Hirvonen, "Some Aspects on Death in the Cold and Concomitant Frostbites," *Int J Circumpolar Health* (2000, Apr, 59:2): 131-6.

22 PEEP and CPAP are more aggressive forms of ventilation resulting in higher arterial oxygen partial pressures.

23 An example of hyperbaric oxygen treatment is 2.82 atmospheres, 2.82 times the pressure at sea level, which is equivalent to 43.25psi.

24 D. von Heimburg, E.M. Noah, U.P. Sieckmann and N. Pallua, "Hyperbaric Oxygen Treatment in Deep Frostbite of Both Hands in a Boy," *Burns* (2001 Jun, 27:4): 404-408.

25 UHMS, 9650 Rockville Pike, Bethesda, MD 20814, USA.

26 K.J. Prommersberger, J. van Schoonhoven and U. Lanz, "Treatment of 3rd Degree Fingertip Frostbite in a Mountain Climber with Semi-Occlusive Dressings," *Handchir Mikrochir Plast Chir* (2001 Mar, 33:2): 95-100.

27 J.F. Socks, *Use of Contact Lenses for Cold Weather Activities,* (Groton, CT: U.S. Navy Submarine Medical Research Laboratory, 1981) Rep. no. 969).

28 C.V. Brown, *Skin Diver* (1978, 27): 23.

29 A. Zaferes and B. Hendrick, *Field Neurological Evaluations* (New York: Lifeguard Systems, 1989).

30 M. McMillan and R.S. Pozos, "The Warning Signs of Hypothermia," *Offshore* (1981, 41:14): 150.

31 D.A. Youngblood and G.H. Egstrom, eds. "The Immediate Management of Thermally Unbalanced Casualties in the Field. Thermal Problems in Diving," (Seminar proceedings; Wilmington, CA: Commercial Diving Center, 1977).

32 R.F. Goldman, R.W. Newman and O. Wilson, "Effects of Alcohol, Hot Drinks, or Smoking on Hand and Foot Loss," *Acta Physiol Scan* (1973, 87): 498-506.

33 T. Fazekas, G. Liszkai and L. Rudas, "Electrocardiographic Osborn Wave in Hypothermia," *Orv Hetil* (22 Oct 2000, 141:43): 2347-51.

34 M. Okada, F. Nishimura, H. Yoshino, *et al*: "The J Wave in Accidental Hypothermia," *J Electrocardiol* (1983, 16): 23-28.

35 R.F. Edlich, K.A. Silloway, P.S. Feldman, *et al*: "Cold Injuries and Disorders," *Current Concepts Trauma Care* (1986): 4-11.

36 F.S. Southwick and P.H. Dalgish, "Recovery After Prolonged Asystolic Cardiac Arrest in Profound Hypothermia: A Case Report and Literature Review," *JAMA* (1980, 243): 1250-1253.

[37] S.M. Schneider, "Hypothermia: From Recognition to Rewarming," *Emergency Medicine Reports* (1992, 13): I-20.

[38] A.K. Ream, B.A. Reitz and C. Silverberg, "Temperature Correction of Pco2 and Ph in Estimating the Acid Base Status: An Example of the Emperor's New Clothes?" *Anesthesiology* (1982, 5): 41-44.

[39] J.B. Reuler, "Hypothermia: Pathophysiology, Clinical Settings and Management," *Ann Intern Med* (1978, 89): 519-527.

[40] A.D. Weinberg, "Hypothermia," *Ann Emerg Med* (February 1993, 22, Pt 2): 370-377.

[41] Res-Q MC-2000 Temperature Controlled Medical Case™.

[42] A.D. Weinberg, "Hypothermia," *Ann Emerg Med* (February 1993, 22, Pt 2): 370-377.

[43] D.H. Olsen and I.H. Gothgen, "Treatment of Accidental Hypothermia," *Lakartidningen* (1 Nov 2000, 97:44): 4992-7.

[44] G.G. Giesbrecht, "Emergency Treatment of Hypothermia," *Emerg Med* (Fremantle: Mar 2001, 13:1): 9-16.

[45] C.A. Piantadosci, D.J. Ball, M.L. Nuckols and E.D. Thalmann: *Manned Evaluation of the NCSC Diver Thermal Protection (DTP)Passive System Prototype.* (Panama City, FL: US Navy Experimental Diving Unit, 1979) Rep no. 13-79 or rep no. 4-80.

[46] J.M. Lockhart and H.O. Kiess, "Auxiliary Heating of the Hands During Cold Exposure and Manual Performance," *Hum Factors* (1971, 13:5): 4576-465.

[47] P. Schmicke, "Rewarming from Accidental Deep Hypothermia by a Short-Wave Therapy Apparatus: Case Report on Three Patients," *Anasth Intensivether Notfallmed* (1984, 19:1): 27-29.

[48] J. Dehn, A.J. Christensen and N.K. Schonemann, "Treatment of Severe, Accidental Hypothermia with a Warm Air Bed," *Ugeskr Laeger* (4 Sep 2000, 162:36): 4817-4818.

[49] J.A. Stedha, "Efficacy and Safety of Prehospital Rewarming Techniques to Treat Accidental Hypothermia," *Ann Emerg Med* (1991, 20): 896-901.

[50] P. Webb: "Afterdrop of Body Temperature: An Alternative Explanation," *J. Appl Physiol* (1986, 60): 385-390.

[51] T.T. Romett, "Mechanism of Afterdrop after Cold Water Immersion," *J Appl Physiol* (1988, 65): 1535-1538.

[52] T. Tveita, "Rewarming from Hypothermia: Newer Aspects on the Pathophysiology of Rewarming Shock," *Int J Circumpolar Health* (2000, 59:3-4): 260-266.

[53] Brandon Bravo L.J., "Thermodynamic Modelling of Hypothermia," *Eur J Emerg Med* (Jun 1999, 6:2): 123-7.

[54] A.D. Weinberg, "Hypothermia," *Ann Emerg Med* (Feb 1993, 22, Pt 2): 370-377.

[55] K.N. Hall and S.A. Syverud, "Closed Thoracic Cavity Lavage in the Treatment of Severe Hypothermia in Human Beings," *Ann Emerg Med* (1990, 19): 204-206.

56 E. Visetti, M. Pastorelli and M. Bruno, "Severe Accidental Hypothermia Successfully Treated by Warmed Peritoneal Lavage," *Minerva Anestesiol* (Oct 1998, 64:10): 471-5.

57 Extracorporeal blood rewarming has proved to be a reliable method for treating patients suffering from accidental hypothermia (core temperature < 28°C). M. Kilgus and H.P. Simmen. *World J Surg* (Oct 2000, 24:10): 1282.

58 M. Farstad, K.S. Andersen, M.E. Koller, K. Grong, L. Segadal and P. Husby, "Rewarming from Accidental Hypothermia by Extracorporeal Circulation: A Retrospective Study," *Eur J Cardiothorac Surg* (Jul 2001, 20:1): 58-64.

59 A Owda, S Osama, *Hemodialysis in management of hypothermia,* Am J Kidney Dis, 2001, Aug;38(2), p E8,

60 K. Farbrot, P. Husby, K.S. Andersen, K. Grong, M. Farstad and M.E. Koller, "Rewarming of Patients with Accidental Hypothermia with the Help of Heart-Lung Machine," *Tidsskr Nor Laegeforen* (20 Jun 2000, 120:16): 1854-7

61 L.A. Kuehn: *Rewarming Strategies: Evaluation of Techniques. AAAS Symposium on Human Performance in the Cold.* (Fairbanks, Alaska: American Association for the Advancement of Science (Arctic division), 1982).

62 I.P. Buckingham and L.A. Kuehn, "Diagnosing and Treating Hypothermia" (letter), *Can Med Assoc J* (1982, 126:11): 1276-1277.

63 L.A. Kuehn and J. Zumrick, "Thermal Protection Afforded by Hot Water Diving Suits,"*Undersea Biomed Res* (1979, 6, Suppl 1): 28.

64 I.M. Light and J.N. Norman, "The Thermal Properties of a Survival Bag Incorporating Metallized Plastic Sheeting" *Aviat Space Environ Med* (1980, 51:4): 367-370.

65 P. Marcus, "Laboratory Comparison of Techniques for Rewarming Hypothermic Casualties," *Aviat Space Environ Med* (1978, 49:5): 692-697.

66 J.B. Morrison, M.L. Conn and J.S. Hayward, "Thermal Increment Provided by Inhalation Rewarming from Hypothermia," *J Appl Physiol* (1979, 46:6): 1061-1065.

67 M.L. Collis, A.M. Steinman and R.D. Chaney, "Accidental Hypothermia: An Experimental Study of Practical Rewarming Methods," *Aviat Space Environ Med* (1977, 7): 625-632.

68 M.L. Conn, P.A. Hayes and J.B. Morrison, "Contribution of Metabolic Can Respiratory Heat to Core Temperature Gain after Cold Water Immersion," (Underwater Physiology VII, UHMS, A.J. Bachrach, ed. 1981): 509-515.

69 J.S. Hayward, "Thermal and Cardiovascular Changes During Three Methods of Resuscitation from Mild Hypothermia," *Resuscitation* (1984, 2183): 21-33.

70 R.M. Harnett, E.M. O'Brien, F.R. Sias and J.R. Pruitt, "Initial Treatment of Profound Accidental Hypothermia," *Aviat Space Environ Med* (1980, 51:7): 680-687.

[71] J.B. Morrison, J.S. Hayward and M.L. Conn, "Effect of body temperature and composition on recovery from hypothermia," paper presented at the Seventh Symposium on Underwater Physiology, Undersea Biomedical Society Annual Meeting, Bethesda, UHMS (1980:19).

[72] M.B. Strauss and W.S. Vaughan, Jr., "Rewarming Experiences with Hypothermic Scuba Divers," *Undersea Biomed Res* (1981, 8, Suppl 1): A30.

[73] T. Vassal, B. Benoit-Gonin, F. Carrat, B. Guidet, E. Maury and G. Offenstadt, "Severe Accidental Hypothermia Treated in an ICU: Prognonsis and Outcome," *Chest* (Dec 2001, 120:6): 1998-2003.

Chapter 9 Ice Diving Procedures

The previous chapters in this book have discussed specific skills, such as tethering and site setup, training, rescue, and recovery. This chapter discusses all the other qualities required to make ice diving a safe operation.

Are Divers and Tenders Mentally and Physically Ready?

Before divers begin dressing, they should be asked if they are mentally and physically prepared to do the dive. This question also implies, "Do you feel that you have the necessary training and equipment to perform the dive?" Even when divers say "yes", officers and tenders should monitor them for their overall state of mind and identify any potential problems.

We have found no correlation between diver certification level and performance ability under the ice. In our ice classes, we have had newly certified divers go into ice holes with comfort and confidence. We have also seen dive instructors who find every excuse in the book for not being able to make it out to the hole. It is critical to recognize the signs of nervousness, impending anxiety, and potential panic. Self-recognition of symptoms is just as important. Do not allow divers with these symptoms to go under the ice. Panicked divers typically rip their masks off underwater, spit out their regulators, and bolt to the surface. If this occurs with ice overhead, the risk of injury and death are greatly increased. Under the ice is not the time to find out something is wrong. It must be made clear that it is perfectly acceptable, and actually commendable, for divers to tell a tender or officer that they are not fully comfortable performing the dive, if such is the case.

The safety officer should watch for "excuses," because excuses are often red flags that the diver is not mentally ready to perform the dive. Excuses take many forms, and it is sometimes difficult to distinguish excuses from actual problems. One difference is that excuses normally come in multiples. For example, because winter is the primary season for colds and congestion, it is not uncommon for divers to have equalization problems that prevent them from ice diving. However, when a diver has equalization problems and no visible signs of congestion, in addition to a too-tight hood, a missing glove, and a headache, it is time to take the diver aside for a chat. The first step is to tell the diver that it is perfectly okay to not dive today, and that serving as a tender would be most helpful. If the diver readily accepts this suggestion, you have just prevented a diver who was not mentally ready from diving. When a diver cannot find a piece of equipment, you can offer to replace it. If the diver says "great" and continues dressing, then the missing equipment was probably not an excuse. If, on the other hand, the diver finds another problem after you provide a replacement glove, consider it a red flag.

Be very watchful of divers who are moving far slower in their equipment preparation and dressing than usual. This is particularly true for public safety divers, who are supposed to move with professional rapid deployment motion. Talk with these divers and find out what is going on in their minds. If you are an instructor or dive team leader, do not just give up on divers who are nervous about going in. Talk with them privately and find out what can be done to make them more confident. Sometimes it may mean putting a buddy with them, if the dives are normally solo. Sometimes it may mean telling them that all they need to do is hang just below the ice roof at the edge of the hole, that they do not need to go any farther out unless they choose to, and that staying at the edge is perfectly acceptable. Often, once they are under and have become relaxed in their breathing and thoughts, they will want to explore farther out.

Asking divers to "kiss the water" prior to submerging functions not just to acclimate their faces, but also serves to gauge how comfortable they are in the ice hole. If a diver cannot comfortably breathe without a mask with a submerged face, then that diver absolutely should not be allowed to dive under the ice. Divers should be made aware of this requirement prior to taking the class or participating in a day of ice diving.

Other diver red flags that tenders should watch for include: frequent and repetitive checking of personal equipment, out-of-character behavior, being unusually quiet or talkative, and repetitively asking their tenders procedural questions. Another check for mental readiness is the equipment check that is performed when the diver is fully dressed. The checklist provided in this book not only ensures that the equipment is fully ready to go, but is designed to also discover how comfortable divers are with their equipment. This check is performed twice, first by the tender and again by the safety officer at the demarcation line. If divers become flustered when they are asked to reach for and touch particular pieces of equipment without visually looking, think about what will happen when they need to reach for that equipment under the ice after the bottom has silted up. Watch how their hands move. If their hands move with rapid, jerky motions, or if they are shaking, it is necessary to find out what may be going on with them.

Physical readiness is equally important. Some divers do not feel comfortable admitting that they are becoming chilled or that their ears feel stuffy, because they very much want to do the dive and do not want to be pulled out of the lineup. It is up to tenders to watch for signs of cold stress. Backup and 90%-ready divers are the most likely to become chilled while they remain relatively motionless at the ice hole during the primary diver's dives. Watch for divers hugging themselves, rubbing their hands together, wiggling their feet, blowing into their hands, pulling blankets tighter around them, and general squirming around. Monitor breathing rates and quality when the divers are on the ice and immersed in the hole.

Tender readiness is also important. The safety officer should watch out for tender cold-stress signs, as well as signs of fatigue. Sometimes public safety diving operations have limited numbers of personnel, which means tenders work extra hard moving and picking up gear, pulling lines, tending, and crossing the ice if there are no transport devices set up between shore and ice holes. Tenders also need high mental alertness and functioning. They need to notice immediately if a tether line is caught under the ice, if a diver's movement changes, if a piece of equipment is missing or incorrectly used/donned, or if anything else is potentially wrong. If tenders did not get enough sleep the night before, have their minds on a personal problem, or do not feel well, small problems can be missed and can quickly evolve into serious situations. Extra care must be taken to monitor public safety diving personnel during operations in the middle of the night, when fatigue and naturally lower core temperatures exist.

For both tenders and divers, we suggest a "three strikes you're out" rule. If a minor mistake is made—well, that can happen to anyone. If a second mistake is made—it could simply be bad luck. If a third mistake is made—the red flag goes up! After three minor problems, the fourth minor problem is far more likely to become a major problem. Accident analysis clearly shows that few injuries or deaths occur because of one major problem. Go back to the fatalities described in chapter 1 and write out the progressive series of errors, while keeping in mind that only the largest mistakes were described and that there were probably at least several other minor mistakes that occurred first that were missed.

If someone who normally does not make mistakes now makes three mistakes, that person is signaling you that something is not right today. That person is not at his or her peak efficiency. There is no sense in risking a serious problem. Talk with the person to find out what might be going on. Have that person move to a job with less responsibility and fewer potentially serious consequences, or just call it a day.

Tender Duties and Procedures

Upon arriving on the site, tenders should:

1. Put on PFDs and other personal protective equipment and tending equipment (*i.e.*, timekeeper, profile slate for the backup tender, and sunglasses as appropriate).
2. Receive plan-of-action briefing from the dive coordinator or the instructor.
3. Assist setting up the staging areas, if that is not already completed.
4. Take air cylinders, tending lines, and all other necessary equipment out of the vehicles and lay them out with pre-planned organization.
5. Turn the air on when any cylinder and pony bottle assembly is touched so that a diver *never* dons a cylinder or pony bottle that has the air turned off or the cylinder less than full.
6. Make sure divers are mentally and physically fit to dive. Ensure that divers have their pre-dive blood pressure checked if appropriate, and make sure they have not recently been drinking alcohol.
7. Assist divers into their exposure suits and equipment.
8. Complete a tender-diver gear check.
9. When divers are standing by, assist them by: offering drinking water and appropriate thermal protection, reviewing dive objectives, and monitoring them for stress. Get divers off their feet as soon as possible.
10. Review line signals, safety procedures, and contingency plans.
11. Walk with the diver to the point of stepping on the ice. Tenders should use the deployment line to belay divers up and down inclines and declines or, if necessary, use a tethered ice board or sled.
12. Follow appropriate procedures for deploying tender and the diver to the point at which divers will enter the water. These procedures could range from walking

with the diver, if the ice is good, to using a human piton system with an ice board that is secured to a pulley system for weak ice.

13. Check that a contingency 80ft^3 (or larger) bottle and a contingency pony bottle are close to the site. Ensure that other safety procedures are being followed, such as securing the diver and tender lines from a stationary object (such as an ice screw) back to the shore.

14. Check that backup and 90%-ready divers are ready and in place before allowing primary divers to enter the water.

15. Ensure safe roof turn entries of divers, and document their times in and down and their starting cylinder pressures. Know what the divers' cold-water surface air consumption (SAC) rates are as well as the divers' average breath per minute rates.

16. Ensure that divers acclimate their faces to the water.

17. Give and receive line signals or use a communication system.

18. Monitor the tether line. Call out each diver-distance-out to the profiler. Alert the profiler when any slack in the tether line occurs.

19. If a communication system is used with a tender headset, the tender should count the diver's breathing rate for one minute with the backup tender giving start and end times. With this system, the primary tender never has to look away from the diver's line. Do this every five minutes and have the profiler record the rates. If a communication box with a speaker is used, the profiler can monitor and record the diver's breathing rate. If no communication system is used, stop and ask the diver to signal a "1" for every inhalation for one minute. After ten minutes, perform estimation calculations for how much air the diver should have left based on SAC and breathing rates. If a communication system with a speaker box is used, the backup tender can take and record the diver's breathing rate.

20. Ensure that the deployment line is always neatly bagged and ready for rapid deployment, especially when using a hardwire underwater communication system. If possible, keep the line off the ice on a tarp.

21. If using underwater communications and there is visibility, periodically ask divers to check and report their air pressures.

22. Be alert to what is happening at the surface as well as underwater. Maintain a global view of the entire scene, and be prepared to act in the event of an emergency.

23. Always maintain direct line access to the diver with the tether line. This may require moving around the hole.

24. Notify the dive coordinator of any problems.

25. Have someone notify EMS personnel and command if the diver signals for help or indicates that the drowning victim/search object is found. EMS can then be prepared.

26. *Never* take your eyes off the diver's tether line!
27. Tell the diver when it is time to return to the hole, and assist the diver out of the hole using proper procedures. Report the diver's ending air pressure and time, transport the diver to shore, undress them, and take the diver to the rehabilitation staging area for blood pressure, hydration, and cold stress checks. If divers were performing searches and they did not find the object on their dive, ask them if they feel fully confident that the area can be secured or whether they think they may have missed the object.
28. Record ending dive time and cylinder pressure.
29. Have a tender-to-diver debriefing so that the diver can report anything found that may be noteworthy, such as obstructions.
30. The backup tender/profiler of a search operation reviews the diver's pattern quality, rate of search (number of running feet covered in twenty minutes), breathing rate, degree of cold stress, ability to find the test search objects, and the diver's post-dive response to the question: "Is the item not there or do you think there is a chance you could have missed it?" Use this information to make a decision about whether the area can be secured or if it should be re-searched.

Tenders must always focus their attention on their divers. If someone is talking to a tender, the tender should listen, but not look at that person. A tender's eyes are always on the water, where the diver is. A look away for even three seconds is enough time for a diver to pop up under the ice roof and sink back under without being noticed. The change in the tether line angle is the only way the tender will know this has happened, and that requires constantly and actively watching the line.

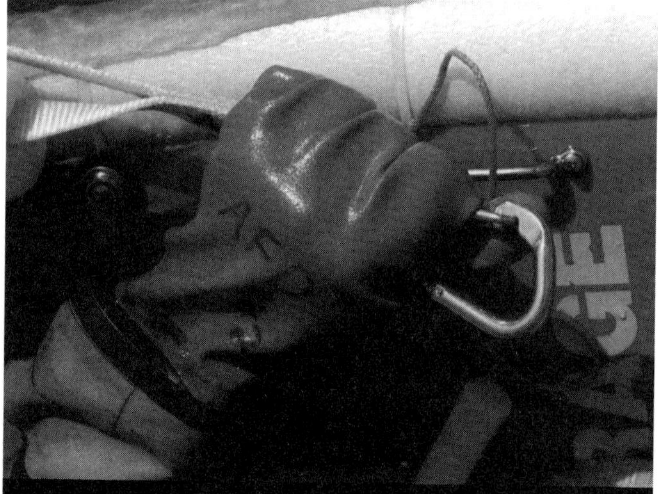

Photo 9–1 It is important to handle equipment the same way every time so it becomes reflexive. Acting spontaneously is important during stressful times. Tenders should grab and hold carabiners in a ready-to-use position.

It takes more training and skill for a tender to know when a diver is snagged under ice than when not under ice. Tenders have to learn how to feel the line for any changes. Because small problems can escalate into larger ones so quickly under the ice, tenders should be left to do their jobs without any unnecessary conversation or interruptions.

Staging and Scene safety

Safety officer

The position of safety officer should be designated in both recreational and public safety diving operations. Safety officers are an integral part of all public safety operations. Fire, police, and EMS personnel should have at least a rudimentary understanding of the safety officer's often difficult and very important job. Because ice diving is an operation, a safety officer is equally important for sport ice dives. A safety officer does not have to be a diver but does have to be at minimum an experienced, certified ice diving tender. It is recommended that sport instructors send one or more people, who will serve as their safety officers on sport ice diving sites, to a public safety ice diving operation program to gain valuable training and certification.

Photo 9-2 The safety officer has to understand the procedures for all levels of the operations to make sure they are conducted safely, from creating the plan of action and staging platforms to the final debriefing.

Photo 9-3 Part of the safety officer's responsibility is accountability, to log who is in the hot zone and to make sure that only the necessary personnel are in the hot and warm zones. The safety officer also has to occasionally step back to view the scene to make sure all rules are being followed such as PFDs and no divers without tenders.

Safety officers must have excellent global awareness and unwavering detail-observation skills. They also need to have confident leadership capabilities, because they are in fact making sure everyone does what he or she is supposed to be doing. Shortcutting is unacceptable. The most important quality of a safety officer is a genuine concern about the health and safety of everyone on the site.

The safety officer's job is to make sure all standards and guidelines are followed, including:

1. Hot (water/ice), warm (shore staging area), and cold zones are established and maintained.
2. A demarcation line is set up in the warm zone. No personnel cross this line without being checked from head to toe by the safety officer.
3. A vehicle route is kept clear to allow EMS to enter and exit.
4. Communication systems are checked to make sure that EMS can be contacted if a problem occurs (if EMS is not already on scene as part of the operation).
5. The EMS personnel on the scene are trained to manage cold stress, hypothermia, immersion, submersion, and diver gas-bubble-injury problems.
6. Shelters or windbreaks are set up.
7. All personnel are wearing appropriate PPE.
8. All personnel in the hot and warm zones are wearing PFDs or other proper flotation devices.
9. Equipment is properly laid out, such that all full cylinders are pointing their valves toward the water and all used cylinder's valves are pointing away from the water; no items are too close to heaters; cylinders are not left standing where they can fall over and break equipment; dive gear is not laid directly on snow or the ground; lines are bagged or reeled if they are not being currently used; lines are not being stepped on or crushed under equipment; etc.
10. All personnel in hot and cold zones are accounted for on a command board or by collecting tender and diver certification cards at the entrance to the warm zone. The latter method also serves to verify the minimum required training level to be in the warm and hot zones.
11. The incident command system is in full operation and has expanded enough to handle the situation.
12. A sufficient risk/benefit analysis was performed.
13. A safe and effective plan of action was created based on ice quality, environmental conditions, and available personnel and equipment resources.
14. Personnel are effectively briefed and understand the plan of action.
15. Sufficient personnel, equipment, and health-related resources are available or are en route to handle the incident.

16. Witnesses, surface support, and divers are continuously monitored for signs and symptoms of cold stress and mental and physical readiness.
17. Diver blood pressure is checked prior to dressing as well as post-dive.
18. Lines from the warm to the hot zone are properly secured with straps to stationary objects.
19. If holes need to be cut in the ice, proper procedures are followed by trained personnel.
20. The route and procedures to transport personnel between the holes and shore are safe and effective and require the minimum personnel effort.
21. Tested and practiced contingency plans are agreed upon and are ready to be put into action should the need arise.
22. Necessary safety equipment is available: a contingency pony and main cylinder at or near the ice holes, one or more rescue throw-rope bags, a thermal recovery stabilizer or similar device, an ice transport device, oxygen (preferably with warm, humidifying capability), and first-aid supplies. Personnel trained to use this equipment are also present.
23. No primary divers are immersed in ice holes without tenders, backup divers, 90%-ready divers and the proper rigging in place.
24. Proper ice diving procedures and guidelines are being followed, with proper record keeping.
25. Divers and tenders are effectively rehabilitated when they come off the ice.
26. A debriefing and, if necessary, a critical incident stress debriefing are conducted at the conclusion of the operation.

Dressing

Once their gear has been laid out, divers should dress with the aid of their tenders, using practiced, deliberate, conservative motions to avoid unnecessary exertion. Professional, rapid deployment dressing is demonstrated in *Blackwater Contingency.*[1] Exertion can lead to peripheral vasodilation and perspiration with subsequent increased cold-stress risks, increased heart rate and breathing, increased nervousness, and an increased chance of making mistakes.

How to properly dress and don a drysuit

The following dressing procedures can be viewed in *Drysuit Dressing.*[2]

Make sure that all divers who plan on diving dry have drysuit diving certification.

1. Stand on a clean dry tarp.
2. Make sure the suit seals are properly powdered with talc and that the zippers are coated with paraffin wax and are functioning smoothly. Make sure that the inflator valve is pointing downward at a 45° angle toward the side the hose will come from, to allow for the most effective quick-disconnect maneuver should the inflator begin to free-flow underwater.
3. Remove watches, earrings, shoes, badges, and other items that can be pressed into the body or damage the suit, should suit squeeze occur.
4. Don appropriate undergarments.
5. Roll the top of the suit inside-out and down to the level of the chest inflator valve. If suspenders are attached, put them to either side.
6. Don the lower extremity part of the suit.
7. Streamline fingers on one hand and slide the hand through the wrist seal. Once the wrist seal is at the first knuckles, assist it with the other hand by inserting at least two fingers to push the seal down over the hand. One finger can act as a knife on cold latex. Do not grab the sleeve and attempt to pull the wrist seal into place, as this can tear the seal. Use the other hand to assist as described. Repeat with the other sleeve. The tender can assist with sliding the seal over the diver's hand by standing to the side of the diver with the diver's elbow in the tender's chest. The tender then reaches into the seal with two fingers on either side to stretch open the seal while pulling it down over the diver's hand.

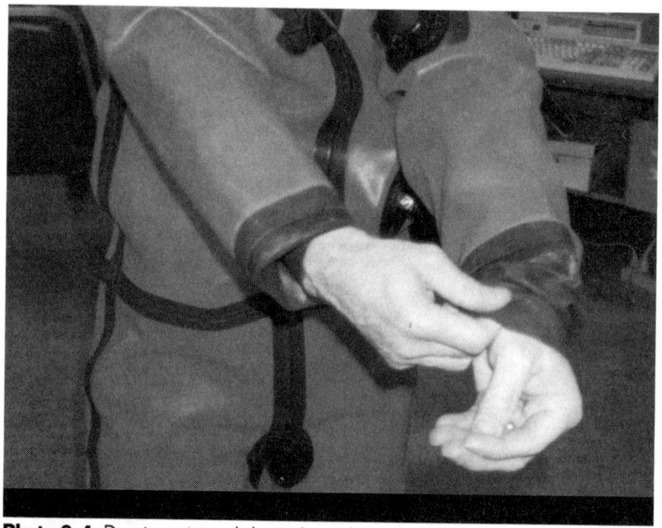

Photo 9–4 Donning wrist seal. Assist the seal over the hand with at least two fingers.

8. Set the wrist seals; make sure they are flat against the wrist and forearm and that the seal starts just below the wrist bone. If the seal covers the wrist bone, the diver loses some wrist mobility, and hand circulation is more likely to be compromised.
9. Warm up the neck seal by slowly stretching and relaxing it ten times. If the seal is latex, fold it like a reverse turtleneck and pull it down over the forehead. Do not jam it down over the top of the head. Resulting neck compression can cause many problems, among them equalization problems.
10. The tender checks that the neck seal is sitting properly.
11. It is the tender's job to close back or front entry zippers. This prevents zipper damage from possible stress of a real call, finger dexterity loss due to cold, or undergarment material in the way of the zipper that cannot be seen by the diver. For back zipper suits, divers raise their arms to shoulder level and slightly bend them forward at the elbows while rounding the shoulders forward. The tender then picks up the bottom part of the zipper with both hands and pulls the suit up gently into the diver's crotch. This serves to bring the top and bottom parts of the zipper together. The tender places the left hand at the start of the zipper and pulls the tab to where the zipper starts to bend in the area of the diver's left shoulder blade. The tender repositions the left hand over the closed zipper just before the tab and continues to pull the tab closed to the right shoulder blade. Repeat the

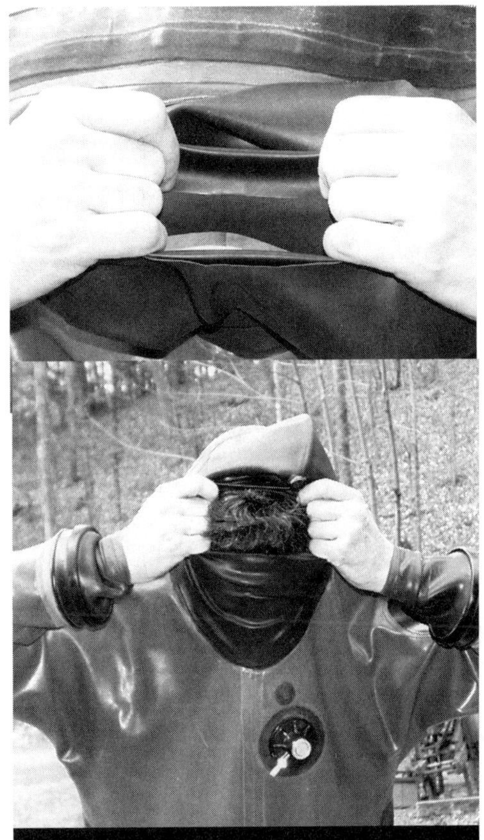

Photo 9-5 and 9-5a Drysuit neck seals can be worn a little looser if they are worn with a reverse turtleneck fold. The diver places this reverse fold in the neck seal, gently stretches it open and closed ten times, and then pulls the seal over the forehead and down the head. If a diver needs assistance, a tender can use two hands to open the seal vertically while the diver maintains the horizontal stretch as shown.

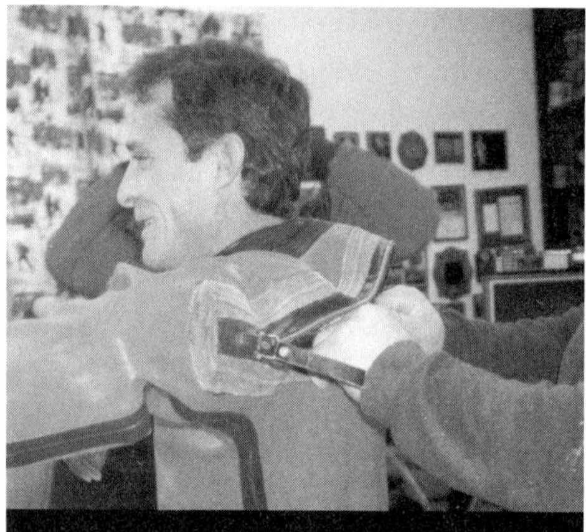

Photo 9-6 The tender first picks up the bottom zipper and gently pulls it upward.

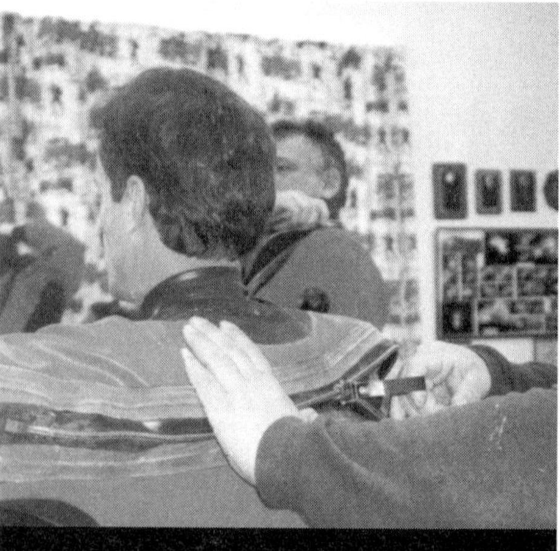

Photo 9-6a The tender closes the zipper in three stages, moving his left hand forward at each stage.

Photo 9-6b The diver performs a final check to make sure the zipper is fully closed.

process until the zipper is fully closed. Tenders can keep one finger between the diver's undergarments and the zipper tabs when pulling the tabs closed to prevent material from becoming stuck in the zippers. Divers then reach up and check that their zippers are securely closed. If the zipper begins to stick when being closed, do not force it. Rather, back it open a little and carefully apply paraffin to the outside of the zipper (the external components) and again try to close it gently. If there are threads coming out of the zipper, cut rather than pull them.

12. Check drysuit seals and hood to see if they need to be cut.
13. Note that if this is the first time the diver is wearing a neck seal, it may feel tighter than it actually is. If it is not obviously too tight, let the diver wear it for a few minutes to see if it becomes more comfortable.
14. Without visually looking, the diver should adjust the exhaust valve all the way open.
15. Burp the suit by going down on the right knee, opening the neck seal with the left hand while rotating and pushing the torso into the upward leg, with arms compressing the chest while the right hand pushes on the exhaust valve.
16. Adjust the exhaust valve all the way closed and then open a quarter to half turn.
17. Stretch arms, torso, and legs to remove wrinkles and trapped air and to put the suit in the right position on the body.
18. Don the hood undergarment and hood.

19. The tender assists the diver with gloves using the same method as the wrist seal assist. The diver's bent elbow is braced against the tender's chest. The tender opens up the glove's wrist seal and eases it over the diver's hand as the diver pushes downward into the glove. This technique supports the diver's balance and allows the tender to use strong bicep muscles.

Photo 9-7 Burping the suit to vent the air.

20. The tender puts appropriately weighted ankle weights on the diver.

21. The diver performs a series of full-body stretches.

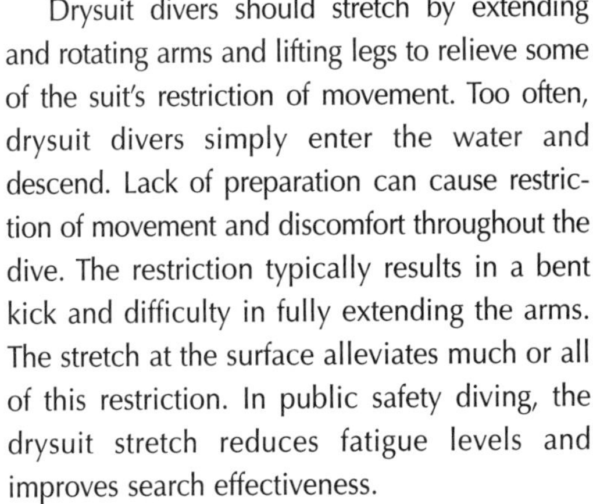

Drysuit divers should stretch by extending and rotating arms and lifting legs to relieve some of the suit's restriction of movement. Too often, drysuit divers simply enter the water and descend. Lack of preparation can cause restriction of movement and discomfort throughout the dive. The restriction typically results in a bent kick and difficulty in fully extending the arms. The stretch at the surface alleviates much or all of this restriction. In public safety diving, the drysuit stretch reduces fatigue levels and improves search effectiveness.

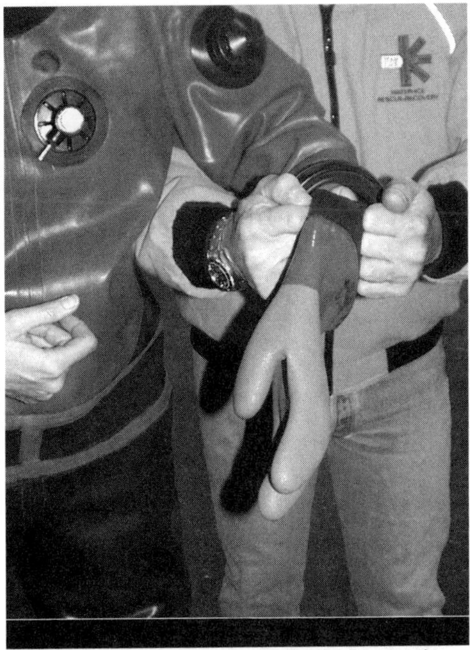

Photo 9-8 The tender places the diver's elbow into his chest and pulls the gloves on while stretching the wrist seal over the hand.

Proper weighting is imperative. Over-weighting is particularly hazardous for drysuit divers in ice diving. Among other problems, the over-weighting increases the probability that too much air will be in the drysuit and BCD. This increases the risk of an uncontrolled ascent to the surface. Public safety divers who recover a body or other heavy search object should *not* add air to their BCD or drysuit to compensate for the negative buoyancy of the body or search object.

Case history 1. In 1987, Mark McMillan, an Antarctic research diver, added air to his drysuit to compensate for a heavy array of lights he was carrying. For whatever reason, he dropped the array and found himself "ballooned up under the ice" with no regulator in his mouth.[3] His partner was not able to vent the air from McMillan's suit and had great difficulty dragging him back to the hole. Cause of death was deemed a carotid sinus reflex caused by the increasing volume of air in the neck seal during the uncontrolled ascent, which prevented his ability to slow the ascent. The final result was an arterial gas embolism.

If a recovered body is more than -5lb or -6lb, or if you feel the need to add air to your BCD to raise the body, have surface support assist with a tow rope instead. You can also have the backup diver assist or use a lift bag if you have been properly trained in under-ice lift bag use. Do not add air to your drysuit to compensate for your own negative buoyancy. Make sure you are neutrally weighted. Only add enough air to the drysuit to stay warm and comfortable, with full extremity movement. Unlike a drysuit, a BCD can be fully purged in seconds in case of an unexpected or uncontrolled ascent. And, unlike a drysuit, an over-inflated BCD cannot cause a carotid sinus reflex.

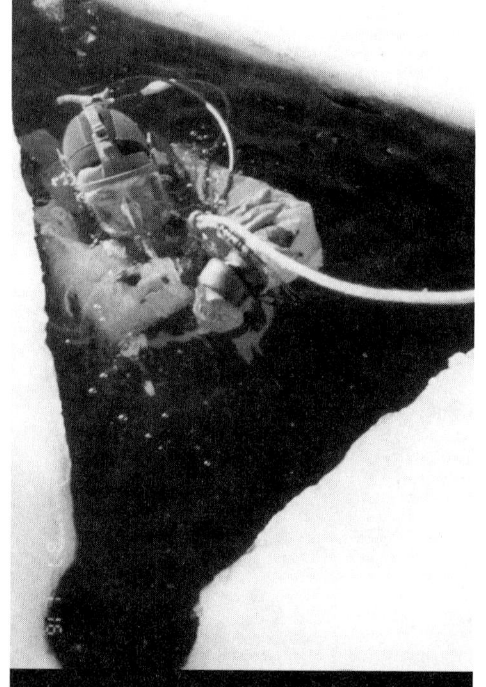

Photo 9–9 Venting a BCD is a simple one-handed, two-second procedure. Drysuit venting requires a specific vertical posture, the use of both arms, and more time.

BCDs are designed to adjust buoyancy; a drysuit is not. A drysuit can hold much more air than a BCD, and that air can end up in places, such as the feet, where its presence increases the risk of injury. A BCD allows simple and fine control of the amount of air purged from it, while a drysuit does not allow this to be done as effectively. These points are important to note because many instructors, and even books, teach divers to use their drysuits instead of their BCDs to control buoyancy. The reasoning is that divers cannot manage to control air sources in two places simultaneously. The people who promote this reasoning do not address the above concerns and do not address the fact that the majority of drysuit divers are severely over-weighted, which means they are adding tremendous amounts of air to their drysuits in futile attempts to become neutral. These same people typically also teach the myth that an over-inflated drysuit can be purged by pulling away the neck or wrist seal before the diver hits the ice roof. They often do not address the fact that increased air in the drysuit increases the risk of carotid sinus reflex.

Case history 2. A public safety diver from a dive team in upstate New York experienced a classic case of carotid sinus reflex. He had recently completed a drysuit certification course offered by a local recreational instructor who followed the certifying agency's course outline. The outline stated that a diver should not add air to his BCD; rather he should use the drysuit to adjust for buoyancy. This diver was definitely overweighted. To adjust buoyancy, he added air to his drysuit, which caused him to ascend. As he ascended, the air in the drysuit expanded and created a pressure against his neck on the carotid arteries. This created a snowball effect—as he ascended, the air expanded and caused a more rapid ascent that in turn led to more air expansion. This created a greater pressure on the neck area. The rapid ascent caused lung over-expansion injuries to the diver and put his buddy at risk as he attempted to slow his ascent. The diver recovered from his arterial gas embolism, but the buddy's family "suggested" that he drop out of the team, which he did.

Case history 3. This victim was a Coast Guard diver who previously dove in a Unisuit™ (a thick neoprene drysuit) and was diving in a vulcanized rubber suit for the first time.[4] There is a strong possibility that he was over-weighted by as much as 20lb (9kg). A vulcanized rubber suit requires far less weight than a Unisuit™. The diver and his dive partner were placed on a T-shaped tether system, and they descended in an inverted position. Because of the inversion and in the over-inflation of the drysuit, and probably because of the ankle weights or fins that were not duct-taped on, the diver lost a fin. The buddy attempted to replace the fin, with the result that they became entangled in the tether line. The tether line then knocked off one diver's mask and regulator. There is a disagreement about whether or not there was a rapid rise to the surface. The diver surfaced without symptoms. Later, in the hospital, he exhibited confusion and the inability to make simple decisions and reported having a serious headache. He was treated for an arterial gas embolism. In mid-treatment, he developed pain in his right leg and numbness in his foot. After further treatment, all symptoms were resolved.

What can be learned from this case? Emphasize the dangers of over-weighting and over-inflation of the drysuit and the potential problems of diving more than one diver in a hole at a time. Inverted descents, fins not strapped or duct-taped on if necessary, and lack of sufficient training with a pony bottle (if they even wore one) are some other errors that were part of the chain of events leading to the accident. The recognition of the patient's inability to make simple decisions was commendable. The example given in the case history was that the patient could not figure out which side of the hospital corridor was the correct side to walk on.

Another problem caused by over-weighting is the increased risk of inflator valve free-flow. The more an inflator valve is used in very cold environments, the greater the probability that it will free-flow. The more over-weighting there is, the greater the use of inflator valves. As the diver rises, the large amount of air expands, and consequently, air has to be purged. Every pound of lead is equivalent to a pint of gas (one kilogram of lead is approximately equal to one liter of gas). This means that eight pounds of over-weighting requires 8pt of air at the surface, 16pt at 33ft, and 24pt of gas at 66ft to achieve neutral buoyancy. Recreational divers and public safety divers in water with visibility should be neutral, while public safety divers in blackwater who need to search the bottom should be at most -2lb. Also, the great compensation chase, as we call this repetitive inflation-deflation process, makes it very difficult to be neutral and hover. Divers are continually physically working to maintain their position in the water column. Exertion increases breathing, which again can change buoyancy by as much as 6lb to 8lb on every inhalation and exhalation. This downward spiral causes other problems, such as increased air consumption and breathing rates, fatigue, and increased risk of lung barotrauma.

Another point to consider in the BCD versus drysuit inflation controversy is that a drysuit inflator valve free-flow is a more serious event than a BCD inflator valve free-flow. Both should disconnect with ease to prevent further introduction of air into the unit, but there is no comparison between purging air from a BCD and from a drysuit. When possible, choose drysuit exhaust valves with a domed rather than a flat top. The dome allows divers to raise their left arms and push the exhaust valves against the side of their heads for hands-free purging. Their right hands can simultaneously disconnect the inflator valve.

Dress to be properly weighted. Only add enough air in the drysuit to have enough mobility and thermal protection as is necessary, and use the BCD to make the necessary minor buoyancy adjustments if there is enough water depth to warrant them. If you have to add air to your BCD by the time you reach fifteen feet, you are over-weighted and should make the necessary weight belt corrections. If you become chilled on a dive, make sure to wear more thermal protection on the next dive rather than adding more air as insulation.

Donning dive equipment

If BCD cylinder assemblies are stored already set up in public safety dive vehicles, tenders turn the air on the main and pony cylinder as they touch the assemblies to remove them from the vehicle. They are then laid down on the staging tarp with valves pointing toward the water, indicating that the tanks are full. Keep regulators and full face masks warm and protected if possible.

The tender checks the BCD-cylinder assembly while it is still on the ground. If there are quick-release BCD shoulder straps, make sure they are fully loosened. Is there at least one cutting tool mounted on the BCD in the area between the diver's shoulders and waist? Place the gauge hose through the BCD armhole. Most divers leave the gauge hose on the outside of the BCD and merely secure it with a clip at the gauge end to the outside of the BCD. This increases the possibility that something could become entangled between the gauge hose and the diver's back or side, which are difficult places to reach. If, on the other hand, the hose runs snugly under the arm and under the BCD, the only place entanglement can occur is at the center of the chest where the gauge comes out from under the BCD. This is a much easier place to free an entanglement. Public safety divers should already be trained to set their gear up to decrease entanglement risks, but far too few of them are, and few recreational divers have even considered it. Make sure that the pony regulator hose streamlines downward and along the BCD backpack so that it does not stick out. Also, place the drysuit inflator hose through a BCD armhole so it can run under the diver's arm, under the BCD, and directly to the inflator valve. The tender should use the equipment checklist to give a quick ten-second check of the BCD-cylinder assembly.

The dive harness goes on next. The diver puts his/her arms back in a handcuff position as tenders slip the harnesses up the diver's arms and shoulders. The tender comes to the front of the diver as the diver put one hands flat over the solar plexus. The tender closes the girth strap while asking the diver to take the largest inhalation possible, and then snugs the strap over the expanded chest and hand. Divers should not close their own harnesses, especially on rescue sites, because they may tighten them too much and hence restrict their ability to inhale deeply should the need arise. Another function of putting the hand over the solar plexus is to make sure that the harness girth strap sits in the safest position, at the level of the solar plexus. Before moving on, the tender makes sure the diver has a cutting tool, such as a pair of shears, mounted upside down on the harness. A locking carabiner with a 3in to 4in (9cm) piece of duct tape wrapped around it is attached to the harness tether point.

Next, the diver turns his/her back to the tender and bends forward as the tender picks up the weightbelt with the free end in the right hand ("you have the right to be free") and gently aligns the belt in place across the diver's back. As the diver secures the weightbelt closed, the tender stands the BCD-cylinder assembly up behind the diver, who gets down on the right knee with arms reaching back and together. The tender slips the BCD assembly up the diver's arms. The reason for having the right knee down and the left knee up is to position the BCD assembly closer to the left arm, where it can more easily be slipped over the drysuit exhaust valve. When the BCD is almost fully on the diver, the diver counts to three, and both diver and tender stand up, with the tender supporting the cylinder from the valve and the bottom. As the diver stands, the cylinder slips right on. The tender comes around and holds the gauge and drysuit inflator hose above the BCD cummerbund as the diver secures the cummerbund. The diver snugs up BCD shoulder straps if necessary, while the tender secures the pony mouthpiece to the neck strap and secures the gauges. This BCD donning method is efficient and reduces the likelihood of tender back strain, improper gear placement, and frustration as a diver stands and tries to get an arm back into a BCD armhole. Frustration impairs mental fitness. The diver secures the drysuit inflator hose to the inflator valve.

The tender now performs a complete tender-diver gear check before approaching the demarcation line. This check should take less than twenty seconds and, if possible, should be done in the warm shelter. Once divers are dressed, do not leave them standing for any unnecessary time. Too often divers are forced to stand while tenders are given instructions or while decision makers figure out where to put the diver. If there is a delay of more than two minutes, sit the diver down on something off the snow and ice.

Tender-Diver Gear Check

Tender check of diver equipment

There are four functions of the tender-diver gear check. One function is to make sure all equipment is in the proper place. If dive gear is not worn in the same place on every dive, the diver cannot become reflexive with it; and reflexivity can make the difference between life and death in a crisis. A second function is to make sure divers are comfortable with their equipment and can reach for each item accurately without looking. Remember that even clear water ice dives can become black if a diver hits the bottom and causes a silt-out. A third function is to make sure that tenders fully comprehend the location and function of all the equipment worn by their divers. Lastly, the check procedures are designed to provide divers with hands-on reinforcement of self-rescue procedures, such as: drysuit inflator valve quick-disconnect, exhaling first through regulator mouthpieces, tipping chin to pony regulator, accessing cutting tools, and finding harness carabiners and weight-belt release mechanisms.

Photo 9–10 The tender performs a gear check with a checklist card to prevent missing a step.

This check should be conducted year round before every dive. Practice and perform it in the same order every time to make it reflexive. Putting the checklist on a slate or laminated card is highly recommended. It should not be done from memory unless it is done several times a week all year long.

This check is performed by the tender and then subsequently by the safety officer at the demarcation. If possible, perform the check in a mildly heated area to decrease the incidence of equipment freeze-ups.

1. Know where to go to assess the scene and stage the operation.
2. Don PFD, gloves & other necessary PPE.
3. Perform Rapid Scene Assessment.
4. Identify life & health hazards.
5. Interview witnessess and document information on the profile map.
6. Assess no. and ASAPS status of victims.
7. Establish communication w/ victims.
8. Mark a spot on shore in front of each victim if possible.
9. Implement IMS. Establish command post, Incident Commander, Staging Areas, Staging Officers, Safety Officers, and other officers and modules required by the size and complexity of the incident.
10. Direct positioning of rescue vehicles to form windblocks & exits.
11. Setup demarkation line.
12. Call for appropriate agencies, including the dive team, EMS, and law enforcement.
13. Secure the scene & determine & mark Hot, Warm, and Cold Zones.
14. Secure witnesses for interviewing by law enforcement & check their conditions.
15. Draw a profile map of the most likely victim location. Find a safe place to deploy divers.
16. Reassess ASAPS.

A = Agressive
S = Self-rescue capable
A = Alert
P = Passive
S = Submerging

Divers Alert Network: For Divers Emergencies 919-684-4326

Make sure the diver is physically, mentally, & emotionally prepared for the dive.

SAFETY OFFICER; ______________________
INCIDENT COMMANDER: ______________________
NOTES: ______________________

THE TENDER SHOULD HAVE AND CHECK:

1. Appropriate PFD closed properly.
2. Appropriate footwear and exposure protection.
3. Proper tending gloves.
4. An appropriate time keeper.
5. A water activated flasher.
6. Eye and sun protection.
7. Back tethered harness if working from a steep embankment.
8. If acting as the back up tender, a profile slate and writing utensil.

WATER SITE MUST HAVE:

1. Contingency reg. on full cylinder.
2. Contingency reg. on full pony bottle.
3. At least one rescue throw rope bag per diver-tender pair.
4. BLS, or better, first aid personnel and supplies.

TENDER TO DIVER:

Right: **Diver's**: Left

- 1. Stop/face & tighten line/ you OK?
- 3. go Right (Divers) ⬅
- 4. go Left (Divers) ➡
- 2+2 search immediate area.
- 3+3 Stand by to leave bottom.
- 4+4 Come up.

There is only 1 job you have to do every time.
—GO HOME. Team LGS.

TENDER CHECK OF DRIVER:

1. Air on.
2. Hood in place.
3. Mask ready, gloves on.
4. Primary regulator, exhale first, breathe 3 times, check tie-wrap and mouthpiece.
5. Pony reg: same as #4, hose under arm, secured w/ mud mouthpiece cover and holder.
6. Drysuit hose on functional.
7. Diver, press BC inflator button 3 times, while tender checks pressure gauge, gauges secured, all hoses trim to body and secured.
8. Diver, without looking, find 1st shears, 2nd shears, & 3rd cutting tool.
9. Diver, find carabiner at harness tether point, tether line attached.
10. Diver, find weightbelt release– Is it right hand? 10" extra webbing?
11. Fins ready, w/ ankle wts. if needed.
12. Back-ups have contingency lines.

DIVER TO TENDER:

- 1. Diver is okay.
- 2. Tender, make notation.
- 2+2+2 Tangled but okay, alert back-up diver.
- 3+3+3 Okay but needs help from backup-diver.
- 4+4+4 Needs immediate help!
- 6+6 Found the object

Lifeguard Systems
P.O. Box 548 Hurley, NY 12443
Tel/Fax: 845-331-3383
www.teamigs.com
www.wateroperations.com
www.rip-tide.org

Figure 9–1 Slate. Because most teams do not run dive operations more than two or three times a month, a complete tender and diver equipment checklist should be used for every check. Laminated cards can be kept in wallets, worn on necks, and clipped to PFDs. Use lists, not memory, especially in a rescue mode.

TENDER CHECK OF DIVER:
1. Air On!
2. Hood **in place.**
3. Mask **ready,** gloves **on**
4. Primary regulator: **exhale first,** breathe **3 times; check** tie-wrap and mouthpiece
5. Pony reg: **same as #4**; hose under arm, secured **w/mud** mouthpiece cover **and** holder
6. Drysuit hose **on/functional**
7. **Diver**, press BC inflator button **3 times while tender checks** pressure gauge; gauges secured; **all hoses trim to body and secured**
8. **Diver, without looking, find** 1st shears, 2nd shears, & 3rd tool
9. **Diver, find** carabiner **at harness tether point; tether line attached**
10. **Diver, find** weightbelt release–is it right-hand? 10" extra webbing?
11. Fins **ready, w/** ankle wts **if needed**
12. **Back-ups have** contingency lines

THE TENDER SHOULD HAVE AND CHECK:
1. **Appropriate** PFD **closed properly**
2. **Appropriate footwear and** exposure protection
3. **Proper tending** gloves
4. **An appropriate** time keeper
5. **A water-activated** flasher
6. Eye **and sun** protection
7. **Back-tethered harness if working from a steep embankment**
8. **If acting as a back-up tender, a** profile slate and writing utensil.

WATER SITE MUST HAVE:
1. Contingency **reg. on full** cylinder
2. Contingency **reg. on full** pony bottle
3. **At least one** rescue throw rope bag **per diver-tender pair**
4. **At least BLS** first aid **personnel and supplies**

Lifeguard Systems
P.O. Box 548
Hurley, NY 12443
845-331-3383
www.teamigs.com

Figure 9–2 Tender check laminated card. Because most teams do not run dive operations more than two or three times a month, a complete tender and diver equipment checklist should be used for every check. Laminated cards can be kept in wallets, worn on necks, and clipped to PFDs. Use lists, not memory, especially in a rescue mode.

1. Air on—main and pony. If the tender stands behind the diver, both air valves should be turned all the way toward the tender and a quarter-turn back.
2. Hood in place (with skullcap if it is a dry hood), no skullcap material sticking out, and the hood and mask sitting properly together. Check diaphragm-type hood vents to make sure they are not damaged. If wearing a full face mask, do not put it on until ready to go into the water.
3. Mask ready, gloves on. If a drysuit is worn, check that wrist seals lay flat just behind the wrist bone. If a full face mask is worn, a standard facemask should be carried in a reachable pocket. When this is the case, ask the diver to reach for the spare mask without looking.
4. Primary regulator: If the regulator is not already wet, gently exhale first, gently breathe, and repeat twice. Check the mouthpiece for damage and check that the tie-wrap will not move when you push it. The "exhale first" is to train divers to

always reflexively exhale whenever putting regulator mouthpieces in their mouths. The majority of divers have a habit of inhaling first because they generally put mouthpieces in their mouths before submerging. The result is that if it ever becomes necessary to retrieve and put a mouthpiece in their mouths underwater, there is a good chance they will inhale first, especially if stress is involved. Inhaling water can lead to choking or even a full laryngospasm, both of which can lead to lung barotrauma or drowning. Tenders should carefully watch every time divers put mouthpieces in their mouths to make sure that the divers exhale first. Next, check the tightness of the tie wrap to make sure that the mouthpiece will not fall off when underwater.

5. Pony regulator, same as primary regulator. Also check that the hose runs streamlined down the pony bottle and BCD backpack, then runs under the diver's arm; that the hose has a hose wrap; that the second-stage exhaust tee is removed; and that the mouthpiece is secured with a neck strap. Can the diver reach the pony mouthpiece by tilting the chin down? If not, snug up the neck strap. The hose runs under the diver's arm to limit the risks of it becoming entangled and pulling the mouthpiece out of the neck strap.
6. Ask the diver to disconnect and reconnect the inflator hose without looking. The diver should be able to disconnect it in one quick motion. The drysuit inflator hose is on and is functional. The inflator valve is pointing downward at a 45° angle in the direction that the hose comes from the regulator. The BCD cummerbund or chest strap does not cover the valve.
7. Without visually looking, the diver reaches for and presses the BCD inflator button gently three times if it is not already wet, while the tender checks the pressure gauge to make sure the needle does not move and to see what the cylinder pressure is. Gauges are secured through BCD armhole and under the diver's arm, and if necessary secured to the BCD with a quick-release buckle that is the same as the BCD chest strap buckle. The gauge buckle and BCD buckle should be the same so that, if an injured diver's gear needs to be removed, the rescuer will not have trouble with it. Do not use metal clips, which can catch fishing line. All buckles that need to be opened to remove equipment must be quick-release and capable of being opened with ice diving gloves. If the BCD inflator is wet and there is no warmer protected area to perform this gear check, do not press the inflator button; rather pull on the hose to make sure it is properly connected.
8. Tender asks diver, "what time are you coming home?" This refers to the minimum pressure that divers should be back at the ice hole with, namely 1,000psi (68bar). In low visibility water or in a silt-out, divers may not be able to read the numbers on their pressure gauges, but they may be able to see where the needle is. Look

at the pressure gauge as if it were a clock, and see what time 1,000psi (68bar) would be. That is the coming home time.

9. Without looking, the diver finds: shears, alternate shears, and cutting tool, in that order. All cutting tools are in the golden triangle region between the shoulders and navel. At least one pair of shears is mounted on the harness pointing downward.

10. Ask the diver to grab the carabiner at harness tether point. Tether line is attached. Lock is secured with duct tape with a section folded over to enable easy removal.

11. Ask the diver to grab the weightbelt free webbing with the right hand. Is it a right-hand release? Are there ten inches of free webbing beyond the buckle? Is the belt unobstructed, and is the free webbing actually free and not tucked in anywhere?

12. Fins are ready, with ankle weights donned if necessary. Fin straps are taped.

13. Backup and 90%-ready divers have contingency straps properly attached to their harness tether points.

14. Is the diver mentally and physically prepared and ready to perform this dive?

It is very important to note that any of the checks that require using an air source should be done when the equipment is dry. For example, divers should avoid breathing from wet regulators when out of the water because, as explained in chapter 5, the movement of air across wet regulator parts can cause freezing and a resulting free-flow. That is one advantage of performing the first equipment check in a warm shelter. If it is very cold outside, it may be better not to breathe on the regulators or not to test the drysuit and BCD power inflator valves at the demarcation line if they were already checked in a shelter. If there is no shelter, divers should proceed slowly and gently when checking the function of air source equipment and be prepared to manage a free-flow problem, should one occur. Remember, long, slow exhalations can help prevent regulator second-stage free-flows.

Transporting Personnel and Equipment to the Ice Hole

Even though recreational divers should only dive if the ice is strong enough to walk on, that does not mean that divers and tenders should walk to the hole. As discussed in chapter 7, unnecessary exertion should be avoided in regards to immersion hypothermia. The

Tender Equipment Checklist

- ❑ Appropriate PFD closed properly.
- ❑ Water-activated flasher (for night operations) and cutting tool mounted on the PFD.
- ❑ Appropriate head and footwear and exposure protection.
- ❑ Ice cleats.
- ❑ Proper tending gloves.
- ❑ Time-keeping device.
- ❑ Eye and sun protection.
- ❑ Appropriate harness with back tether point.
- ❑ Laminated equipment checklist and line signal cards.
- ❑ Knowledge of the diver's personal SAC and breathing rates.
- ❑ If serving as a backup tender, a profile slate and writing utensil.
- ❑ If available, communication system with the hardwire line in a backpack.

Site Safety Equipment Checklist

- ❑ Contingency regulator on full cylinder with carrying harness.
- ❑ Contingency regulator on full pony bottle.
- ❑ Rescue throw-rope bags.
- ❑ Lines properly secured to stationary objects.
- ❑ Shelter.
- ❑ Medical equipment and trained personnel.
- ❑ Hypothermic stabilizer capsule or the equivalent.
- ❑ Communication equipment to activate 911.

most efficient transport device is an ice board set up on a pulley-type system that is fixed to the ice with a mariner's hitch and controlled by shoreline tenders. The mariner's hitch is also where tenders and divers will secure the ends of their tether lines, so that tenders and divers are tethered all the way back to shore. To set the board up, attach one line to the D-ring on the bow underside and another line to the stern cable. Lay the bow line over the top of the board for the tender to lie on and keep out of the way, while moving the board to the designated location at which the ice screws will be placed in the ice.

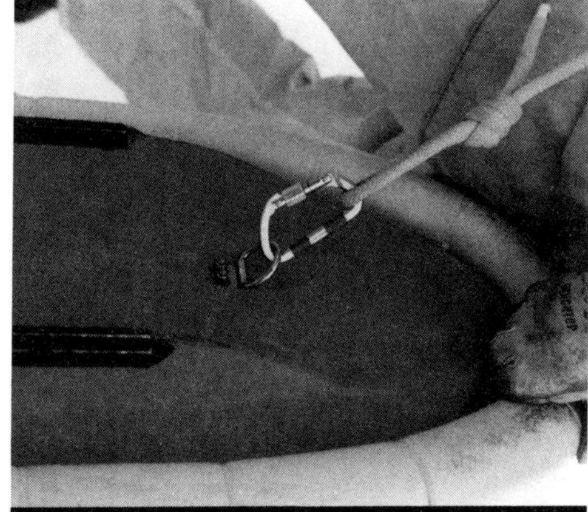

Photo 9-11 First, attach lines to the underside of the board's bow and stern.

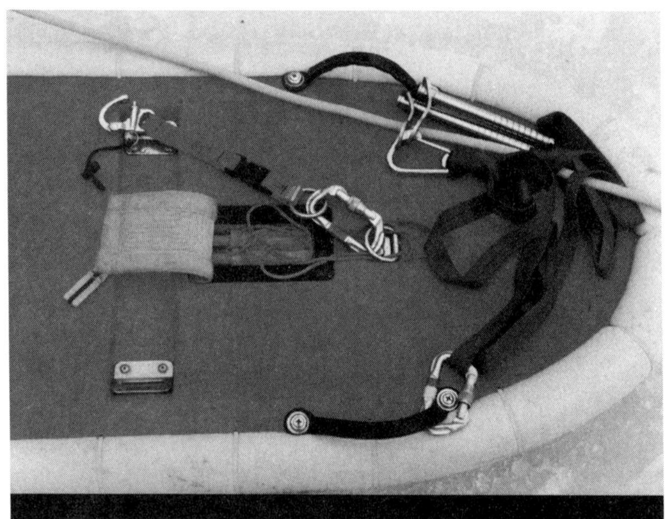

Photo 9–12 The ice rescue board is dressed and ready to go with the tender who will establish the double ice screw mariner's hitch into the ice. The tender will bring the pulley in a bag to put on the board.

Place the following items in a bag on the bow: two ice screws, a section of webbing with a loop on either end, that is approximately 5ft (1.5m) long,, a pulley, and a carabiner at either end of the webbing. Secure the bag to the board's bow. In a rescue mode, simply attach these items to the board handles to save time. Also at the board's bow are two ice awls, tethered into the board, and tether straps with a snap shackle at the loose end. The snap shackle is snapped into the front harness tether point of the rider. A snap shackle is used because it is very easy to release with thick gloves. Simply take tension on the strap by pushing away from the board; pull the snap shackle cord and it will immediately snap open.

Photo 9–13 Students learn how to use an ice pole to test the ice.

The tender lies on the board, snaps into the shackle, pulls the ice picks out (which remain tethered to the board), and pulls his way out to where the diver point-of-entry will be. The lines attached to the underside of the board are each longer than the distance between the shore and the Diver-point-of-entry (DPOE).

If the ice is believed to be strong enough to walk on, the tender can walk out and test the ice with an ice pole, while the board tethered to his harness follows behind him. If weak ice is approached, the tender should get on the board and use the ice awls to progress further.

Shore personnel maintain a hold on one end of each line. The tender should screw the ice screw into the ice as follows: Angle the screw with the top slightly away from shore, away from the load that the screw will be taking up. A good ice screw properly placed can pulley 300lb (136kg) or more back and forth all day without moving at all,

Photo 9-14 As the screw enters the ice, ice is pushed up through its center.

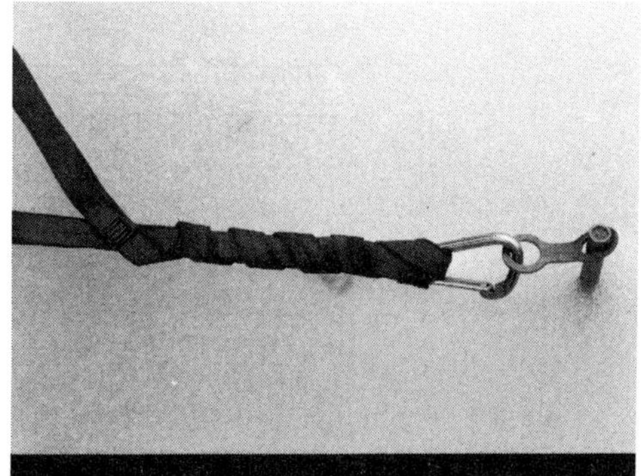

Photo 9-14a Once the screws are in the ice approximately 30in to 36in (76cm to 91cm) apart, tie the webbing between the screws and secure it.

Photo 9-14b Attach a strap with a pulley to shore-side ice screw.

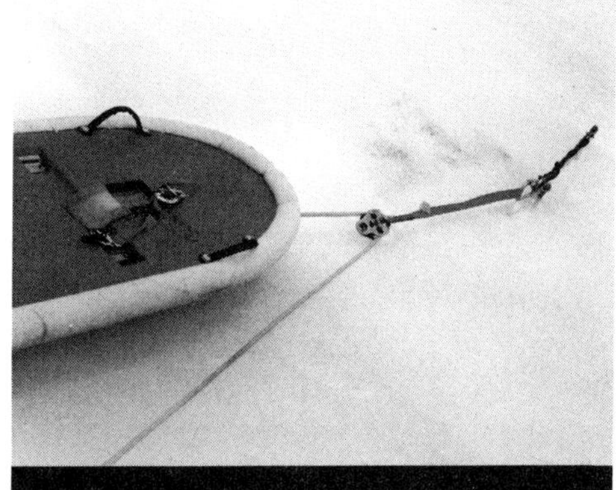

Photo 9-14c Attach the bow-line to the pulley.

when placed in as little as three inches of ice. Screws are hollow inside, and good screws will force ice up inside their core, replacing the air as they are screwed in. Therefore, when screwing into the ice, make sure your hand does not cover the top of the screw and prevent air from escaping. If that happens, the ice cannot push up into the core and the screw will not descend into the ice. Do not bang ice screws in, as that will fracture and therefore weaken the ice that keeps the screws in place. Placing screws in with a screwing motion not only does not weaken the supportive ice, but may cause a very slight melting that then refreezes and provides the screws with even greater support. If

Photo 9-14d Board pulley system complete. For long distances, or if there are not enough shore-line tenders to control two separate lines, use this system. Attach a third ice screw with a pulley nearer to shore. Run the bow-line through the pulley, then through the carabiner on the board's stem cable, and then run the line to the shore-line tender.

the ice is thin or soft, use two inline ice screws tethered together with webbing and carabiners. Use a mariner's hitch to provide extra stability.

When the mariner's hitch is in place, snap a carabiner into the shore-side screw handle and run the board's bow line through it. This will enable shore personnel to pull the board back and forth between the shore and the holes with ease. For distances greater than 75ft (23m), or if there is snow on the ice, use a pulley to make the shoreline tender's job easier. Instead of running the line through the mariner's hitch carabiner, attach webbing with a pulley to the carabiner and then run the line through the pulley.

The primary tender should go out first, after being checked at the demarcation line by the safety officer. The tender snaps a strap with a snap shackle to connect the board to a front D-ring on his harness. Once at the hole, the tender disconnects from the ice board and connects a section of webbing from the back of his harness to the DPOE-side ice screw. The strap should be just long enough for the tender to be able to move completely around the ice hole. Webbing is used because it is easily distinguished from diver tether lines and because it is easier to grab for self-extraction from an ice hole should accidental immersion occur.

The next person to go out is the fully-dressed diver, after being checked out at the demarcation line. The diver lies down on the board with a tether-line-deployment bag, snaps into the shackle, and is pulled to the hole. The recommended rate of pull is a maximum of 2ft/s (.6m/s). If the ice breaks, the board needs to be moving slowly enough that the passenger will not end up banging into, or submerged under, the ice roof.

Photo 9–15 An incoming diver is attached to the tether line and disconnects from the ice board.

Putting Divers in the Hole

If the ice is thick enough to kneel on, help divers sit at the edge of the hole by holding their tank valves. Have the divers kneel and then rotate toward their tether lines to sit on the edge of the ice hole with lower legs immersed. Move slowly to prevent anyone from slipping and to prevent damaging the ice. Once they are seated, have the divers splash water on their faces to begin acclimating to the cold water.

Photo 9–16 Diver entering water. (1) The tender snaps the tether line twice on the diver's left side to signal "turn to your left".

Photo 9–16a Diver entering water. (2) The diver responds by performing a left roof turn entry using the triangle corner. The tender holds the diver up with the tether line.

When the divers are ready to enter the water, have them perform a roof turn entry, also known as a controlled seated entry. As in any tethered diving situation, to keep divers from becoming tangled in their tether lines, the tender should gently pull the line against the diver's body. This tells the divers which side to turn to. After placing the regulator in their mouth, divers should place their inside hand on the ice (hand on the same side as the tether line), roll their body toward their tether line, place their outside hand on the ice, and gently slip into the hole. The tender should gently help support some of the diver's weight during the rotation to help lower the diver into the water slowly.

Photo 9–17 Board entry. Another option is to lower a diver into the water with the board.

A proper roof turn involves the upper portion of the body (not the stomach) leaning on the ice shelf; this helps prevent losing a weightbelt. A controlled roof turn keeps the diver's head out of the water.

Divers should not perform any type of violent or aggressive entry, such as a giant stride or sitting-forward entry. There are several concerns with these types of entries. Once a regulator is wet, it needs to stay wet. If divers perform an entry in which their faces submerge with their mouthpieces in, and then lift their heads and mouthpieces above water, the potential for freeze-up is great. Also, divers should not enter in a way that might cause them to surface under the ice and potentially hit their head. The last concern deals with the sitting-forward entry, which consists of the diver simply sliding forward off the ice, with or without the tender's assistance. This procedure puts the tender very close to the ice lip in an off-balance position, which reduces the tender's safety. Without tender assistance, there is a risk of the tank valve and regulator first stage hitting a diver in the back of the head.

With any fast entry, the temperature shock to the system can be particularly uncomfortable. Such a shock can result in reflexive gasping and panic. A slow, properly performed roof turn significantly reduces these problems.

An important safety function of the roof turn is to prevent the diver from becoming rotationally entangled in the tether line. Such entanglements require the tender to untangle the diver, with pull-start effect. This type of entry entanglement can create chaos if it occurs to a backup diver responding to a primary diver's call for immediate help.

A controlled, slow entry is particularly important, because the diver may be unaware of underwater obstructions close to the surface. Another advantage of the roof turn is that, as divers immerse slowly with support from the tender, they can reach up with one hand and point their pony mouthpiece downward to prevent it from free-flowing as it immerses into the water. Once the pony second stage has filled with water, it can be released.

If a diver is wearing a full face mask, it can either be donned prior to the roof turn, placed on the top of the diver's head, or be held out of the water by a tender as the diver enters the water.

Once in the water, the diver can perform another full-body drysuit stretch to give it a final place positioning.

Kissing the water hello

To reduce the chances of a shock to the system when the diver submerges, the next step is to kiss the water hello. This involves removing the mask, placing the regulator into the mouth, slowly lowering the face into the water, and breathing for twenty to forty seconds, or until breathing becomes comfortable. Once acclimated and breathing comfortably, signal the divers to remove their regulators from their mouths, keeping the regulators upside-down and submerged, and then pick their faces up out of the water.

As described in chapter 7 regarding immersion hypothermia, this process of kissing the water hello is particularly important in overhead environment diving. In the case of sudden mask flooding or dislodgement, divers have a greater chance of being physically shocked by the sudden rush of cold water against their faces. This shock can cause a reflexive gasping that can result in aspiration of water into the upper airways. Such aspiration can lead to panic and a reflexive shutting of the epiglottis over the trachea, called a laryngospasm. Panic can produce a bolt to the surface, which in the case of ice diving is the ice roof. Surfacing quickly creates not only a risk of lung barotrauma, head injury, and drowning for the diver, but also a risk for the diver's buddy if they are using a yoke line or if two separate tether lines become entwined.

Photo 9-18 A backup diver splashes his face to move into ready-status on weak ice. Notice he holds the mask out of the water.

Kissing the water hello acclimates the face and helps ensure that divers can comfortably breathe underwater without a mask or with a flooded mask. If divers cannot comfortably breathe without a mask, they should not be allowed to dive. This skill should be worked on in warmer months. Divers who know they have trouble with this skill should develop a routine of splashing their faces with cold water in mornings and evenings, and practice breathing through a snorkel with their faces submerged in cold water in a sink or bathtub. These at-home practice sessions are helpful for all other divers as well.

Once the diver is ready, the mask is donned. The tender holds the diver up to avoid too much use of the BCD power inflator, as the communication system is checked and as the backup tender records the diver's in pressure and time. The tender should ask the diver if he feels prepared for the dive.

Checking for proper weighting

Every aspect of public safety diving equipment should be set up in advance of an operation, including checks that all divers are weighted properly. Therefore, during a drill, conduct diver weighting checks. The best time is when a diver has 1200psi to 1500psi in their cylinders.

1. Completely empty your BCD on the surface while you are wearing full gear. Weight checks conducted with only partial gear will on average result in 2lb to 4lb of over-weighting.

Photo 9–19 If bone mic pouches are not available, wrap the wire around the bottom mask strap twice and secure it under the strap against the skull behind the ear.

Photo 9–19a The tender holds the diver up while the communication system is checked and the profiler records pressures and time.

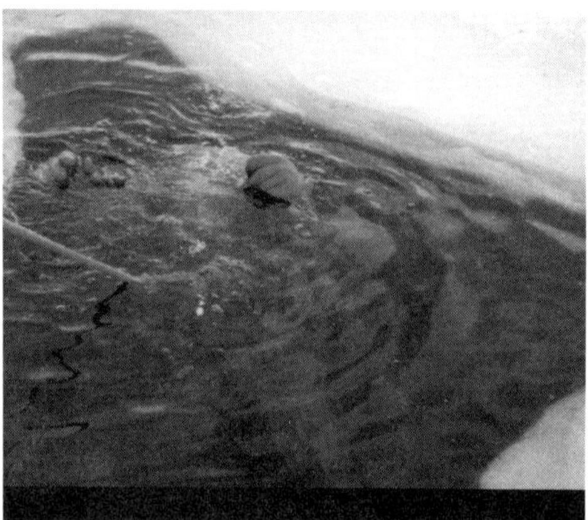

Photo 9–19b Perform a weight check before every dive. Vent the BCD and drysuit and with normal breathing and no kicking, hang with the top of the head just at the surface.

Photo 9–19c When the diver feels the weighting is good, he gives the OK signal to let surface personnel know everything is good and that he is ready to descend.

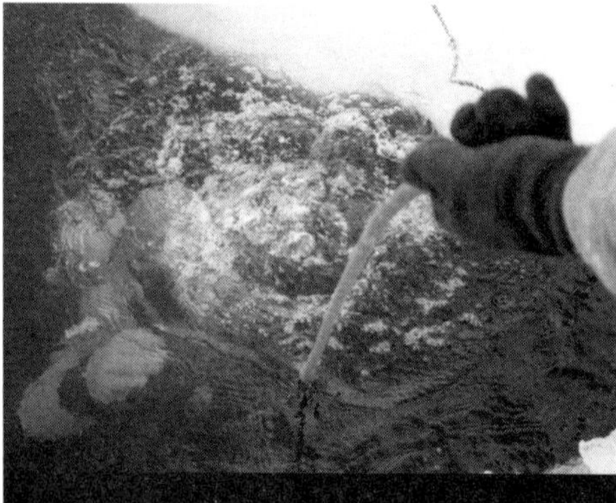

Photo 9–19d Let the diver hang and breathe for one minute just below the surface to acclimate and check for possible free-flows prior to commencing the dive.

Photo 9–19e Tenders wear one headset earpiece over an ear and the other off the ear so they can hear what other surface personnel say to them.

2. Ensure that all of the unnecessary air is out of your drysuit. If you are wearing a wetsuit, pull the suit out and stretch while in the water to release any air trapped in the crotch or armpits of the suit.
3. Stop kicking and sculling, and remain motionless.
4. With the BCD empty, breathe normally, and begin adjusting your weights in small increments. You have the proper amount of weight when you are floating so that you slowly rise and descend between eye level and the crown of your head when you inhale and exhale.
5. To descend, simply raise one arm over your head and gently, slowly exhale a little extra air. Once you have descended a couple of feet, gently bring the arm back and continue the descent with normal breathing. You can also cross your arms gently over your chest and tense your arm muscles if you wish to make the descent a little quicker. Remember, every movement will have a delayed reaction in the water, so be patient. Never fight to descend, as fighting will only make you more buoyant as you take in more air and spread your arms outwards.

If you need air in your BCD on the surface to keep from sinking, you are over-weighted by the amount of air in the BC. If you need 5pt of air to keep from sinking, you are over-weighted by 4lb to 5lb. Because you cannot easily measure the amount of air in the BC, simply remove weight in small increments until you can perform the above procedures.

Descent and diving

The diver should make a slow descent to the bottom, directly below the hole. Slow descents enable acclimation to the cold water, proper equalization every 2ft to 3ft (.6m to 1m), and reduction of the amount of disturbance of the bottom. Equalization may be more difficult in cold water because of hoods and possible wintertime congestion, so equalize gently and more frequently than during warmer months. Contact the bottom gently to prevent a silt-out, and if one happens, simply do not move. Wait for the particles to settle down. Once on the bottom, divers should kneel facing the line.

To give divers line signals, tenders need to create an angle in the line so that it is not vertical. There has to be at least a 10° to 15° angle for the diver and tender to understand controlled directional diving. If the line is vertical, and therefore directly up and down over the diver's head, the diver cannot figure out how to face the line and tender, and therefore cannot differentiate between left, right, out, or in. To create an angle, tenders may have to move a few feet to the left or right. Another option is to send divers in one direction while giving them a little slack, then take tension on the line and give the stop/face line signal. This causes the divers to swing out, creating an angle in the line. It

Photo 9–20 If the line is vertical over the diver, the diver cannot know what direction he is facing.

does not matter whether the divers move left or right, only that they move a few feet in one direction to create an angle.

Once divers are kneeling comfortably on the bottom, they should do a simple gear check to ensure that their breathing is under control and that their equipment is comfortable and secured. When everything is ready to go, the lead diver should signal a "1" to tell shore that everything is okay and that they are ready to go. The tender returns the signal. Decide before the dive whether the dive is going to be a tender-directed dive or a diver-directed dive. If tender-directed, tenders tell divers where to go, when to stop, and when to surface. Tender-directed diving is used when visibility is low or zero, and when search patterns are the dive activity.

Diver-directed dives are conducted by recreational divers whose dive activity is to swim around looking under the ice roof. On a diver-directed dive, the divers tell the tender they are ready with a "1" that the tender returns. The divers then head off slowly in their exploration. If the divers begin to go to an area outside the planned dive profile, the tender should stop them with a "1" and then, after the "1" is returned, send them in the correct direction with a "3"(right) or a "4"(left). For example, based on the direction the divers are moving, the tender realizes they will soon be between the shore and the hole. This is an area designated as a non-diving area. The tender should stop the divers and send them back to the planned diving area. On a diver-directed dive, if the tender sends a "1", requesting them to stop/face the line and check whether they are okay, the tender would then send them a "2" to signal the divers to resume what they were doing.

If divers are swimming in shallow water to explore the ice roof, tenders need to lie on the ice to bring the tether lines as low as possible. This will maintain a direct line access by preventing the ice roof from compromising the line (no bends or obstructions in the line).

When the pre-planned dive time is almost up, and the divers have not already returned, the tenders should give them a "1" signal to stop, face the line, and take up slack. When that signal is returned by the divers, the tenders will give the "time to return" signal "4+4". Remember: all signals are returned verbatim.

Time and air

The following is a list of useful definitions relating to time and air:

- SAC rate—surface air consumption rate
- DAC rate—depth air consumption rate
- bpm—breaths per minute
- psi—pounds per square inch

Wreck, cave, and recreational technical divers often use the "rule of thirds" for air management: one third out, one third back, and a minimum of one third back on the surface. In addition, each diver typically has a true alternate air source, such as a pony bottle. In ice diving, a minimum of air to bring back to the surface is 1,000psi (68bar) in an 80ft³/3,000psi cylinder.

If you are using cylinders larger than 80ft³, you can continue to use the 1,000psi minimum standard to return to the surface. If the working pressure of a cylinder is 3,500psi, use 1,200psi as the minimum to return with.

Divers should only dive with full cylinders, especially under the ice. Our standards call for a minimum of 2,800psi as full for an 80ft³/3,000psi cylinder. Sometimes public safety diving teams end up with slightly lower fills. When this happens and there is a need to perform the dive, the rule is to remove one minute of dive time for every 100psi less than a full cylinder. If the cylinders have less than 2,300psi, the dive should not be performed.

Along with air management, time management is very important. The maximum time for safe ice diving is fifteen minutes with a possible five minute extension for a total maximum of twenty minutes. This estimate is based on the average diver's air consumption, exposure equipment, depth, and distance out. Divers who have high SAC rates or who become chilled quickly should use shorter maximum dive times. These short times allow divers to rehab quickly so that they can assume surface duties and dive again if necessary.

Air and time rules are critical to ice diving because there is no direct access to the surface; divers must return to the point-of-entry. Also, the risks of entanglement and silt-outs may be greater, as is the risk of equipment free-flows. Cold stress and just the stress of diving under an ice roof can significantly increase air consumption.

Since recreational ice divers dive where they can see, they can monitor their air to make sure they meet the minimum amount of air with which to surface. Both divers and tenders can monitor the time to surface and call it when it is reached. Public safety divers, on the other hand, often dive in low or no visibility. If there is one or more inches of visibility they can use the "time coming home" technique described above in the diver equipment checklist. The "time" for most gauges is between 9 o'clock and 11 o'clock. If there is zero visibility or blackwater, divers can use a trick of duct taping a Ziploc® bag filled with clear water over the gauge, which is placed against the facemask with a penlight shone sideways into it. This enables divers to read their gauges in the dirtiest of waters. Ultimately, it is the public safety diving tender's job to have a good idea of how much air the diver has. This is determined by knowing the diver's SAC rate and average breath per minute rate. See the Appendix to learn how to calculate SAC rates.

In warm-water diving, public safety diving backup tenders should count and document diver breath per minute rates every five minutes. For standard, non-moving, blackwater search dives, the average rates are as follows:

- Experienced diver (fifty or more blackwater search dives): 8bpm to 12bpm
- Intermediate diver (twenty to fifty blackwater search dives): 13bpm to 15bpm
- New diver (fewer than twenty blackwater search dives): 16bpm to 17bpm

To count a diver's bpm in non-overhead diving, each set of exhalation bubbles seen hitting the surface is counted for a full minute. Unfortunately, the ice roof prevents us from using this method in ice diving. If communication systems are used, tenders can easily count the breaths by hearing them. The backup tender gives the primary diver a minute start and end time, so that the primary tender can count breaths without taking his eyes off the diver's line to look at a watch. If a communication system is used, the backup tender can count and document the bpm, as would be done in non-overhead environment dives.

If a communication system is not available, after every five minutes, the tender gives the diver a "1" to stop and face the line. When that signal is returned by the diver, the tender gives a "2" pull telling the diver to give a "1" on every inhalation until the tender signals "stop". The tender does not return the "I inhaled" signals. This method is not nearly as accurate as the other two techniques listed above, because anytime we think about our breathing we change it. EMS personnel often pretend to be taking a patient's pulse when in fact they are counting breaths. Nevertheless, getting a diver's bpm using this line-pull system is far better than not getting one at all, and if the diver does consciously or unconsciously slow the bpm down during the count, it is a plus for the diver.

Let us look at an example of how backup tenders/profilers can calculate a psi estimate for a primary diver that is generally correct within 200psi (13.6bar). Drysuit diver John normally breathes 10bpm with an SAC rate of 20psi/min. He is diving to 33ft (10m), so his depth air consumption rate is expected to be 40psi/min. As long as the backup tender sees that

John's breathing rate stays within 9bpm to 11bpm, it is safe to assume that his DAC rate is 40psi/min. This means that ten minutes into the dive John should have used:

40psi/min x 10min = 400psi .

If he started with 3,000psi, he should have 2,600psi in his cylinder at the ten minute mark. At the fifteen minute mark, he should have used:

40psi x 15min = 600psi

so his gauge should read 2,400psi. With that amount of air, John should be allowed the five minute extension if all else is going well. At the end of twenty minutes, he should surface with 2,200psi.

With this system, there is no need to bring blackwater divers up to check their air. This is not a good procedure anyway because it wastes time, reduces the area searched, increases the probability of diver equalization and equipment problems, and wastes effort.

Sadly, most divers have no idea what their personal SAC rates are, or their average bpm. Training programs provide an excellent time to log and learn these important rates. Warmer weather pre-ice dive training sessions can provide excellent baseline numbers that can be used for the first ice dives with a 20% increase until actual cold-water rates have been logged. If there is no opportunity to get these baseline rates, then the following rule of thumb can be used for the first couple of ice dives:

- Experienced diver: 8bpm to 12bpm, less than 30psi/min SAC
- Intermediate diver: 13bpm to 15bpm, 30psi/min to 40psi/min SAC
- New diver: 16bpm to 17bpm, 40psi/min to 55psi/min SAC

It is important to note that SAC rates can double during actual rescue-mode search dives and if exertion is involved a doubling of the bpm rate can result in a tripling or quadrupling of the diver's SAC rate. If a diver is using a full face mask with a positive pressure system, expect SAC rates to increase by 10% to 20%; this is especially true for divers with little full face mask experience. Cold water can also increase SAC rates, especially if the divers are wearing wetsuits. Monitoring diver bpm every five minutes is done not only to know when it is time to end the dive, but to monitor the diver's physical and mental state. Any medically trained person will tell you that one of the key signs that something is wrong is a higher or lower than normal breathing rate.

If a diver has a bpm more than twenty, the diver is physically or mentally stressed. The tender should give the diver a "1" to stop and face the line and then "3+3" to stand by to see if the diver can get control of, and lower, the bpm rate. If the diver's breathing does not slow down to a reasonable rate, the dive needs to be aborted. If the diver is on a communications system, the tender can ask questions to find out what is wrong. The tender can then work with the diver to make gentle, slow, long exhalations lasting three or four seconds. Concentration should be on the exhalations rather than the inhalations, because the goal is to lower the diver's carbon dioxide level, which in turn will slow the heart rate and the bpm. When divers are stressed, they tend to hypoventilate (not venting enough carbon dioxide, resulting in hypercapnia) rather than hyperventilate (venting too much carbon dioxide, resulting in hypocapnia). If the bpm returns to a safe rate, the diver can continue the dive. Do what it takes to get the diver's bpm a little lower before giving the "4 + 4" to ascend, because ascending while breathing heavily increases the risk of lung barotrauma. A high bpm is not only a safety risk while diving, it also signifies a decreased quality of searching . Public safety diving teams should not secure an area that was searched by a heavily breathing diver. That area should be re-searched by a different diver. Make sure to find the cause of the rapid or heavy breathing and fix it.

Once divers have documented their average ice SAC and bpm rates, the instructor or team coordinator can determine which divers can have the maximum allowable ice dive time and which ones need shorter times. Divers with high air consumption might have a maximum dive time of fifteen minutes with no possible five minute extension. In summary, ice divers should leave the bottom with at least 1,200psi in their main air sources. They should be back on the surface with a minimum 1,000psi, and they should have a maximum dive time of fifteen minutes with a possible five minute extension as per the dive coordinator. The minimum psi at the surface and the maximum dive time should be adjusted if conditions are strenuous, the diver has a large SAC rate or becomes chilled quickly, or if, during a dive, the diver's breathing rate is higher than normal for that diver.

Removing the Diver from the Hole

With a board

An ice board makes it easier to get divers out of ice holes in all conditions, but its value is best appreciated when the ice around the hole is weak or broken. Public safety divers using an ice board for diver recovery do not need to remove their weightbelts before exiting the water. When there is bad or broken ice, removing the belt at the hole may cause the belt to drop through the ice and be lost.

The tender pulls or pushes the board into the hole to the diver. The diver grabs the lower right handle and upper left handle and pulls himself up and onto the board. With practice, a

Photo 9–21

Photo 9–21a

Photos 9–21, 9–21a, and 9–21b Shore personnel send the board to the diver who slides it down in front of his body in the water. He raises an arm and waves it in a circular motion to tell shore to take up the line to pull him out of the water and back to shore.

Photo 9–22 When the diver reaches shore, he gets into a pushup position as tenders rapidly move from his crotch to head, opening all straps and disconnecting all equipment. Another tender simultaneously holds the weight of the tank and then lifts it off the diver.

diver should be able to mount the board by grabbing the forward center strap. The tender signals the shore to pull the board and diver out of the hole. The diver may need to counterbalance the board by shifting weight toward the stern if it breaks through the ice. Once the diver and board are on good ice, the tender gives a stop signal and the backup tender releases the diver's tether line, as the diver snaps the board's shackle into his/her harness front tether point. Upon reaching shore, the diver moves into a pushup position on the board as surface personnel release the board's snap shackle, remove the diver's mask, open all chest and shoulder straps, disconnect the drysuit inflator valve, lift the BCD-cylinder off the diver, and remove fins. The process looks like what happens to a car that pulls up to a maintenance stop during a NASCAR race. As was learned in chapter 7, blood can

Photo 9–23 The first piece of equipment to remove and pass off is the weightbelt.

Photo 9–24 The diver gives a big kick as the tender pulls the diver onto his belly at the triangle corner.

Photo 9–25 It is everyone's responsibility, tenders and divers alike, to clean, dry, and maintain the equipment after the operation.

pool in the core and blood volume can decrease during immersion. Therefore, divers should not stand up on their own with all their gear still on immediately after cold-water submersion. Once the gear is removed, the diver is then assisted up and taken to the rehabilitation staging area.

Without a board

When the diver is ready to leave the water, remove the weightbelt first. With a coordinated effort, the diver should kick as the tender pulls the diver horizontally out of the hole to good ice. If the ice is breaking around the hole, it may be easier to remove the diver's BCD assembly and take that out separately. Let the diver sit for a moment before standing up. Make sure the diver feels all right before standing. Divers should not carry anything other than their fins and mask back to shore after a dive. The gear should be transported on a sled or by having the tender carry it in stages. Never let divers walk back to shore on the ice with weightbelts on.

Night Diving Under the Ice

Not many recreational divers experience the serenity of diving under the ice after dark, nor the spectacular light show created by dive lights under and over the ice roof. With a few exceptions, night ice diving is not very different from a daytime ice dive.

Only conduct night ice diving operations after plenty of ice and night diving training and experience.

Surface lights should light up the entire area. The best choice is soft yellow lights. Yellow lighting is best because it:

- reduces glare off the ice.
- reduces the likelihood of blinding a diver who happens to look into the light while entering the water or hanging at the surface.
- reduces lamp heat that causes the ice to melt.

Mount lights on plastic crates, keeping them up and off the ice or on ten-foot poles. This further reduces the possibility of ice melting, while increasing the size of the area being lit.

Hang a flasher light in the far corner of the ice hole, enabling the divers to see the hole during ascent. Place the flasher in the far non-diving corner of the hole to reduce the possibility of entanglement. The light should be approximately 3ft to 6ft (1m to 2m) beneath the surface. Each diver and everyone on the ice should wear a flasher or other type of light. Avoid using strobes as they can cause epileptic seizures in some people.

Each diver should have two lights. In non-tethered, non-overhead situations, three lights are recommended. One difference is, with tethering and the use of the "1" stop and face the line signal, divers always know which direction is home. Divers should use head-mounted or wrist-mounted lights rather than hand-held lights, because the latter decrease hand circulation and therefore increase the risk of hand cold stress and the resulting loss of dexterity and strength. Keep hands and fingers free when possible.

Prepare the hole during the day. This reduces the chance of an accident with the cutting tool in the dark.

Public safety divers often have the need for night ice diving operations and should practice night operations before the actual situation arises. When shore lighting is insufficient to safely light the operations area, do not dive unless the ice is strong enough for staged lighting.

Spidering

Spidering is a technique of moving from one place to another under the ice roof quickly and efficiently in a direct straight line. It is a technique developed by author Hendrick that enables divers to cross from one hole to the victim's point-of-entry to search directly under the ice roof for possible floating victims or an accidentally disconnected diver. When spidering, divers are upside-down with hands and the front of their bodies facing the ice roof and their tether lines. Divers move backwards by crawling with their hands against the ice or by using wrist-mounted ice awls. In the beginning, spidering can be a little disorienting, and it takes practice to become skillful. Spidering is purely tender-directed and is the only situation in which divers do not return line signals; they simply react in the proper direction.

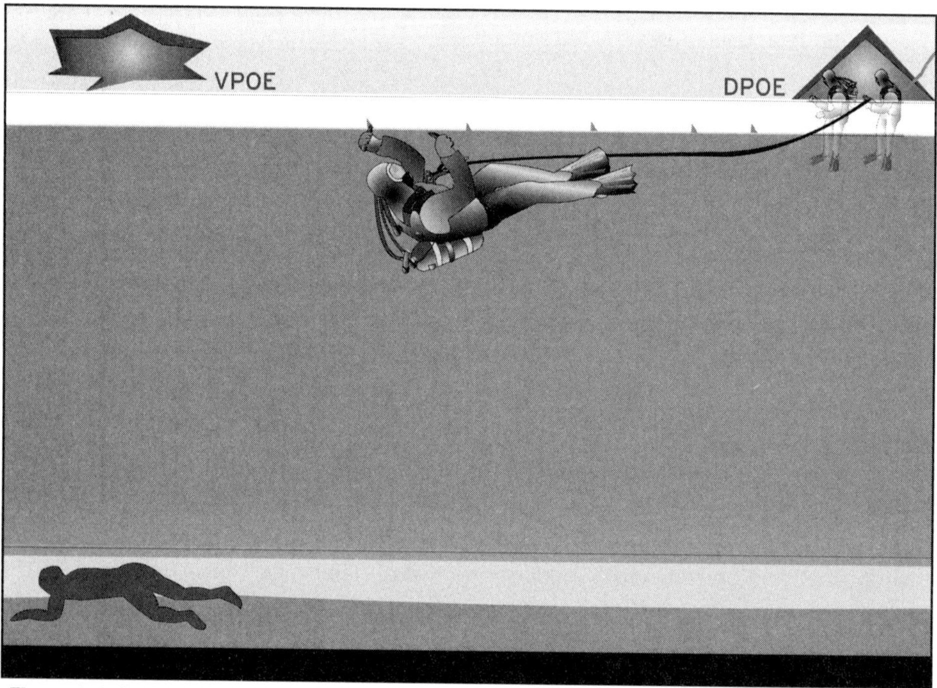

Figure 9–3 Spidering. The diver is inverted, slightly buoyant, with ice picks in hand, and finely tuned to his tenders every signal. With gentle kicks, he pulls himself from point A to point B.

A diver can move much more freely inverted under the ice roof than in a standard prone dive position. In the supine spidering position, the tank and first stage are not banged against the ice and the tether line does not end up between the diver's legs. Spidering improves the divers' field of vision because they can look toward the line, turn their heads to look to either side, and can bend their head backwards to see what is in their path. Spidering enables divers to maintain a taut line for a longer time and at a greater extended distance. Very importantly, spidering also makes it much easier for divers to swim a direct course, because they can easily maintain the line at the same angle to their bodies. If there is visibility, they can see the line's angle and know whether they are veering off to the right or the left. Without visibility, simply feeling for the angle of the carabiner to the harness tether point gives line-angle information.

Photo 9–26 The first step of spidering is for the tender to hold an arm out to indicate direction while the diver works backward to get under the ice roof.

The diver does not return left or right signals while spidering. He/she makes small incremental changes of no more than two feet and returns all other signals.

It is very important to note that like all ice skills, spidering requires hands-on training with a qualified instructor. Divers should not attempt spidering or any skill after simply reading this book. The same is very true for ice instructors, who absolutely should not attempt to teach these skills without being properly trained and certified by a qualified instructor trainer.

Photo 9–27 Reaching the VPOE. The line must be kept taut by the diver to prevent entanglement. The diver spidered straight to the VPOE hole in less than forty seconds.

When training tenders and divers to perform the spidering technique, create three triangular holes. Begin with holes that are 50ft (15m) apart. Place the tender down on one knee with one arm straight out pointing in the direction the diver needs to go to get to the target hole. The divers should think of themselves as a spider upside down on a ceiling, looking upward at the ice. It is important that the tether line be correctly held by the tender. If the line is held too snugly, the diver is forced to pull the line out of the tender's hand, which causes unnecessary exertion. If the line is held too loosely, it will be slack, and the diver will not have the line tension needed to provide directional guidance. If slack does develop in the line, tenders need to gently take it up. During each debriefing, divers and tenders should discuss how the line felt to make the appropriate adjustments for the next dive.

This technique requires practice for divers to make it in one straight shot to the targeted hole. Four or five spider runs are usually what it takes to no longer need tender corrections. Before this is perfected, divers may veer off a little to the right or left. When this happens, the tenders need to make an immediate correction by giving a left or right signal. When the divers receive this signal, they move three to four feet in that direction and resume going straight back. If that correction was not enough, the tender gives another left or right signal and the process is repeated.

It is important to understand that the farther out the divers are from their tenders, the larger the corrections need to be to get divers back in line with the target hole. For this reason, tenders should make corrections immediately, and if the divers are far out and need a large correction, it will reduce diver fatigue if the tender pulls the diver in a bit to make corrections. If the diver does not begin in a straight line, bring the diver back and begin again.

When divers are in the process of making corrections, the tenders should tighten up on the lines so that the divers only progress left or right without continuing to go backward. How hard the lines should be tightened up depends on how large the correction needs to be. The larger the correction, the tighter the line should be held. Once the correction is made, the line is loosened back to normal and the divers continue on their way. These and other "tricks" should be taught during hands-on spidering training.

Divers should move at a slow, reasonable pace. They should not give more than a very gentle kick as they use their hands to crawl under the ice roof. They should also make themselves a touch light in buoyancy. If they have to kick at all to stay under the ice roof, they are too negatively buoyant. The use of wrist-mounted, handheld ice awls helps divers move quickly and in a straight line. Positively buoyant awls with retractable points are preferred. If pre-ice dives are conducted as part of ice training, divers should practice removing and replacing the ice awls from their holders. If such dives are not conducted, divers should practice this just under the ice at the hole prior to using them during spidering. The wrist mount should be easy to remove by the diver in case the diver needs to ditch the awls for whatever reason.

Photo 9–28 Wrist mounts are the most effective way to wear ice awls for both surface rescuers and spidering divers.

Once the diver can consistently move from hole to hole, the next step is to have the diver surface at the target hole, give an okay signal out of the water, and then descend to the bottom. As the diver descends, the tender needs to give line so that the diver ends up at the bottom in the far edge of the hole to find a search item. To fine-tune the accuracy of spidering, insert a pole about a foot down into a weep hole and hold it there. Spider the diver right to the pole and have the diver push the pole up out of the hole.

Spidering applies to necessary ice diving procedures. Public safety diving teams should use it for searching the VPOE when sending a diver from a hole closer to shore. In this situation, the diver can bring a drowning victim right up to the VPOE to an awaiting surface person with an ice board on a shore-controlled pulley system. Both public safety divers and recreational divers should use spidering to go to the hot zone in lost-diver scenarios. The hot zone is the last documented place the diver was believed to be before the disconnect occurred.

With a little experience, divers can connect holes 70ft and 80ft (21m and 24m) apart in a matter of thirty or forty seconds and search everything under the ice roof in their path. They can then descend and search the best hot zone very quickly and effectively under the VPOE.

Diver-Controlled Swim

A diver-controlled swim means that divers choose where to go, rather than the tender giving the diver directional instructions. This method is more common in the recreational community and should not be used in public safety work. Two conditions that should be met before considering this technique are that visibility be good and that divers be very familiar with the underwater site.

Before the dive starts, it must be decided how far out the tender can allow the divers to swim. The tender should be asking the divers every three to five minutes if they are okay.

This technique requires an experienced and very attentive tender who can keep written records and who knows and understands designated area diving (DAD) as explained below. Planning is the key to a safe diver-controlled swim. The tender must continually adjust the line distance to prevent slack line, to prevent loss of communication, and diver entanglement in the tether line.

Tender and Diver Positioning

Whenever a diver is deployed, the distance between diver and tender must be monitored in case of an accidental disconnect, in case the diver requests the backup's assistance, and when a search is being conducted. If tenders move their location, the diver distance-out could change. Every time the tender moves, a search pattern can lose or gain a foot or two. On a shore-based operation, it is easier to keep the tender in one place, but on ice the tender may have to move, depending on where the divers are, to maintain direct line access. When it is possible, tenders should attempt to maintain their feet and tending hand at the same distance from the edge of the hole no matter what side of the triangle they are on. This reduces the change in tether line length as tenders move around the hole.

It is very important to maintain constant direct line access to the divers such that there are no bends in the tether lines. Tether lines should never go down under the ice and back under the tender; rather, they must be straight and forward of the tender. Straight tether lines become extremely important for the public safety diver, to maintain accurate search patterns. It is critical for all divers to be able to effectively receive and give line signals. If a diver does end up under the quadrant the tender is standing on, the tender needs to move immediately to the opposite quadrant to restore direct line access. Do not attempt to keep the tender in place and use line signals or verbal communication to move the diver to the proper quadrant. Line signals are compromised by the bend, and the diver may lose sense of direction.

It is important to know how far out the diver is at any time. All distances are measured from the tender's hand. Whenever tenders are required to move, they need to adjust hand positions to keep their hands at the same distance from the edge of the hole. If the ice is quickly eroding, thereby enlarging the hole, that should be taken into account, and larger overlaps of distance should be taken when putting in new divers.

For safety reasons, tenders should always call out line markings as they pass in and out, to ensure that distance is monitored and to help remain alert. The backup tender should then look at the tether line to see if the line is actually at the distance the tender called out. This is even more important in overhead environment operations. It is imperative to know how far out the diver is in ice diving operations in case of an accidental disconnected-diver emergency.

Designated-Area Diving (DAD)

Designated Area Diving (DAD) helps improve safety and speed up diver rescue in the worst-case scenario—a disconnected diver. It is extremely important in an overhead environment to always know where the diver is.

DAD is a method of sectioning the area around the hole into four separate sections or quadrants. Starting at the top of the hole, divide the area into four quadrants always moving in a clockwise direction. Consistency is imperative. The idea behind DAD is to improve the overall safety factor in sending assistance to the primary diver or divers.

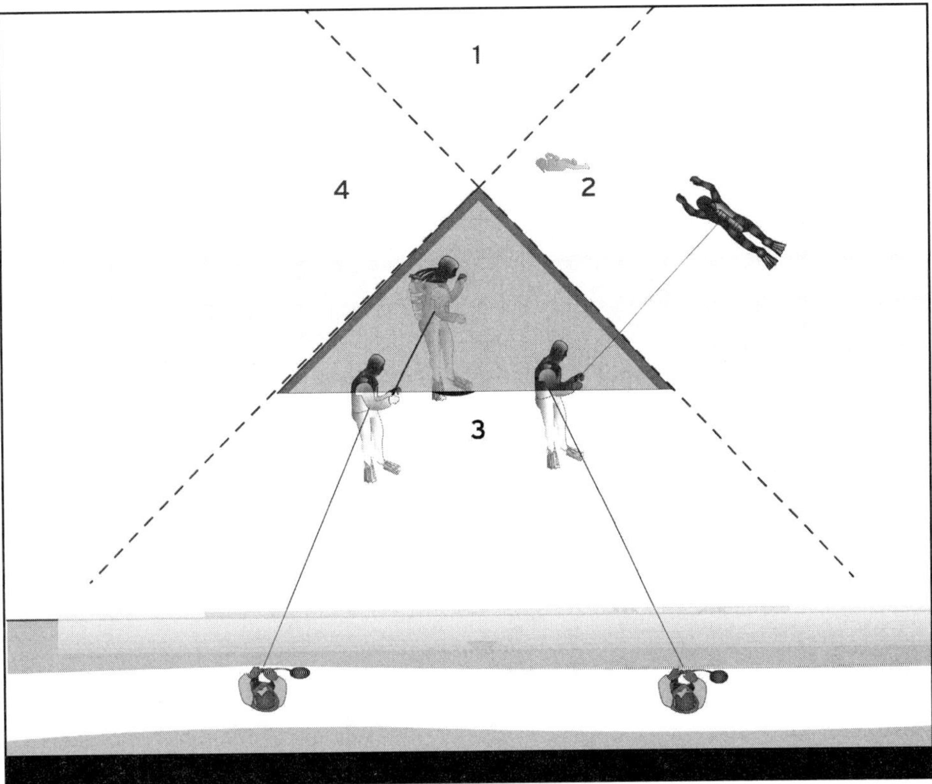

Figure 9–4 The DAD system greatly increases safety and search effectiveness.

When the divers are in quadrant one, the tender knows exactly where they are and how to get to them in case of an emergency. As the divers move into quadrant two, the tender now has a new primary dive area. The tender knows how far out the divers are from the measured tether lines. Combine this information and the tender has a good idea where the divers are. The responsibility of maintaining DAD or the last-known whereabouts of the divers is in the hands of the profiler, the backup diver's tender.

Since tenders know which quadrant the disconnected divers were last in, command has a starting point for a diver-disconnect search. Tender-directed diving makes it easier to control the divers' direction and maintain records of the quadrant and distances the divers are working. In recreational diving, divers are for the most part free swimmers. While tethered, they are free to move in their own chosen direction, so DAD is a good tool for tenders to keep track of diver locations.

Whenever possible, keep tenders standing in three, the quadrant closest to shore, and keep divers out of quadrant three. This decreases erosion from divers' bubbles under the area of ice where personnel and equipment are transported back and forth to shore, and prevents divers from ending up in too shallow water.

Ice Diving at Altitude

Ice diving is mostly a freshwater activity, and since many lakes are at altitude, it is not unreasonable to expect to ice dive at altitude. The specific techniques for ice diving at altitude are the same as for ice diving in general, the techniques needed for diving at higher altitudes apply to ice diving as well. It is recommended is that you not have to learn both at the same time. Gain proficiency with altitude diving during the summer months, and then have the time to focus solely on ice diving. Do not forget to move over one dive-table time-group to compensate for increased decompression sickness due to cold water.

Summary Questions

1. What is an important question for tenders to ask divers before a dive?
2. List one or two red flags to distinguish excuses from actual problems.
3. List at least five signs that divers are not mentally ready to perform a dive.
4. What personnel are the most likely to become chilled in ice diving operations?
5. Explain the "Three strikes and you're out" rule.
6. List the first eight tender duties when arriving on an ice dive site.
7. What should the tender check immediately prior to assisting the diver into the water?
8. Explain how to monitor diver breathing rates with communication systems.
9. How often are diver breathing rates checked if only line signals are used?
10. Tenders should never take their ______ off their diver.
11. If a search object is not found, what criteria does the backup tender use to determine whether a search was performed accurately or needs to be redone?
12. Explain the importance of the statement in question 10.
13. All ice diving operations, from recreational to search-and-recovery, should have a safety officer. True or False?
14. A safety officer should have diver certification. True or False?
15. List at least nine shore staging-area concerns that the safety officer needs to monitor.
16. How should a drysuit be prepared before donning?
17. What should drysuit divers do after they have donned the suit and entered the water to make sure the suit will be comfortable?
18. PSD divers who recover a body or heavy search object should add air to their drysuit or BCD to help them bring the object to the surface. True or False?
19. How much air should be put in a drysuit?
20. Give at least six reasons why drysuits should not be used as a BCD.
21. An over-inflated drysuit can cause a diver to lose consciousness because of a _____ ______ ______.
22. Purging air from a neck or wrist seal is an effective self-rescue technique to manage an inflator valve free-flow. True or False?
23. Describe an effective way to secure the gauge hose.

24. How are harnesses closed?
25. Describe how tenders assist divers with donning their BCD assemblies.
26. List the four functions of the diver equipment check.
27. State the steps in the diver gear check.
28. List the equipment needed to set up an ice board pulley system.
29. What is a safe diver entry technique into the water?
30. What does it mean to kiss the water hello?
31. When divers are ready to search or begin their dive, they give a ___ line-pull signal.
32. Tenders give a ___ line-pull signal before every other signal.
33. _____ psi is the minimum air to be back on deck with, or use the rule of ____ with a cylinder with at least a ____ ft^3 volume.
34. Tenders can use the diver's current ______ _____ _____ rate and the SAC rate to make a good estimate of how much air the diver has at any point during the dive.
35. Describe spidering.
36. What is the VPOE?
37. What is a diver-controlled swim?
38. Describe the DAD quadrants.
39. Which is the best DAD quadrant to keep tenders in if possible and why?

Notes

1 Walter Hendrick, Blackwater Contingency, LGS Productions,1986.

2 Walter Hendrick, "Rapid Dressing and Flooding of a Drysuit", Delphin Productions,1989.

3 J. Stewart, "Review of Antarctic Diving Accidents," AAUS Polar Diving Workshop, San Diego, eds. Lang and Stewart (May 1991): 44.

4 Ibid., 45.

Chapter 10

Spring Pre-Planning for Safe Ice Diving Operations

We frequently ask fire, police, and EMS personnel who have worked on ice rescue/recovery incidents, "What was the most important thing you learned from the operation?" The most common response is, "We were not prepared! We didn't have the right equipment or procedures. We didn't have sufficient standard operating procedures and guidelines, and incident command fell apart."

Water operations typically stand alone as compared to other fire services when it comes to preparedness. No fire department in this country would allow someone without fire training to enter a building and fight a fire as part of its team. No EMS organization would let a basic EMT, let alone a person with no medical training, insert an IV into a patient. What chief would allow a firefighter with no high-angle training to step off a high-rise during a rescue? Those same departments and chiefs would more than likely allow their members to go out onto the ice to save or recover someone even if those members had no ice training. Would your chief allow you to enter a fully involved fire without turnout gear or SCBA? Of course not. Would your chief allow you to dive under the ice without a quick-release pony bottle or with nothing more than a recreational diving certification? Very likely yes, if yours is like most departments.

We as a society and as public servants do not respect the power of water and cold. It is not uncommon for ice incidents to claim the lives of would-be rescuers. If performed with the right SOP/SOG, equipment, certified personnel, and training, sub-surface ice search-and-rescue/recovery is one of the safest types of operation that any public safety person can perform. If conducted improperly, the results can be fatal. The first step in pre-planning is to get into the mindset that ice operations require training, certification, drill time, standard operating procedures and guidelines, and the right equipment—just like confined space, hazmat, SWAT, high angle, and all other public safety disciplines.

The result of not being prepared is to put the lives of the rescue/recovery personnel in danger as well as to severely reduce the probability of saving the original victims. When the responding departments are unprepared, an ice incident scene typically contains the following:

- A weak or non-existent incident management system (IMS).
- No real plan of action.
- Lack of safe, functional staging areas.
- Personnel scrambling around trying to find a way to cross the weak ice to reach the victim's hole.
- Poor, or no, scene control; personnel without personal flotation devices or other personal protective equipment (PPE) on or near the ice.
- Little or no accountability for personnel in the operations zone or on the ice.
- Unprepared, unprotected, untrained, hypothermic personnel.
- Victims, if found and recovered, are banged, yanked, and dropped at least three times en route back to shore and the waiting ambulance.

Sometimes the victim is recovered, sometimes not. Sometimes the divers make it back to shore, and sometimes they do not. In almost every case we have studied, the danger to divers' and tenders' lives could have been easily prevented with proper pre-planning. Pre-planning includes discovering the needs of the community in regard to ice incidents, obtaining the necessary trained personnel and equipment, and writing a safe and effective SOP/SOG.

Pre-planning is critical even for well-trained and equipped teams. Pre-planning can greatly reduce the time it takes to reach the victim, which can make the difference between life and death. Let us take a look at the first step in pre-planning: reviewing past incidents and potential problem situations.

To be prepared, teams need to know what to be prepared for, so they can pre-plan responses for a variety of potential incidents.

All responding agencies (fire, police, park rangers, EMS) should be involved in the pre-planning process to:

1. Create the safest and most effective standard operating procedures and guidelines possible.
2. Facilitate better inter-agency cooperation and organization during actual incidents.
3. Develop an effective IMS, so that each responding agency understands that there will be only one incident commander (IC) and one command post (CP). Agencies should understand how the IMS is established and managed.
4. Increase inter-agency communication and pooling of resources.
5. Ensure that all necessary agencies are dispatched. For example, dive teams are immediately dispatched even if the incident is currently a surface one.

The importance of inter-agency pre-planning and follow-up drill sessions with debriefings cannot be stressed enough. Without effective and knowledgeable leadership and IMS, a simple ice incident can become a disaster. For example, consider a simple ice rescue incident of a boy lying on the bottom in nineteen feet of water under the ice fifty feet from shore. How difficult is it to plan, command, manage and perform such a rescue? Not very. With a well-trained and equipped team, the rescue should take less than fifteen minutes from time on scene to the time the victim is packaged in the ambulance. How often is that accomplished?

Unfortunately, the necessary organization is not often seen in the real world. Who shows up for an ice call? Either no one or the entire world. Turf wars, battles for the IC position, lack of equipment and trained personnel, difficulties accessing the site, freelancing rescuers, uncontrolled media and bystanders, lack of inter-agency coordination, and a variety of other problems often arise. Inter- and intra-agency preplanning and drills can prevent almost all of these problems.

The first step is reviewing prior problem sites:

1. Where are the most common problem locations?
2. What are the best access and exit points to these locations?
3. What hazards are known and might be expected in these areas?
4. Typically, how far from shore do incidents take place in each area?

5. What have been past causes of ice incidents (*e.g.*, motor vehicle accidents, snowmobiling, ice fishing)?
6. What month is most likely to have the most incidents and at what time of day, evening, or night are incidents most common?
7. What agencies responded to past incidents, and what were the results?
   - Were there too many responders, or too few?
   - Did responders have the right training for the operations at hand?
   - If they had training, did they use it, or was the operation chaotic?
   - Was the necessary equipment available?
   - Were any personnel ever put at unnecessary risk? If yes, why, and could the risk have been prevented?
   - Was there an effective debriefing?
   - How gentle and effective was the patient handling?
8. Did the IMS work? If not, why not? What needs to be done to prevent the same problems from recurring?
   - Was an effective risk/benefit analysis done?
   - Was a certified ice diving SAR technician assigned to the scene as an advisor to the incident commander?
9. Are there any measures the community can take to prevent future ice incidents, such as putting up signs in known problem areas, educating the public, erecting fences, making problem areas shallower in the winter, conducting ice accident prevention programs in schools, or passing laws imposing fines for driving on the ice?

You may find no consistency that can be tracked, and may have to pre-plan based on the unpredictable. On the other hand, you may discover a consistency, such as the same general location year after year or the same causes being repeated. For example, if you discover that two ice incidents in the last decade involved submerged vehicles with passengers caused by drivers deliberately driving on the ice, the following steps should be taken:

1. Overhead-environment underwater vehicle extrication operation training and equipment must be obtained.
2. An underwater vehicle extrication ice operation SOP/SOG must be written.
3. Since a submerged vehicle creates a hazardous materials situation, divers and tenders need to be hazardous materials trained and capable, and a hazardous materials team needs to be onsite to clean the divers post-dive.

4. Plans for vehicle removal need to be made.
5. The county may want to pass laws that impose fines on drivers who cause these rescue-and-recovery efforts, which cost taxpayers thousands of dollars and damage the environment with fuel spillage. Lawmakers may want to consider automatic manslaughter charges on surviving drivers if child passengers are drowned.

An example of the value of reviewing past incidents is demonstrated by members of the Saranac Lake Fire Department in New York. They discovered that they had enough long-distance operations, because of snowmobilers and large lakes, to invest in an airboat. A review of past incidents showed that they needed to apply gunwale reinforcements to their airboat, because an incident with a large victim caused the gunwale to blow out when it came into hard contact with broken ice blocks. They also realized they needed some type of thermal recovery capsule to contain the hypothermic victims in the relatively long trip in the airboat back to shore.

Let's take a look at a sample review process.

1. Review past ice rescue incidents to find the potential ice problem areas in your district.
   - Where do people most commonly congregate on the ice? Where do they ice fish, snowmobile, ski, and skate? Do they ice fish in shacks, and do they drive their vehicles to these shacks?
   - Where might cars accidentally or purposely end up on the ice?
   - What ice areas are known to be potentially weak? Where do geese and other waterfowl commonly congregate in the winter? Where are bubblers installed? What areas of the lake have natural springs? Where does the water have a current in the winter?
   - Are there areas of hidden ice? Where might cross-country skiers or snowmobilers unknowingly end up on a snow-covered body of water?
   - Are there ponds or lakes near schools or in restricted wilderness areas where children or teenagers hang out?
   - Is the first ice in November or December a common time for ice accidents? What time of day is most likely to have ice incidents?
2. Once the potential sites are located and marked on maps, determine the potential problems and hazards these sites might present.
   - If there is a major roadway near waterways, could a truck with hazardous materials be a potential threat? Are the guardrails designed to keep trucks on the road should an accident occur? If hazardous materials are spilled into the water, where will they go? What parts of the road are typically icy?
   - Are there biological contaminant hazard concerns, such as from geese, farm manure, or sewage? Are there chemical hazards such as pesticides

and PCBs? Golf course ponds, for example, are often toxic enough to require anyone entering them to wear dive gear with at minimum full face masks, blocks for pony bottle air access, and hazardous materials tested drysuits.

- Could a plane, bus, or train end up in the ice, causing a massive casualty incident? Think about the Air Florida crash into the frozen Potomac River in Maryland. Consider how pre-planning between and within responding agencies could have dramatically changed that event.
- Is the site difficult to access because of steep embankments, trees, large rocks, fencing, other obstructions, ice, or deep snow?
- Is the problem area a long distance from shore or is the body of water a long distance from road access? Will snowmobiles or other transport modes be necessary to reach the shore from the road access?
- Where are the closest helicopter landing zones to the water area?
- What are the potential water currents and depths? Currents over .5knot (50ft/m) require advanced training and additional equipment.
- Are there any potential snow avalanche problems that may affect the staging areas?
- Are there any underwater obstructions, entanglement hazards, or shallow water areas to consider?
- Are there known or potential suspended ice roof hazards caused by the water table being lowered after ice has formed? Dive teams near ski slopes that drain ponds for snowmaking are particularly vulnerable.
- Can there be moving blocks of ice? During a surface ice rescue incident, Harrison Township Fire Department in Michigan discovered the severe hazards of being caught on a moving ice block, including the length of time it could take to rescue the rescuers and the potential trauma that can occur.
- Time of day can be important. Will calls occur at night, causing the need for artificial lighting? Do calls occur in late afternoon when kids are home from school and happy hour is in effect, and when it may be difficult to assemble a volunteer crew? Dive teams in Alaska are faced with more dark hours per day during winter months than their southern peers.

3. Next, figure out what equipment, training, personnel, standards, and guidelines are required for these operations to be safe and effective.
    - Incidents involving a vehicle in the water require a hazardous materials operation and the necessary training to safely approach vehicles on/in the ice.
    - Weak ice requires that dive teams have the training to use ice boards, pulleys, and other equipment, with procedures that allow tenders and divers to work continuously in a prone position on the ice. Tenders may have to operate while lying in open water.

- Ice rescues of snowmobilers may be compounded by trauma, and head and spine stabilization transport procedures may be required. Some type of backboard-capable transport device will be necessary, as will the shortest victim-recovery time possible.

Vehicle involvement

Do dive team personnel have the training to understand and manage the specific hazards of vehicles in the water? These hazards include diver entanglement and entrapment, objects shooting air at divers, jagged metal, fuels, bio and cargo hazards, and patients with trauma and/or entrapment problems. If the vehicle is only partially submerged, do personnel fully understand that if they are close to a submerging vehicle, they can easily be pulled under with it? Do they understand that aspiration of fuels can result in lipoid pneumonia and other problems? The dangers of immersed vehicles are often based on ignorance of the potential consequences. Partially or wholly submerged vehicle incidents require a dive-team response—a dive team with submerged vehicle, hazmat, ice, equipment, and certification training.

Serious hazards

The following hazards should be written in an SOP as "no-go" for the majority of ice search-and-rescue/recovery (SAR) teams. If they are not written as "no-go" incidents, the department is more likely to be held liable for not responding. If the community wants the department to be able to respond to such incidents, they need to provide the necessary funds for training, equipment, drill time, and personnel.

Hazmat. If the water has biological or chemical contamination, technicians should be hazmat-certified ice SAR divers with similarly certified surface personnel as the tenders. The shore and medical personnel who handle the patient(s) need to be protected as well. It is likely that would-be rescuers will attempt to immediately get out on the ice to save someone submerged in contaminated water, because the threat of contamination is not as easily perceived as more visible threats, such as fire or a collapsing trench. The need for secure scene zones is imperative to keep this from happening.

Suspended ice roofs. If a victim falls through ice that is suspended above the water level, there is little chance that the victim will be alive by the time responders arrive. The dangers to anyone attempting such a rescue from the surface are immense. There is also a possibility that the victim is alive under the ice and that rescuers could crash the remaining ice roof down upon them. This is a job for dive teams trained for suspended ice roof operations.

Currents. The maximum current acceptable to dive under the ice should be .5knot to 1knot (one knot is approximately 100ft/min), depending on the level of training. An operation involving currents .5knot to 1knot requires extensive training beyond standard public safety diving SAR. Currents can significantly change ice stability. Shore personnel duties become critical because they need to make sure on-ice tenders are not pulled under the ice roof as the tenders work to keep their divers in place. A diver who normally breathes 9bpm to 10bpm with a 20psi/m SAC rate in no current can easily increase to 20bpm to 22bpm with a 100psi/m SAC in a 1knot current. Now think of regulator free flow risks. Currents increase the likelihood of equipment freeze-ups and diver heat loss, make contingency diver assists more difficult, and increase the probability of tender accidental immersions. Currents compound existing problems to create greater ones.

Moving ice blocks. This problem is not uncommon for communities with very large bodies of water with thick ice or with rivers that freeze over. Technicians should not enter the water when there is the possibility of their heads being crushed between moving ice blocks. A small (1ft^3) block is large enough to cause serious trauma if it is moving. Victims can be stranded on a block of ice that has broken away and is now moving with the wind or water current. A block may also break free during a rescue attempt.

If there is water current potential, estimate the most it could be (in knots) and then calculate how long a block could drift in the time it would take to get a boat or helicopter response. For example, a 2knot current could push a block 2,000ft (610m) in ten minutes. That will determine where a boat should be launched and where a helicopter should begin a search. Approaching boats should understand that a floating iceberg could be twice as wide underwater as it is above water and that the submerged section could have debris such as tree trunks frozen into it. Wind can also have an effect on the block, so estimating wind speed will be important as well.

Ice fishing shacks. What problems could ice shacks with heaters present? Ice shacks can puncture the ice roof with the occupants still inside, creating a more complex confined-space problem.

Weather and low-visibility conditions. If a call comes through during a blizzard, what tactics may need to be deployed? Are tow trucks part of the standard dispatch to ice scenes? How about sand/salt and plow trucks? Is the guideline written that the EMS duty crew stays in the ambulance to remain warm, dry, and functional while other responders bring the victim to the ambulance or at minimum to the shore? Blizzards can dangerously lower visibility above water, cause equipment problems, task loading, and hypothermia. They are often accompanied by high winds. Deep snow not only weakens ice, but hides the ice edge. Pre-plan how hot and warm zones can be obviously divided with visual markers once the shoreline is found.

If severe winds are known to occur over a body of water in the district, what special equipment and procedures are in place to work in such conditions? Are enough ice cleats available for rescuers' boots? Is there enough personal protection equipment to protect rescuers from the chilling effects of rain, wind, and snow?

How does the operation change during nightfall? Is there a sufficient lighting capability for the staging areas, the command post, and the different sectors of the operation? What is the plan for large-area surface ice searching during nightfall? Can a helicopter with a spotlight be deployed? Do tenders have waterproof hood lights mounted on their suits to use if shore or other lighting cannot reach the ice hole in the rescue time frame? If the operation is within the recovery time frame and lighting is a problem, the operation should most likely be postponed until the next morning. What policies are in place for water-related incidents during electrical storms?

Shore access locations and staging areas. Decide on planned shore access points and have the necessary tools to make the location usable. Find the shortest or safest route to the middle of every body of water. Sometimes rescuers respond from the shore area that is closest to the victim, but this may not be the best choice if that shore area is particularly steep, rocky, or full of trees. An area farther away with a friendlier embankment may be the better choice. Make sure all personnel are aware of the preferred routes, access, and staging locations. Draw and label these sites on maps so that an incident on the lake in zone "1" will be accessed by shore area "A".

If a body of water is completely surrounded by thick forest, chainsaws may be needed to clear a staging area. Be prepared to manage deep snow or icy shore conditions. Ambulances may need chains on their tires. Perhaps only four-wheel drive vehicles can be used. If embankments are steep, a rope team should be deployed.

Bodies of water far from road access. Special problems must be considered when water sites do not have access to good roads. How do rescuers reach the site? Do they even know how to get there? Is there a common rendezvous site? What is the best route? Are snowmobiles with sleds accessible to transport equipment and personnel if necessary? Be familiar with all these bodies of water and make sure they are clearly documented on maps. In the event that a child or adult is reported missing in the winter, these areas may need to be searched. Is a helicopter accessible for assisting with the search?

Once routes and transport devices are found, figure out how many personnel are needed and what equipment must be brought to the scene. Then figure out how to load these resources on the transport devices and go for a test run. How long does it take to load up the gear and personnel, travel, and reach the destination?

Learn how to avoid fatiguing and stressing the divers before they get to the scene. Let support personnel and transport devices carry the gear to save diver strength and equipment and to prevent them from perspiring in their drysuits. Tenders who need to wear surface ice rescue suits because immersion is a possibility should don them just prior to stepping on the ice.

Lastly, a plan must be made and tested for transporting the victim(s) from the remote site. Is the transport device sufficient for a patient with head or back injury, trauma, hypothermia, and/or other problems? At least one rescuer with EMT or higher status should be on the scene. Similar plans must be made and tested for injured divers and tenders. Ensure that all personnel and equipment are ready to do the job.

Personnel. What time of day are calls most likely to occur? If most of the rescuers are volunteers, how many can be available during those times, and where will they be coming from? Do they keep warm clothing, boots, gloves, and hats in their vehicles at all times?

What mutual aid teams are capable of assisting during large operations? Are all personnel trained and certified at minimum to the Ice Rescue Awareness level? Are sufficient numbers of the right rescuers trained at the Ice Operations and Technician (diver) levels, and will they be the personnel available during the high-risk times? By "right" personnel we mean those who are not overly large, heavy, unfit, older, hypertensive, insulin-dependent diabetics, poor swimmers, and/or people with asthma who do not belong out on the ice performing ice dive operations. Are a sufficient number of EMTs and paramedics trained to the Awareness Level? Do EMS and hospital emergency department staffs have written protocols for long-term drowning, near drowning, immersion hypothermia, cold stress, fuel contamination, gas bubble injuries, and other water/cold related problems? Do personnel know what their responsibilities and duties are, and have they redundantly practiced them? Are police trained to recognize homicide, neglect, and/or abuse by drowning? Are officers sufficiently trained to manage ice incidents?

Immersion can cause a rise in blood pressure. Is a system set up to check the blood pressure of all technicians both before they suit up and after they leave the ice? Do dive team members have training, equipment, and certification in surface ice rescue so that they will be capable of assisting their team members during accidental immersion? Do they have ice diving search-and-rescue/recovery certification, rather than merely recreational ice diving certification? Do all team members have a copy of the ice operation SOP/SOG, and do they take annual or biannual written competency examinations on its contents?

Equipment. Late autumn is a good time to prepare equipment and rescue vehicles for ice operations, and to make sure all personnel know where everything is and how to use it. It's also a good time to make necessary equipment purchases, identified during the pre-planning meetings.

Make laminated cue cards of communication signals between tenders and divers, diver/tender equipment checklists, and first-on-scene duty checklists, and mount them on PFDs with lanyards. Practice the procedures outlined on these cards.

If you have sized ice rescue suits and drysuits, make sure sizes are clearly and boldly written on both the suit and the suit bag. Make sure all suit zippers are paraffin waxed, and both sides of seals are powdered, and check that the suits are in good working order without leaks. Have profile slates and other documentation paperwork ready and easily accessible, along with writing instruments. Make sure rope bags are clearly marked for length and that the line itself is marked in five-foot increments and is in good working condition. Make sure personnel know which bags to use and how to read the line markings. Use equipment check log sheets to regularly ensure that all equipment is functional and in its place.

Set up the gear for easy and rapid access. Store each complete set of tender-diver's gear in individual bags or bins that can quickly be pulled off the truck and used. A common mistake is to put all like-gear together. That means for a tender-diver pair to dress, they must go to the PFD area to grab a PFD, pull out the harness box to get a harness, reach for a suit in the suit locker, etc. This wastes time and increases the occurrences of lost and forgotten gear. Bins can be designated by gear size or by the name of the diver who will fit the size. Make sure all equipment requirements in the SOP/SOG are met. If your department follows NFPA standards, check that all the standards in Document 1670 are met.

There's only one job you must do every time—Go Home! Pre-planning will help make sure you, and hopefully the victims, can do that most important job.

Summary Questions

1. What is the first step in pre-planning?
2. Ideally, who should be involved in the pre-planning process?
3. List seven questions to ask when reviewing agencies' response to past incidents and the response results.
4. If there is a potential for submerged vehicles, what must be planned for?
5. What time-of-day factors should be considered?
6. List six dangers of submerged vehicles.
7. List examples of hazards that will or may warrant a no-go operation.
8. What can be done to prepare drysuits for the winter season?

Chapter 11
Recreational and Public Safety Diving Training Sites

Choosing an ice diving site is a luxury afforded to the recreational diver and trainer. In public safety diving, circumstances choose the site. The techniques presented in this chapter help ensure a safe dive for the recreational diver and a safe training site for the public safety team.

Choosing a Site

Begin site selection before the ice forms. Instructors can develop interesting ice diving programs and fun marketing techniques with ice diving programs that begin in warm weather. Summer planning and dives integrate very nicely with advanced courses. Teaching underwater navigation, mapping, surface work, and site setup is a year-round possibility. Pick the location for the ice dive and build a dive task for it. A dive task involves completely mapping and outlining the area for the winter dive season. The warm-water planning dives should not count toward the minimum required dives under the ice for ice diver certification.

Choose an area

- that is familiar, or make it familiar during warm weather dives.
- with depths less than 40ft (12m).
- with few or no underwater obstructions.
- with easily recognizable underwater landmarks.
- with easy parking access.
- that typically has a minimum of 6in (15cm) of ice in the winter.
- that does not require a long walk from the water line to the intended diving area.

To help avoid a long distance between the shore and the proposed hole, find an area where the underwater slope line from shore reaches a minimum 15ft (4.5m) of depth within a relatively short distance from the shoreline. Avoid shallower water to prevent possible diver entrapments between the bottom and the ice roof.

If possible, choose an area that does not host much ice fishing. Monofilament line and hooks can pose significant entanglement threats. Schooling fish can cause areas of unexpected weak ice; and ice-fishing activities can create both conspicuous and inconspicuous weak spots. Also, parking and rapid exit in the event of an emergency may be limited if fishermen arrive after divers are already set up and on the ice. Once public safety divers have achieved an entry-level Ice Diving Search Operation certification, they should dive where people fish, to be prepared in the event of someone falling through while fishing. Of course, you want to dive these sites during the summer, and be sure to include lots of blacked-out diver monofilament and fishhook entanglement management training.

Photo 11-1 Rescue/recovery teams do not have the option of choosing a site—the victim chooses it for them. Consequently, the more sites they dive during drills, the better off they will be.

The site should have easy access and exits for vehicles that reach as close to the water as possible. The distance that divers and tenders must travel between the vehicles, the water line, and the ice hole should be minimal. If divers and tenders must walk far out to the ice hole, problems of fatigue, equipment freeze-ups, and general discomfort are far more likely. Unless there is a specific reason for a longer distance, plan for a maximum of 200ft (61m) between the water line and hole. Public safety divers may need to be able to perform operations in hard-to-reach places; therefore, second-level ice training should be obtained to learn safe and effective procedures for such sites.

Remember to always plan for possible emergencies. Do cellular phones work at the site location, and, if not, are radios capable of contacting emergency personnel? Never dive a site without being able to contact emergency personnel.

Unless tenders and divers are already trained and certified for blackwater diving, avoid sites with low visibility and very silty bottoms. In the winter, particles may settle to the bottom, providing better water clarity with a siltier substrate. Divers planning on diving such a site should first make enough silty-substrate, warm-water practice dives to be capable of moving 1ft (30cm) off the bottom without stirring up silt. During these practice dives, divers should practice gently settling on silty substrates to learn how to minimize the mess, become comfortable with sudden losses of visibility, learn how to patiently "sit the silt out", and gain skill and confidence performing such tasks as gear adjustments, entanglement problem solving, and sharing air while settled on very silty bottoms.

It is important to keep in mind that ice reduces light levels. Sites with relatively good visibility in the summer may provide low visibility under semiopaque ice covered by a foot of snow in the winter. Recreational or recovery divers have the time to shovel snow off the ice so that they can experience the improved visibility caused by the protective ice roof.

Water level changes

Talk to local government agencies such as the department of Environmental Protection or local watershed police, to determine how far the water level rises or falls during the winter season. The height of the water level can be measured by the amount of beach area covered by water. These changes in water depth can occur from human interference. For example, a mountain in New York has an excellent swimming pond during the summer months, but it becomes significantly shallower in the winter because it is one of the three main drawing pools for the local ski resort's artificial snow machines. In this case, the level of water usually drops dramatically.

This drop is of concern for several reasons. If the drop occurs before ice forms and is not taken into account when examining site maps made in warmer months, divers could end up in shallow water, at risk of becoming entrapped between the bottom and the ice roof. The most serious danger of lowered water levels, however, is a suspended ice roof. This occurs when ice forms and then the water level drops, creating an air space between the ice and the water. Ice that would normally be considered safe (6in to 8in, 15cm to 20cm) may now be dangerous. Suspended ice lacks the support of the water below it, and, as a result, when weight is put on the ice, the ice will bellow and be much more prone to fracture. A second problem is that people who fall through suspended ice are immersed in water at a level below the ice and find it difficult or impossible to get out of the hole unassisted. Such victims cannot easily be seen or heard, as they are hidden below the ice and are yelling under the ice roof. Their voices are lost in a cavern of ice and water. Because these immersed victims cannot easily gain support from the ice roof, full submersion is far more likely. There is also the danger that would-be rescuers can cause the suspended ice roof to come down on top of a victim, thereby causing injury and increasing the risk of drowning.

If the watershed falls more than 6in to 8in (15cm to 20cm) after icing, do not plan dives in that area because of these problems. If the water level drops before the first freeze, make a notation on the site map.

Site map

The reasons for making a site map include:

- Being able to find the planned site during the winter.
- Discovering whether the water level changed between map-making and ice formation.
- Figuring out the approximate extent of water-level change, if one occurred.
- Gathering information necessary for a safe and enjoyable dive site and noting where underwater entanglement hazards and obstructions exist, where there may be too much current to dive, where contamination could be a problem, and where shore access/exit problems may occur.
- Practice making effective, court-ready scene document profile sheets for evidence recovery operations.

From a permanent stationary object, such as a tree, rock, or building, measure the distance to the water line. Next, plan a depth and area for the dive. Plant a stake at the water's edge and set a weighted buoy approximately where the planned ice hole will be. Maintain the buoy on a relatively taut line so it watches (remains) directly over the planned hole. Measure the distance from the permanent object to the buoy. Draw in permanent ranges that are approximately 90° apart, or use compass bearings to show directions. To take a range, line one object up directly behind another one. Find and document a second range 60° to 120° away. If two or more sets of ranges are recorded, it is easy to return at any time in the future and stand on exactly the same spot. Record all ranges, distances, and directions on the site map.

Next, add or subtract the approximate estimated rise or fall of the water level occurring before the first freeze. If unknown, compare the summer measurements with the winter measurements to ensure that there is no significant change. Know the distance between the stationary object and the water's edge in the summer to be able to see a water level change in the winter. If there was 10ft (3m) of water 100ft (30m) from the water's edge and now the ice is unexpectedly back 10ft (3m) past the previously measured water line, the water may be too shallow to dive from the previously planned hole.

Photo 11-2 Tenders work from DAD area 3, the shore side of the triangle where the ice block is pushed under for extra support. Divers are kept out of area 3 if the ice is thin to prevent bubble erosion under the path of transport between the hole and shore.

Example. The distance between a large oak tree to the stake at the water's edge is 85ft (26m). From the stake to the buoy is 130ft (40m). If these two measurements are in a straight line, the future hole is 215ft (66m) from the tree. However, if the water level drops 15ft (4.5m) prior to the first freeze, there would be an additional 15ft (4.5m) on the shore side, with 100ft (30m) between the tree and the water's edge. For the moment, these numbers may not seem important. They are critical, though, for they show the potential for suspended ice or dangerously shallow water. Another problem of shallow water is that rocks and other objects on the bottom can become obstacles to both the diver and the tether line. If the water level has dropped 15ft (4.5m) for example, and the depth has changed by 6ft (2m), the divers could find themselves in difficult straits.

The next step in planning is to completely map the bottom, starting at the buoy and working a 150ft. (46m) circumference. Even though the divers should only dive a maximum of 75ft to 100ft (23m to 30m) from the hole, backup divers, depending on experience, may need to extend that distance to 125ft (38m) if searching for a lost diver. It is necessary to know on paper what the maximum safe distance from the buoy toward shallower water is, and where the diver (upon reduction of the water level) will have a minimum clearance of 3ft to 5ft (1m to 1.5m) above his head in a swim position. Ice divers should not dive in less than 6ft to 8ft (2m to 2.5m) of water. Ensure that divers can work in a 360° circumference of that maximum safe distance from the buoy.

Use the map information to delineate the DAD. For instance, from the hole to the shore, divers cannot safely go more than 65ft (20m) out, because the water becomes too shallow. However, the open water side of the hole allows a full 100ft (30m) distance out. Quadrants 1 and 2 are clear of all obstacles, while 3 and 4 are restricted line-length areas. In this situation, there are four basic options:

1. Move the buoy and proposed ice hole farther out by a minimum of 35ft (11m).
2. Do not use lines longer than 60ft (18m) for backup divers, and make primary diver lines even shorter.
3. Train tenders to thoroughly understand and be skilled with line-handling procedures and DAD.
4. Change locations.

If option 3 is used, make another dive to the same place with a boat, using the boat as a platform representing the ice hole. Train the tenders from this mock ice hole to work with the lines and DAD. This is a good dive to practice during warmer months. Use a three-anchor system to keep the boat in place when one or more tethered divers are working off it. For more information on dive boat anchoring see *Public Safety Diving* or such boating books as *Chapman Piloting: Seamanship and Small Boat Handling.*[1,2]

Often divers prefer to put the hole over a specific object to help divers orient. Divers can pre-set an environmentally safe, easily identifiable object such as a colored cinder block.

An area with submerged trees is not a good area to learn how to ice dive and should not be used by recreational divers even after ice diving certification. Severely tree-infested underwater sites require specific training and direct overhead access to limit the probability of entanglement. If there are obstacles safe enough to dive around, record their depth, location, and distance to the proposed hole. To map these obstacles, take a series of small buoys and attach them to the objects. If there are more than two or three objects, the buoys can be numbered or colored. Then measure the distance between the object buoys and the hole buoy. A better method is to do these dives with tethered divers, just as they will be done during the ice season. Dive off a boat placed right next to the proposed ice hole location. Use the tether line markings to calculate the distance of these objects from where the tenders will be standing on the ice at the hole. This system has two advantages: the distance measurements will be both more meaningful and more realistic because they are based on actual distances from tenders to divers.

If measurements are taken only at the surface from the tender to the object buoy, depth will not be taken into account. If this system is followed, use the chart below to make the corrections for depth. In the winter, tenders will then be able to look at the tether line distance marking and know how far the diver is from any particular obstacle.

Surface Distance

Depth	10	20	30	40	50	60	70	80	90	100	110	120	130	140	150
10	14	22	32	41	51	61	71	81	91	100	111	121	131	141	
20	22	28	36	45	54	63	73	82	92	102	112	122	132	142	
30	32	36	42	50	58	67	76	85	95	104	114	124	133	143	
40	41	45	50	57	64	72	81	89	98	108	117	127	136	146	
50	51	54	58	64	71	78	86	94	103	112	121	130	130	149	
60	61	63	67	72	78	85	92	100	108	117	125	134	143		
70	71	73	76	81	86	92	99	106	114	122	130	140	143		
80	81	82	85	89	94	100	106	113	120	123	136	144			

Figure 11-1 Depth-Distance Out.

Whenever possible, measure underwater obstacles with tender-directed, tethered divers working from a three-anchored boat that serves as the ice hole location. The divers' tether lines are marked every 5ft (1.5m) with tape so that the backup diver's tender can draw the location and distance of each obstacle between the primary diver and primary tender. Divers can give tenders a signal of two line-pulls to tell the backup diver's tender, or profiler, to make a notation. During the debriefing, the diver can tell this tender what each recorded notation was. Having an underwater communication system, whether hardwire or wireless, facilitates mapping.

Photo 11-3 Divers mapping an area from a boat that is placed right in front of a proposed ice hole location. This method will provide accurate line distances from the ice hole to underwater obstacles.

If setting up a boat is not possible, take these measurements underwater by using compass bearings, kick cycles, and arm spans, all of which enhance navigation skills. Besides being an exceptional way to practice underwater mapping skills, this system discovers the location of potential hazards. Remember, this may be the first time some divers are diving with lines, so obstacles are very important issues. It's important to note that divers should never be allowed to assume they can use these navigation procedures to dive under the ice without being tethered properly to a trained tender! Finding a hole without being tethered is not an option, especially if the water silts up and a life depends on it—yours. This is equally true for expert compass users and experienced underwater navigators. Never dive under the ice without being properly tethered.

Be aware that the dive location never looks the same in winter as it did in summer from the shore. Snow and ice confuse the perception of distance, and without pre-planned line measurements, it is very difficult to make the hole in the planned location.

Second or third safety holes should be pre-planned using the same techniques as used for planning and documenting the location of the first hole. At least one safety hole should be cut and set up.

The day before

It is advisable to check the planned site the day before the dive to ensure safe ice thickness. Check the ice to ensure that it is thick enough to work on safely and to make sure that your chainsaw is at minimum 2in to 4in (5cm to 10cm) longer than the ice is thick. Do not forget to check the weather forecast as well.

Testing the ice

Always test the ice before sending people onto it. The simplest way to start testing is to send one tethered person wearing appropriate immersion PPE (thermal and flotation) onto the ice with an ice pole. Banging the end of the pole onto the ice gives a good audible indication of the condition of the ice. Good ice should sound solid. If the pole goes through the ice, so will people. A sample of ice thickness can then be taken with an ice screw or a longer tool.

What is good ice?

All ice is potentially dangerous. In public safety diving, "no ice is safe ice". The reason is obvious. When a public safety dive team is called, someone has already gone through the ice. True, there may be a few cases in which a person fell through a pre-cut fishing hole, so the ice may be good, but they are rare. More often than not, the victims have punctured weak ice and are now submerged.

The recreational diver and trainer can pick when and where they are gong to dive. A reasonable understanding of what constitutes "safe" ice is still needed, however, so that the dive can be planned around good ice diving conditions.

The salinity of the water will make a difference. Salt water freezes at a temperature 2° or 3° colder than fresh water. The exact difference in temperature depends on the salt concentration. Consequently, salt water ice dives require colder temperatures than freshwater ice dives. Water movement also interferes with ice formation: the faster the movement of water, the weaker the ice. Other ice-weakening factors include a high density of waterfowl and the resulting excrement, underwater springs or human-installed bubblers, rain, previous human destruction such as that made by snowmobiles, water pollution, weather changes causing alternate melting and freezing, and snow.

Keep in mind that ice can change hourly and that some of these changes can be dramatic. Ice can go from safe to unsafe in a couple of hours. Given all the contributing factors, diving conditions should be better than just "okay".

Ice scale

A common question is: "How thick should the ice be to be considered safe?" Ice thickness is just one of several variables. For example, 2in (5cm) of good-quality black ice would last longer and hold more weight than 5in (13cm) of ice made of frozen snow and slush. As a rule-of-thumb for black ice:

- **1in (2.5cm)**—May be unsafe for crawling and is dangerous to small animals. For rescue/recovery operations, a platform-transport device or dive vessel is mandatory.
- **2in (5cm)**—Capable of holding one small adult on foot, moving very slowly and gently.
- **3in (7.6cm)**—Safe for two or more people on foot moving slowly at least 3ft to 4ft (1m to 1.2m) apart.
- **4in (10cm)**—Capable of supporting two people standing 1ft to 2ft apart, minimal ice diving possible.
- **5in to 6in (13cm to 15cm)**—Capable of supporting a dive operation, three to four people standing 1ft to 2ft (30m to 61m) apart, plus dive equipment.
- **7in to 8in (18cm to 20cm)**—Capable of supporting small vehicles, snowmobiles, or four-wheelers, less than 1ton (907kg) gross weight.
- **9in to 12in (23cm to 30cm)**—Capable of supporting small passenger vehicles, less than 2.5ton (2,268kg) gross weight.

Add an inch or more to the ice scale whenever there is a question as to the overall safety and supporting capability of the ice.

To better understand how to determine ice quality, consider how ice is formed. During the warmer months, colder water is typically found at deeper depths, because cold water is denser than warm water. As cooler weather begins in the fall, the surface layer of water, the epilimnion, loses heat to air by convection, and as it cools it becomes heavier and sinks downward. This is sometimes called "fall turnover" because the epilimnion water (upper layer of water) switches places with the hypolimnion (bottom layer of water), as the epilimnion becomes colder than the hypolimnion.

When water reaches 40°F (4°C), it is at its densest state, and as it becomes colder the density actually decreases until it reaches the frozen state. Frozen water is less dense than water at 40°F (4°C) and hence ice floats. Water has this unique feature because of hydrogen bonding. When water becomes cooled below 40°F (4°C), energy is lost, and the molecules slow down and settle themselves into a less dense arrangement. Ice can be almost 10% less dense than 40°F (4°C) water.

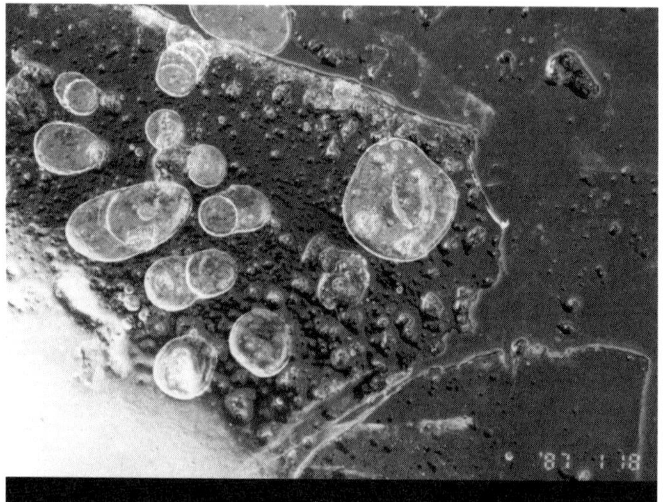

Photo 11-4 Skim-ice forming over diver's hole with diver air bubbles freezing in the ice.

When water drops a little below freezing (32°F [0°C]), ice crystals called *frazil crystals* begin to form. When these crystals accumulate, they become *frazil slush* or *grease ice,* aptly named because the water gains a greasy, shiny appearance. As the crystals further accumulate to form a harder surface, it becomes skim ice, or a very thin clear covering. Frazil and *skim* ice are most commonly seen by ice divers in very cold weather as the water in ice holes begins to visibly refreeze. (

Photo 11-5 Twenty-four inches of ice history. Primary ice = 12 1/2 inches of black ice. Secondary ice = 2 1/2in of rain, 5in of snow, 3 1/2in of rain melting and freezing with snow.

The first layer of ice formed is called *primary* ice. If primary ice forms in one continuous sheet called field ice, it can have the appearance of clear glass, called *black ice,* and is the strongest form of ice. It is called black because, when viewed, it is the same dark color (black, blue, brown, or green) of the water beneath it. If wind and water movement break up the ice as it forms, the primary ice will form as connected pieces rather than as a continuous sheet and will be uneven in appearance and weaker than black ice. If the field ice is anchored to both shores, it is called *fast ice.*

Ice that forms above or below the primary ice is called *secondary ice.* An example of secondary ice is snow or rain that freezes to the primary ice. Ice made of frozen snow is milky in color and weaker than clear ice and is called *snow* or *milk ice.* When using ice thickness charts to determine what kind of activity can be conducted on the ice, add 2in to 3in (5cm to 8cm) to the chart's number if the ice is snow ice.

As cold water contacts the underside of ice, it can slow down and freeze, thereby thickening the ice. A thick layer of snow on top of the ice can actually retard the growth of this bottom ice layer by serving to insulate the ice and water below it from the colder air temperatures. If water below the ice is warmer than 32°F (0°C), the protective snow blanket can actually cause the ice to begin to melt.

Ice formed from layers of snow ice and primary ice that has melted and refrozen develops a layered appearance, and is therefore called *layered ice.* Each layer may be a different density and strength, so be conservative when using the ice-thickness charts.

Hummocks are heaps of broken ice that have been forced upward by pressure. *Pack ice,* or *ice dams,* is formed by chunks of ice pushed up against one another by water currents or wind. When the chunks, which can be several feet thick, meet obstructions, they begin to pile up against and on top of one another, forming ice dams. Freezing may then join the separate pieces. Be aware that the newly formed joints may be weaker than the original chunks. Uneven ice can also produce *pressure ridges* when it buckles under the pressures created by thermal expansion. The formation of a pressure ridge can be accompanied by a loud, thunderous, reverberating, cracking sound. Pressure ridges can force blocks of ice downward at an angle below the water surface to form what looks like the keel on a boat; hence, they are called *ice keels.* Ice keels should be noted because they can snag diver tether lines. If pressure ridges are evident in a dive area, have divers check under the ice to see if any keels exist.[3]

As ice degrades it can reach a point of becoming *rotten ice,* sometimes referred to as *candled ice,* which falls apart when picked up. This ice should not have weight placed on it. When ice begins to form or when it starts to disappear, areas of open water called *polynya* will appear between areas of ice. Ice that has a layer of liquid ice on top of it is called *flooded ice,* and is dangerous because it is very slippery. Ice divers most commonly see flooded ice at the edge of an ice hole after extended use, as water begins to come up onto the ice.

When daytime temperatures are slightly above freezing, and night temperatures are below freezing, the ice can remain in a relatively stable state. If both day and night temperatures are above freezing, the ice will begin to melt and weaken. Strong winds can inhibit the formation of ice, but slow winds at temperatures below freezing can accelerate it.

There are many signs of weak ice. Look for holes, soft spots, areas where water pools on the ice, large dips or low points in the ice surface, changes in type of ice, slush, and cracks. Observe carefully before going onto the ice.

Ice sounds and what else to pay attention to

Often while standing on the ice, or even occasionally while under the ice, you will hear a loud, reverberating, thunderous cracking sound. As ice expands, contracts, or moves, it creates cracks or pressure ridges. The noise this produces is compounded by an echo factor at the air-water interface. The sound can travel through the ice for hundreds of feet. Sometimes after such a sound is heard, a pressure crack can be observed. It is a little unnerving to hear one of these thunderous sounds and realize that 28in (71cm) of ice has just cracked and can now be seen through. Unless the crack is large enough to get a foot caught in, it is not usually of serious concern; this is a normal and continuous process for all solid ice. Do not confuse the thunderous sound with the loud cracking sound that is heard when too much weight is placed on a section of ice. That sound can be followed by the weight falling through the ice, as more than one dive team has discovered a bit too late.

Photo 11-6 A tender tests the ice between two pressure cracks that extend from bank to bank.

Pay attention to the sounds of ice. The more experience you have, the more you will be able to understand what is happening. If you are an ice instructor, teach students how to pay attention to the sounds and show them the corresponding changes in the ice.

Another sign to pay attention to is the development of a layer of water on top of the ice. If there is no rain or melting snow, it is usually a sign that water is coming over the top of the ice from natural or manmade holes. This situation warrants extra attention and care about the ice quality. Not only is the water a sign that the ice roof may be dropping and weakening, but the water on top of the ice can increase the erosion rate. The water makes the ice more slippery, so good ice cleats should be worn if personnel are walking on it. If the water starts rising over the ice around the hole, move all surface support back a few feet to take the weight off the weakening edge. This may also require moving back the ice screws that tenders and divers are tethered to.

If the ice is very thick and vehicles are driving on it anywhere near where you are working, pay attention to how closely the vehicles are following each other. Imagine the ice under a car. Even if the ice is very thick, it will bellow downward a little under the weight of the car. This will send an upward wave of water under the ice behind the car. If a second vehicle following the first car drives on top of this upward crest, it could puncture through even very strong and thick ice. There are some places, such as Canada, where roads for 18-wheel tractor-trailer trucks cross hundreds of miles of frozen lakes that are deemed safe for moving vehicles, but in general it is preferable to not work on sites where vehicles are driving. It is also not recommended in most places to park a vehicle on the ice. Four wheels supporting 2,000lb (907kg) in one place for hours have a different effect on the ice than that same vehicle slowly moving across the ice. If heavy shelters or other such objects are placed on the ice, avoid diving underneath them unless the ice is very thick.

Do not allow snowmobilers to ride continuous circles around the diving ice holes. This may sound obvious, but when it comes to ice, many people seem to lose common sense. Just think about the snowmobilers who have contests to see who can jump over the largest area of open water on a frozen lake. Somebody has to lose.

The Dive Day

The first job upon arrival on the scene is to check for any unexpected hazards and then mark off hot, warm, and cold zones. Next, plan where and how to set up the different staging areas in the warm and hot zones. Locate the landmarks on the site profile map if one exists.

Photo 11-7 When an ice call comes in, often the whole world shows up. It is imperative to set up effective, secure zones to keep all personnel without ice diving operational- or technician-level certification in the cold zone.

Downtime

Because of the cold and wet conditions associated with ice diving, avoiding downtime is very important. Aim for no more than ninety seconds between dives, so that as the primary diver(s) are preparing to exit the hole, the new primary divers, who were most likely the previous backup divers, are preparing to enter the hole. The tenders should make the switch as quickly.

Some of the ways to avoid downtime include: effectively cycling personnel, having redundant equipment, and having enough certified surface personnel and divers. Ice instructors should have a pre-plan containing a dive schedule that lists who will dive and when. This plan should be created before arriving at the dive site to prevent wasting time at the scene. On the other hand, the ability to be very flexible and make quick, yet effective, decisions is imperative, since things can change so quickly on ice sites. Divers should be capable of dressing in less than six minutes for recreational diving and three minutes for public safety diving. Scene organization is key also.

Plan to have a "next" diver ready and waiting in a shelter, in addition to the primary, backup, and 90%-ready divers who are at the ice hole. In case one diver has to abort, the "next" diver can immediately be rolled into the cycle. Without this "next" diver, the remaining two divers on the ice would be forced to wait for another diver to get ready. Waiting and downtime on the ice increase the probability of physiological, mental, and equipment problems. Once the "next" diver is moved to active status, there should be another diver immediately ready to take the "next" diver slot. For this to happen, a new diver-tender pair should begin dressing approximately eight minutes into the primary diver's dive, if the maximum dive times are fifteen minutes, or start dressing thirteen minutes after the primary diver descends if the maximum dive time is twenty minutes. This contributes to efficient cycling.

Photo 11-8 Effective diver cycling. The next 90%-ready diver is pulled to the hole by shore personnel. He will lie on the forward mat. There will be no downtime between divers since all necessary personnel will be ready to go before the primary diver is out of the water.

Always have extra equipment available to prevent having to fix or share, rather than simply replace equipment. Have a few pairs of extra ankle weights onsite to make quick and safe weight adjustments to divers. Instead of removing the diver's belt to add 2lb, simply attach an ankle weight where it is most needed, such as around the tank valve or on the diver's chest harness. Never add weight by putting it in BCD pockets unless the pockets are part of a weight-integrated BCD system. Among other problems, weights in pockets can fall out, which can be particularly dangerous during overhead dives.

Less equipment sharing generally equals less downtime. During public safety diving training sessions, rent gear if necessary so there is enough equipment for almost every diver in the class.

If each diver has a personal quick-release pony bottle and harness in addition to the standard dive gear, much time can be saved. If the budget does not enable each diver to have a pony bottle, at least make sure each diver has a quick-release pony harness system so that when a bottle becomes available, the diver only has to slip the bottle into the harness. Otherwise, time will be wasted donning the harness system to the BCD assembly. When equipment is removed from a diver after a dive, it should be made ready for the next dive by replacing cylinders, opening up tightened BCD shoulder straps, running the equipment through an equipment checklist, and laying it down with the valve pointing toward the water in a warm area if possible.

The need for strong leadership personnel who have ice experience and organizational skills is imperative. Ice dive training classes require enough certified ice instructors and certified ice divemasters that a watchful eye can be kept on all parts of the operation, from shore to hole, that efficient cycling is maintained, and that assistance is provided wherever needed. Leaders need to continually count the number of full cylinders available to ensure that there are always enough to maintain the cycle. Plan how to top off pony bottles if they are used repeatedly, or in case a free-flow occurs. Leadership onsite should ensure that equipment is kept orderly in the equipment staging area, to prevent downtime from having to locate missing items and to prevent picking up broken items. Good organization means less downtime.

Staging areas

Staging areas should be set up with vehicles broadside to the wind, while at the same time direct access to exits should be maintained. Use vehicles as wind shelters. Set up tarps between vehicles to protect divers from wind chill, wind shear, and other weather effects. Tarps can be set up as walls and roofs to create tent-like structures. Be sure to run the tarps from the top of the vehicles, all the way to the

Photo 11-9 Vehicles can make good windbreaks, especially for short operations. The two propane heaters provide warmth. Divers and tenders dress in the heated vehicle.

ground, because wind blowing under vehicles can be sufficient to cause tender/diver feet to chill and quickly become useless. Wind that is low to the ground also increases equipment freeze-ups.

Leave at least one vehicle free of tarps and lines for rapid exit capability. If this is impossible, make sure everyone involved knows the location of shears or a knife capable of cutting the lines in an emergency.

Photo 11-10 Example of a shelter over an ice hole in Winnipeg, Canada.

Photo 11-11 Trucks with heaters can make good dressing shelters.

If available, use large tents or inflatable hazmat shelters. Space heaters also provide comfort. When using heaters, be careful to keep equipment, clothing, and tarps well clear of them. If people are not used to working with such heaters, they may not realize how quickly an object within a few feet of the heater can catch fire or melt. Someone should be assigned to maintain a continuous watch on heaters and the equipment around them. If the gear is overly warmed, there are additional chances of freeze-up, because ice is more likely to form after snow melts on the warm gear and later refreezes when the gear is moved away from the heat. Divers and tenders should be instructed on how to keep their gear orderly in the shelters. Disorganization leads to broken and misplaced equipment and increased downtime.

People on the ice

Once it is determined that the ice is safe, keeping in mind that "safe" is a relative term in ice diving, start sending people out to do the various set-up tasks. Keep the number of people on the ice at any one time to a minimum to decrease the extent to which the ice is weakened from use. At a minimum, two people should be sent to perform any task—one does the work and the other helps in case of accidental immersion.

Everyone on the ice should wear a PFD, or other appropriate flotation device, in case he/she falls in. If these personnel are divers, make sure they do not sweat in their drysuits during this process; if they do, they will become colder faster after the hole set-up operation is completed. It is important to prevent submersion. Hole openers in particular have a tendency to fall in head first because of their posture while cutting the ice. In public safety diving, holes may need to be cut in relatively thin ice. If the ice is relatively thin, the hole cutters should be wearing harnesses as well, so they can be tethered back to the shore or another safety point. Remember, always plan and be ready for contingency operations.

No one belongs on the ice without proper PPE, including hat, gloves, safety glasses, and proper footwear.

Ice and hole preparation

In recreational diving training sessions and during public safety diving operations, access to the water is needed. This normally means cutting two or more holes, unless access is already present from the victim's hole or very weak ice. Before holes can be cut, the location of the pre-planned holes needs to be found. Use a measured line and the range points from the map to find the location selected during the warmer season visits.

At least two safety holes should be cut in addition to the main hole to provide additional access routes to manage problems and provide more training opportunities. Safety holes can be approximately 75ft (24m) apart and 75ft (24m) from the main hole. If divers will be limited to one direction, such as DAD quadrants 1, 2, and 4, cut the main hole and then cut two more holes, one at 11 o'clock and one at 2 o'clock. If divers will be allowed to dive in any quadrant of the main hole, put the two additional holes at 11 o'clock and 5 o'clock from the main hole.

Place all tools to cut and set up the hole on an ice board or sled, and secure the tools with a quick-release system. A board distributes the weight, reduces the chance of breaking through the ice, and in the case of an accidental fall-in helps prevent full submersion. The hole-cutter tethers the ice board to the front of his harness with webbing that is about 15ft (3.6m) long. Shore personnel hold a line secured to the other end of the board. If the tender breaks through, the board will be far enough behind so the tools will not fall in. The tender can simply pull himself out with ice awls or the board's webbing.

Photo 11-12 Brightly colored flags are strung up around the hot zone perimeter. Wood stakes with fire tape and cones mark the area with the holes. When the operation is over, the flags are removed, the last side of the fire tape is put up, and the holes are sealed.

Set up a safety barrier around the proposed hole location. Cut four small "X"s into the ice around each hole and place stakes. Hammer each stake 4in to 5in (10cm to 13cm) deep into the X. These should freeze in place. Tie tape or ribbon from one stake to another. Tie the tape taut enough to prevent it from flapping in the wind, but not tight enough to pull the stakes over. Leave the shore-side section open for divers and tenders to access and exit the hole.

Remove snow from the walkways to the hole to reduce fatigue and heat loss from feet and to prevent uneven substrate caused by trampled snow that later freezes. Uneven substrate not only increases fatigue, but can result in people falling, injuring themselves or breaking through the ice. Unless the ice is very thick, be sure to shovel multiple paths or a very wide path so that the same narrow path is not used over and over, weakening the ice. As simple as this may sound, accidental immersions have occurred because ice was weakened in this manner.

Walkways to the hole should be in direct line access to the holes from the shore, as long as the ice thickness permits. If the ice is thick enough—10in (25cm) or greater—shovel one access out to the main hole, and then shovel accesses from hole to hole. If the ice is less than 10in (25cm) thick, or there are many divers and surface personnel, create an access and exit route from the shore to each hole.

Tethering

Now is a good time to install the ice screws that will tether the surface personnel who will be cutting the hole. Use these same screws later to secure the divers' tether lines. Place one screw 5ft (1.5m) from each corner of the shore side of the triangle to be cut. Tenders stand opposite the sides of the triangles that divers will be diving. Divers should not dive on the shore side of the hole as that is where the majority of tending takes place. Remember, even if shallow water is not a problem, several hours of diving under the entrance/exit path can make even seemingly thick ice weak. The divers' bubbles can quickly erode ice.

Cutting the hole

The standard hole shape is a triangle, which is used for its ease of cutting and for providing multiple, tightly angled corners for efficient diver exit and entry. The optimum size for training is 6ft (2m) per side, which gives ample space for several divers in case of an emergency. At no point should a hole be smaller than what will accommodate a conscious and unconscious diver. Five feet on a side is the smallest for safe operations. All procedures should be based on both standard operations and tested contingency plans. And because bigger is not always better, consider the ice-edge erosion rate when planning what size hole to cut.

There are a variety of ways to cut the hole. Hot- or warm-water cuts are used in the Arctic and other areas where there are many feet of ice. Small explosive charges are also used for extremely thick ice. Neither of these methods is very practical for the recreational diver or public safety team. Handsaws or hand augers used by experienced ice hole cutters like the Eskimos and the Mohawk are very quick and efficient, but are not commonly used by divers.

Chainsaws are the most common choice. If a chainsaw is used, it is critical to have proper training and to always use the safest techniques possible with no shortcutting. Chainsaw manufacturers state that their equipment is not designed, manufactured, warranted, tested, or approved to cut holes in the ice. When used properly, though, chainsaws cut a hole in ice as quickly and safely as in other materials. When used improperly, they are dangerous every second.

Chainsaw safety

The following rules and guidelines should be followed when using a chainsaw to cut a hole in ice:

- Read the manufacturer's safety manual.
- Manufacturers mandate the use of safety equipment whenever using a chainsaw. Wear safety boots with ice cleats when cutting the hole. Afterwards, change into boots that will prevent heat loss. Wear eye protection.
- Have an advanced first-aid kit and a properly dressed, medically trained person onsite.
- Ensure the chain is properly attached and the slack is adjusted.
- Always start the saw and allow it to run for a minute or so before taking it on the ice.
- Never walk on the ice with a running chainsaw.

- Always use the bar lock before pulling the saw up and out of the ice.
- Learn how to set the bar lock with a forward movement of the wrist so as not to push the hand forward and over the top. Practice with the saw off.
- Never stand on the inside of the cut.
- Never tie the saw to the cutter or to anything that restricts total freedom of movement. Lying on the bottom of the lake for a few hours will not hurt the saw. Equipment means nothing in comparison to personal safety.
- Except in an emergency, never put the saw down in the snow where it could freeze.
- If stopping for more than a few seconds, turn the saw off and remove it from the hole.
- Never put the saw down while it is running. If moving it more than a few inches, turn the saw off.

Cutting technique

Mark the outline of the hole on the ice. Once the starting point is established, the cutter places his feet so that the forward foot is at a right angle to and at least 3in to 4in (8cm to 10cm) from the blade. Place the rear foot outside the cut and cantered parallel to the blade. Cut in a direct downward motion, 90° to the ice, with a smooth plunging motion. At no time should the saw blade be at more than a 10° to 15° angle of vertical. If for some reason the cutter slips or loses control of the saw, it should find its natural balance point blade down through the ice. Whenever cutting is stopped, even for a few seconds, leave the saw in the blade-down, 90° cutting position. When switching cutters, leave the saw blade in the cut with the bar lock on.

The sawing and cutting action is not forceful. In fact, forcing the blade slows the cut down and increases the potential of the saw kicking back. Work the saw gently back and forth with minimal angles and only slight pressure. Continue to make the cut as straight as possible perpendicular to the surface. If the hole is cut at an inward angle, the ice block is very difficult or nearly impossible to push down because the bottom portion of the block is smaller than the upper portion. An outward-angled cut is wider and therefore requires more time and energy to cut.

Make a 1ft to 2ft (30m to 61m) long cut, then backtrack until the cut has reached the water and is wide enough not to refreeze. Never move the saw farther than is physically comfortable. Keep your body weight squarely over and on all four corners of your feet.

When cutting properly, the cutter should not become soaking wet. A cutter may become a little damp when cutting thin ice, 3in to 4in (8cm to 10cm) thick. If the cutter is getting soaked, something is wrong. The cutter's body may be angled improperly to the angle of the saw, or the saw may be angled improperly. Be sure to con-

Figure 11–13 Author Hendrick demonstrates the proper way to start a chainsaw—kneeling with the saw securely fixed to the ice between his knees.

Figure 11–13a Right toe should be just behind the blade at right angle to the saw blade. Left foot should be a few inches behind the blade and parallel to it.

Photo 11-13b At no time should your feet be in the way of the blade in either direction! Bite-teeth should remain close to the ice at all times.

Photo 11-13c A simple way to get the cut started in thick ice is to use an ice auger to cut an initial hole at each of the marked triangle corners. These holes allow the blade to enter the water and begin cutting the ice easily. They help prevent corners from re-freezing during the operation.

tinuously warn anyone standing behind the cutting action to move, since the saw can and will throw water and ice chips as much as 20ft (6m). When finished, dry the saw and refuel, so that it is ready to go.

Figure 11-13d Keep your weight squared over your feet with hips slightly behind feet and knees bent. Shoulders are in line with the saw. The left elbow is stabilized against the left knee.

Figure 11-13e Improper sawing position. Be very observant to catch and correct improper footing. The right foot should not be pointing toward the saw blade. The left arm is not stabilized against the left knee and the feet are not far enough apart.

Chainsaws can freeze up, particularly when they are being worked too hard. Let the saw do the work; do not use muscle strength. If the saw blade freezes, allow the saw to continue at idle and dip the blade fully into the water for twenty to thirty seconds. Next, gently power the trigger on and off. If the ice is too thick for the blade to be immersed in water, try immersing it at least halfway, or use a bucket of water.

"Pull it out?" or "Push it in?" That question is often asked about the ice triangle now floating in the hole. In almost every case, it is recommended to push the block after making it easily retrievable with an ice screw. The effort required to lift a block out can be tremendous. Typically, the block is cut into smaller pieces, because picking up a block that can weigh more than a person is not only difficult but can in fact be dangerous, resulting in injuries ranging from back injuries to broken limbs. Cutting the block into smaller pieces is time consuming and requires more skill and coordination than cutting the original triangle, because the pieces now move around.

Once the block is out, the question becomes: "Where is it stored?" On top of the ice, it can be an obstacle to maneuver around and a weight that weakens the ice underneath it. Generally, pulling the block out is a waste of time and energy.

There are two reasons why ice blocks are often removed. The first reason is to avoid dive tether lines becoming snagged on thick ice blocks that are pushed under. There are two ways to prevent this problem: (1) Keep diving out of DAD quadrant 3, under which the block is pushed. If divers do need to enter quadrant 3 for a search,

Figure 11-13f Slowly work the saw back and forth at a 45° angle, allowing the saw and its own weight to do the work.

Figure 11-13g At any point in time, if you stop or slip, the saw blade should remain down by its own balance for safety.

have the tender lie on the ice and put an arm low to or in the water. This will work for ice as thick as 40in (1m) and possibly more. (2) The second option is to work the quadrant-3 area from a different hole.

A second reason for pulling the ice block out of the hole is to build a wind block by stacking thick ice blocks to form a wall. It takes less effort to create wind blocks with tent shelters or tarps stretched across stakes driven deep into the ice.

There are thus several good reasons for pushing the ice block under, rather than pulling it out:

- Saves a great deal of effort and time.
- Keeps unnecessary weight off the ice.
- Does not create a topside obstacle.
- Can be used to strengthen and support the ice under the tenders while they work.

With training, the procedure for pushing the block under is safer than the procedures used to pull the block out.

Most divers pull the block because they do not use a DAD system. Divers are allowed to go wherever they want. With DAD, divers are kept out of the quadrant

Figure 11-14, 11-14a To remove ice blocks, Winnipeg PD dive team members cut multiple blocks that are pulled out with ice tongs.

Photo 11-15 With a little training and practice, three surface personnel wearing good ice cleats can safely, easily, and quickly push an ice block under quadrant-3 using three pipe poles. This triangle has 5ft sides and 39in of ice.

closest to shore, which is where the tenders stand, and which is where the block should be pushed under. So, don't pull it out, push it under.

Once the block is cut, cut the corners off about 1ft (30cm) from the ends and either push them under or remove them from the water. This makes it easier to submerge the block and to later replace it. If the plan is to lift and remove the corners, put an ice screw in each corner with attached webbing before cutting them. Then simply lift the corners out with the webbing.

When the block is ready, screw an ice screw into one corner of the block approximately 12in (30cm) from the end and attach a 3ft to 4ft (1m to 1.2m) piece of tubular webbing. When the block is pushed under the ice, the strap facilitates block retrieval when it is time to close the hole.

Have three tenders with ice poles push down on the end of the block opposite the ice screw until it goes underwater. Then gently push the block under the ice roof to the shore side of the triangle (DAD quadrant 3), where tenders generally will stand and where divers will not dive. If the area of the block is part of the dive site, it is imperative that the tender be extremely observant of the line, to make sure it does not become pinched between the block and the ice roof.

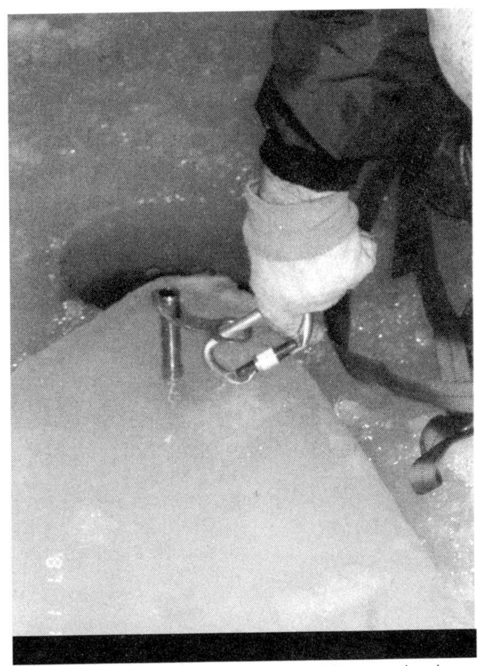

Photo 11-16 Attach an ice screw, carabiner, and webbing to the ice block for later retrieval.

In most cases, the block will not freeze solid to the ice roof. Retrieving the block might require the use of ice poles, to push down on the lip to break the seal under the roof. If divers do not dive in the area of the block, it stays in place, making retrieval easy.

In some cases the ice is so thick (more than 14in [36cm]) that the triangle may have to be cut in half to overcome the buoyancy of the block when pushing it under the ice roof. If using the ice screw system for retrieval, put a screw in each half.

Photo 11-17 Pole placement. To push the block under, place poles at each corner. The person holding the webbing is in charge and steers the block as needed with the webbing and by giving commands to pole holders 1, 2, and 3.

Do not put a buoy in the hole to mark it, because even an experienced tethered diver/tender team can end up tangled with the buoy line.

Thicker ice

If the ice is more than 12in to 14in (30cm to 36cm) thick, the block might be caught when moving it under the ice. To lessen the chance of this, cut the triangle in two stages. Cut the block at an angle to make the top surface smaller than the bottom surface. Make a second cut 1in to 2in (2.5cm to 5cm) inside the first cut to make the block smaller by 1in to 2in (2.5cm to 5cm) around its perimeter. The perimeter ring that is left is crushed and destroyed during the chain sawing of the second cut and as the block is pushed under the ice roof. Make this second cut with a straight vertical angle instead of a slanted angle. A slanted cut is easier to make with one foot on either side. When the triangle is completely cut, it is not safe to stand on it to make the second cut.

What if the ice is thicker than the chainsaw blade? Get a longer blade. If that is not possible, be prepared for a good deal of work. If the blade is shorter, here are a few tips:

1 Cut a pie wedge of ice out to access the chainsaw blade to the deeper ice. To do this, cut a long "V" shape 4in to 5in (10cm to 13cm) wide, angled inward to make a pie wedge that can be removed by chopping it up.

2 Put the chainsaw into the wedge and make a deeper cut.

Another option is to use a 25lb (11kg) ice chisel. In the hands of an expert, 25in to 30in (63cm to 76cm) of ice poses no problems. For the rest of us, such a job is a physical nightmare.

Photo 11-18 At the end of the training day, an infant mannequin is dropped in a weep hole in front of a tape-covered hole for the next day's first search operation.

Weep holes

Weep holes divert exhaust gas bubbles exhaled by the diver, to reduce erosion at the main hole. Cut three to four small holes as wide as the chainsaw bar, approximately 20ft to 25ft (6m to 8m) forward of the main hole in the diving quadrants. Allow 15ft to 20ft (4.5m to 6m) between the weep holes.

Once the two or three primary dive holes and weep holes have been cut, set up the tethering system described in chapter 6.

Weep holes can also be used during training exercises. Place a pole through a weep hole to simulate a lost diver, and have tenders sweep divers under the ice to find it. Public safety dive teams can practice spidering from their hole to a weep hole. Spidering is a technique for rapidly crossing under the ice roof to reach the victim's hole from a hole closer to shore, described later in the book.

Keeping the hole neat and safe

A challenging task on the ice dive site is to keep the area clean, orderly, and safe. "Safe" means reducing the possibility of injury to personnel and equipment. It is not uncommon to see photos of ice diving sites at which tether lines are everywhere, and tenders and divers alike are standing on lines. With too many lines, there is greater risk that someone can be caught in a line and pulled in, or that someone can trip and fall. Working around the ice hole will bring dirt from shore. Dirt combined with snow and ice becomes ingrained into the line when it is stepped on, thereby reducing its overall life. An overabundance of line lying on the snow and ice will tend to become frozen and difficult to work with.

In an emergency, a well-organized dive site reduces overall confusion and disorganization. Ropes and diver tether lines should:

- never be stood on or stepped on.
- be kept in line bags, both for ease of use and to prevent direct contact with snow and ice. Lines running between the shore and the holes may be more than 300ft (91m) long, so it may be easier to store them on non-metal reels with brakes.
- be changed for every diver, checked between dives, and always tied off.
- be cut or tied off at maximum lengths of diver and safety diver allowed distances (100ft [30m] for divers and 125ft [38m] for safety divers). These limits ensure that the diver cannot get farther away than planned.
- never be wrapped around anyone's hand.
- have distance markings every 5ft (1.5m).

If possible, have a small piece of carpet or tarp to place under the line bag to further reduce the possibility of direct contact with the snow and ice. The safety officer should make sure that lines are repacked in bags or on reels whenever they are not being immediately used.

Do *not* allow the safety diver to enter the water as he switches to primary diver without first changing the line bags. *Any* safety plan is destroyed if the primary diver ends up with more line than the backup.

Only the equipment being used and contingency equipment should be on the ice. One of the many advantages of using the ice board for extracting divers from the hole and bringing them directly to shore is that all their gear stays with them and is not removed until they are back in the shore staging area. This means there will not be weightbelts, fins, and other gear left lying on the ice from previous divers, that must be picked up and brought to shore. If an ice board is not available, the safety officer needs to make sure the gear is picked up and put where it belongs.

Maintaining a neat hole also involves keeping it from refreezing and knocking off the thin, overhanging edges that develop from erosion. In very cold weather, it may be necessary to stir the hole with a pipe pole or other object, and sometimes the water surface may need to be skimmed out if filmy pieces of ice form. The safety officer, or whoever is in charge at the hole, should keep a watchful eye on the ice edges surrounding the hole. Tap the edges with a pole; if they break, they are unsafe and need to be removed. If this is not done, a well-intentioned tender who kneels on the edge to assist a diver with an equipment problem may end up in the water.

Ice shelters

Recreational ice divers have the ability to plan their dives well in advance and can therefore build some very effective ice shelters to make their day far more enjoyable.

Michael J. Knorr, an experienced ice diving instructor, created one of the most effective shelter systems and shares its construction process so others can build one.[4] Keep in mind that Mike and his dive operation typically have over 30in (76cm) of good ice to work on. Plans for a very workable design were developed with the help of Ardin and Dale Niemi of S-M Enterprises of Moorhead, Minnesota (a metal fabricating company). The design requirements were such that the shelter is portable, stable, comfortable, dry, and warm.

Start with the floor.

1. Use six 4ft × 8ft (1.2m × 2.5m) sections made of treated lumber that resemble dock sections, with a slotted top so that water introduced by divers will run down to the ice.
2. The supports are 2in × 3in (5cm × 8cm) with the top covering made of 1in × 5in (2.5cm 13cm) treated boards with about ½in (1.25cm) spacing.
3. Supports are made easy to carry by drilling ¾in (2cm) holes spaced about 1ft (.3m) apart at each end and running ⅝in (1.5cm) three-strand rope through to make a loop.
4. Where the ends butt together, leave about a 2in (5cm) overhang on the top board so those ropes are hidden when set up.
5. In one corner of one of the sections, make a triangular 6ft × 6ft × 6ft (2m × 2m × 2m) hole for the diving entrance. A 4ft × 6ft (1m × 2m) rectangular opening may be used with a ladder.

6. A cover is made to drop down over the hole to provide security during setup and breakdown and when diving activities are halted. This is done because the ice hole is already cut when the shelter is put in place over it.

7. The sides of the structure are 4ft × 7ft (1m × 2m) panels. They are framed in 1in (2.5cm) aluminum square tubing welded together with a support running horizontally halfway down the panel. Triangular pieces of aluminum are welded into each corner for additional strength.

8. Insert 1in (2.5cm) insulation sheets with their foil side to the inside of the shelter, and cover the outside with aluminum sheets. Two of the panels have doors of the same construction. The total number of panels for the shelter is 14.

9. The panels are set upright and held in place with steel bases, with 5/8in (1.5cm) steel rod about 2in (5cm) in length welded in place. The tops are held in place with the same type of design, but instead of a base, these pieces have a 1in (2.5cm) steel plate that holds the panels parallel with each other. The top corners are of a like design but have the steel plates set at a right angle to give stability.

10. The roof consists of five expandable bows that drop into the channels provided by the wall construction. Over these bows, a white canvas is stretched (for light penetration) that is held fast with grommets and bungee cords. An option is to integrate a bubble insulation sheet under the canvas to reduce heat loss. With such a design, the structure can be easily enlarged by adding additional panels and extensions onto the roof canvas as well as to the floor.

11. Heat is provided by a free-standing propane heater. The 50,000Btu heater can take the temperature from subzero to 70°F (-18°C to 21°C) in a matter of minutes. Door traffic and air circulating under the floor provide adequate ventilation for fresh air.

The inside design was approached with the goal of keeping all equipment off the floor.

- All attachments are made of aluminum that hooks over the top of the panels and extends down into whatever design is needed.
- The tether lines are anchored to the floor through bolted pad eyes with backing plates and locking carabiners.
- There are three 2in × 4in (5cm × 10cm) treated boards on holders that have coated hooks spaced about 1ft apart on three walls of the structure. This allows all coats, harnesses, cameras, etc. to be hung.
- Shelves keep a radio, beverages, and other objects off the floor and out of the way.

- At the far end of the shelter (away from the hole), a 4ft × 12ft (1m × 4m) section of 1/2in (1.25cm) rubber matting is designated as a dry area where people can dress down in comfort at the end of the dive.
- Getting in and out of the dive hole is made simple by the use of an aluminum ladder that swings down into position. This ladder swings up and out of the water while divers are submerged to prevent the lines from snagging it.
- An 8ft × 16ft (2m × 5m) trailer is positioned next to the shelter and is used as a changing facility and equipment-assembly location. This is also heated and has a large air cascade system for onsite tank fills.

With this shelter, concerns about the weather are greatly diminished, as are cold-related equipment problems. Equipment is kept warm right up to the dive. A large insulated Igloo™ cooler, filled with hot water heated on a furnace, is kept inside the shelter. As the wetsuit divers complete their final check, their tenders flush their gloves, boots, and suit with this water. All seams are then duct taped to minimize circulation. This shelter also works well for night ice dives, since it is easily lit inside.

One of the main advantages of the shelter is that tenders and divers are protected from the weather during the dive. If such a shelter is used, one to three safety holes and weep holes should still be made. These would of course be located outside of the shelter. Make sure to have ice screws set up at each safety hole to tether tenders and divers should it become necessary to use one of the holes.

Closing the hole and making it responsible

Unless a night ice dive is to be conducted, all ice diving should be fully shut down a minimum of an hour and a half before sunset. After all divers are out of the water and the operation is over for the day, pull the blocks back into the hole. Maintain the safety barrier around the perimeter of the site with stakes, orange ribbon, and orange cones, making sure to close up the shore side as well. Then cones can be placed around the hole to tape off the holes themselves. In addition, an "X" can be laid over each hole with barrier tape and a few stakes. It may be necessary to put orange cones over weep holes if holes are large enough to accidentally put a foot through.

Weather

On pre-planned dive operations, check wind conditions before setting up the ice site. As we learned earlier, very strong winds can erode ice, as can light warm winds. Another wind concern is that surface support personnel can be pushed into ice holes. Good ice cleats become mandatory on windy days. The convective heat loss from wind can greatly increase the risk of hypothermia for all personnel, especially divers who are wet after div-

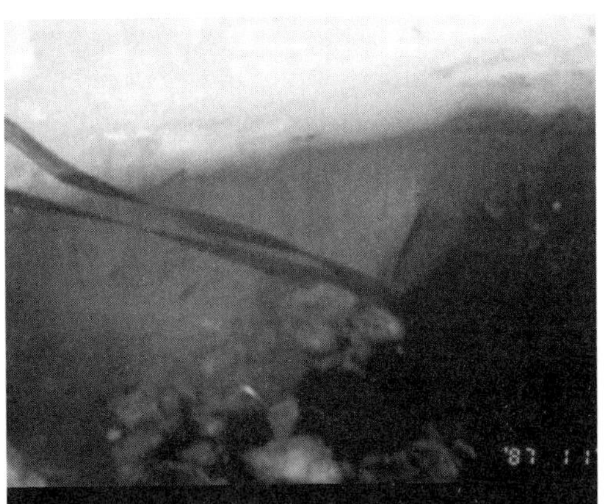

Photo 11–19 Even if the block has moved away from the hole during the operation, it can be retrieved by the webbing secured to its ice screw.

Photo 11–19a To close the hole, the webbing tender pulls the block slowly out from under the ice roof as pole tenders 1, 2, and 3 are ready in place to keep the block horizontal. The webbing tender steers the block into place as the pole tenders allow it to slowly surface.

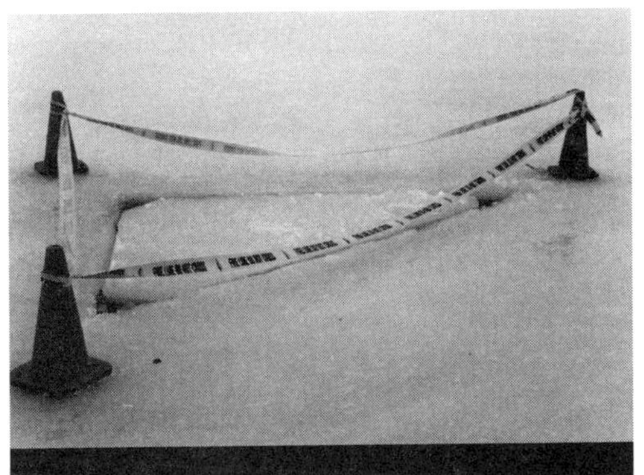

Photo 11–19b To secure the hole, nail cones into the ice and string fire tape around the hole. If desired, tape can be crisscrossed over the ice block and nailed in as well.

Photo 11-19c Orange cones can be placed directly on the ice block in very cold weather. They will quickly freeze to the block and stay in place.

ing. It also increases the likelihood of equipment free-flow problems. Wind can blow snow in the faces of surface support personnel, making it difficult for them to do their jobs properly. It can also blow away anything that is not properly secured down. Make sure you have extra ice screws with plenty of webbing on windy days.

The Beaufort scale is very useful because it provides a means of visually estimating wind speed on land and at sea. To understand how fast a knot is, consider that a nautical mile is approximately 6,000ft Take one nautical mile and divide it by an hour:

$$6{,}000\text{ft}/60\text{min} = 100\text{ft/m} \;.$$

This means that a light 4knot breeze is moving at 400ft/m. A strong breeze of 20knot will move snow across a frozen lake at 2,000ft/m.

If swaying branches are observed on the way to the dive site, a decision must be made about whether or not to initiate the dive operation. Public safety ice diving teams are more likely to deploy when winds are high than other divers are. The ice board becomes even more important to public safety dive teams to transport personnel and equipment during high winds because it is the lowest-profile technique possible.

The original Beaufort scale was created by Rear Admiral Francis Beaufort, British Knight Commander of the Bath, in 1805. The scale was a scale of force; it did not give wind speed. The scale was based on a set of numbers from 0 to 12 that Beaufort assigned to the effects of wind on an eighteenth century fighting ship. Numbers 0 to 4 described wind in regards to how it propelled a ship; numbers 5 through 9 described wind in the ship's ability to carry out a mission and carry a sail; and higher numbers described the ship's ability to survive. In 1846, T.R. Robinson invented the cup anemometer to measure wind velocity. In 1912, the International Commission for the Weather Telegraphy asked for wind-speed equivalents for the numbers on the Beaufort scale. These equivalents were set in 1926, and by 1955 all thirteen of the Beaufort numbers were replaced by wind speeds in knots. For example, a Beaufort 5 is a fresh breeze with wind velocities from 17knot to 21knot.

The wind chill scale should be used in conjunction with the Beaufort scale. As discussed in more detail in earlier chapters, we lose heat by convection when the wind blows. If you look up the actual temperature and apply a reported or estimated wind speed, you will find the wind chill temperature. The wind chill temperature relates to the rate of convective cooling. For example, a person standing in 30°F (-1°C) with a 20mph wind would lose heat at the same rate as the person standing in 4°F (1°C) with no wind.

The Beaufort Wind Scale

Beaufort Number	Wind Speed (knots)	Description*
0	<1	Calm
1	1-3	Light Air
2	4-6	Light Breeze
3	7-10	Gentle Breeze
4	11-16	Moderate Breeze
5	17-21	Fresh Breeze
6	22-27	Strong Breeze
7	28-33	Near Gale
8	34-40	Gale
9	41-47	Strong Gale
10	48-55	Storm
11	56-63	Violent Storm
12	>=64	Hurricane

* World Meteorological Organization

Figure 11–2 The Beaufort scale is a useful tool to predict and understand wind velocities, in order to make safe dive operation plans and risk benefit analyses.

Temperature (°F)

Wind (MPH)	30	25	20	15	10	5	0	-5	-10	-15	-20	-25
5	25	19	13	7	1	-5	-11	-16	-22	-28	-34	-40
10	21	15	9	3	-4	-10	-16	-22	-28	-35	-41	-47
15	19	13	6	0	-7	-12	-19	-26	-32	-39	-45	-51
20	17	11	4	-2	-9	-15	-22	-29	-35	-42	-48	-55
25	16	9	3	-4	-11	-17	-24	-31	-37	-44	-51	-60
30	15	8	1	-5	-12	-19	-26	-33	-39	-46	-53	-60
35	14	7	0	-7	-14	-21	-27	-34	-41	-48	-55	-62
40	13	6	-1	-8	-15	-22	-29	-36	-43	-50	-57	-64
45	12	5	-2	-9	-16	-23	-30	-37	-44	-51	-58	-65
50	12	4	-3	-10	-17	-24	-31	-38	-45	-52	-60	-67
55	11	4	-3	-11	-18	-25	-32	-39	-46	-54	-61	-68
60	10	3	-4	-11	-19	-26	-33	-40	-48	-55	-62	-69

Frostbite occurs in 15 minutes or less

Figure 11–3 Wind-chill temperature is the equivalent windless temperature that would cause the same heat loss as the combination of wind and ambient temperature. It should be considered during dive operation planning.

Summary Questions

1. List at least seven characteristics of a good site for recreational diving or basic ice training.
2. If the water level drops after ice has formed, the lake now has a _______ ice roof.
3. The primary danger of shallow water is diver ______.
4. List three functions of making a site map.
5. List two methods of documenting underwater locations on site maps.
6. For public safety divers, no ice is ____ ice.
7. List at least six variables that will weaken ice.
8. What is the minimum ice thickness to support two adults standing 2ft apart?
9. What is the minimum ice thickness to support a dive operation?
10. Ice with a layer of liquid on it is called ______ ice.
11. As ice expands or contracts it can create a ______, _____ sound that is not usually a concern, unlike a loud _____ sound that occurs when too much weight is put on the ice.
12. List at least five ways of reducing downtime.
13. At least two _____ _____ should be cut in addition to the main hole during training programs.
14. Place an ____ _____ 5ft from the shore side of where each hole will be cut to serve as a tether point.
15. Explain why a triangle is an efficient shape for an ice hole, and provide an effective side length.
16. When using a chainsaw to cut a hole, the forward foot is at a ____ angle at least _____ inches from the blade, while the rear foot is anchored _____ the cut and is _____ to the blade.
17. Cut in a direct _____ motion, with the blade _____ degrees to the ice.
18. If the hole cutter is becoming wet, something is _____.
19. List three advantages of pushing the hole ice block under, rather than pulling it out.
20. Place an ___ ____ in one corner of the ice block with attached webbing, so it can be easily retrieved post-operation.

21. ____ holes divert exhaust bubbles to reduce main hole erosion. Cut __ to ___ of these holes approximately ___ to ___ feet forward of the main hole in the diving quadrants.

22. List at least three rules of tether line handling.

23. Unless night dives are done, all diving should be shut down a minimum of _____ before sunset.

24. Holes should be made ____ before leaving the site.

25. The ______ scale provides a way of visually estimating wind speed over land and water.

Notes

1 Walt Hendrick and Andrea Zaferes, Public Safety Diving (Fire Engineering, 2000).

2 Elbert S. Maloney, Chapman Piloting: Seamanship and Small Boat Handling (New York: The Hearst Corp, 1977).

3 Environment Canada Aeriological Observers Course, Module 2.7 (Ice Survey), 2001.

Chapter 12
Communications

Communication involves numerous primitive and complex skills we take for granted a thousand times a day. Verbal communication is certainly the most common form of human interchange, while body language, physical gestures, and even the look of an eye or the turn of a brow can inflict or create a statement. As sport scuba divers, we were taught basic hand signals: "okay", "up", "down", "ears not clearing", and "where is the boat", with each signal forged out of a need and desire to communicate. Without communication, human interaction is lost. Ice diving adds another dimension to diver communication, namely communication between submerged divers and topside tenders. Another challenge that may be new for recreational divers is blackwater that can occur when bottom sediment is stirred up. Blackwater prevents divers from using the hand signals taught in recreational classes.

A tested, reflexive line-signal communication system between divers and tenders is necessary for all ice divers, no matter what their activity, and for public safety divers to conduct effective searches. This is true even if electronic communication systems are used because such systems can fail, and, by Murphy's Law, the most likely time for them to fail is when communication is especially needed.

Among recreational, public safety, recreational technical, commercial, and military diving, there are a variety of signal systems being used. By far, the most proven signals are those used by commercial and military divers. The signal system presented in this book is a simpler version of those signals; it was modified to remove the signals that would not be needed by recreational or public safety divers and to add a few that might be needed. This system, Lifeguard Systems' line-signal system, has been tested and used internationally for more than thirty years by thousands of divers during tens of thousands of training dives and actual rescue/recovery operations. A few improvements were made over the years, so the signals are slightly modified from the original version first used in the late 1960s.

When using the term "simple", it is important to make a distinction between "keep it safely simple" and shortcutting. Many of the line-signal systems taught today have crossed the line to shortcutting, which can greatly hinder the ability of dive teams to find search objects, and can increase the risk of injury or even death to any kind of diver. Some of the problems with today's commonly used signal systems will be explained in this chapter. Find instructors certified in each system and take their training, then go out and test the different systems hands-on to choose the most effective system. Never take anything that you read or hear at face value.

Electronic Systems

The best option for communicating underwater is an electronic system, backed up by a line-signal system. When functioning properly, such a system reduces the chance of miscommunication and ensures that the tender always knows the status of the diver. Being able to talk to someone is also psychologically reassuring for the diver. Underwater communication systems are especially important where direct line access between the diver and tender could easily be lost, such as when diving under ice. Electronic systems add a greater margin of safety because overhead environment dives hinder the ability of tenders to visually monitor diver breathing rate and quality, and the electronic systems provide continuous auditory monitoring capability.

Electronic systems can be either hardwired or wireless. A hardwired system relies on a cable that physically connects the diver's gear to a unit on the surface. The system includes a microphone that is typically fitted into the diver's full face mask, two bone microphones that are affixed to mask straps against the diver's skull behind the ears, a hardwire tether line, and either a tender headset system or a speaker box.

We prefer the tender headset system rather than the speaker box for several reasons. First, few divers want the whole world to hear them breathing and to listen to what they have to say. Most divers would be more reluctant to admit that they are frightened or experiencing a minor problem if they know that everyone in the immediate area will hear them instead of just their tenders. Secondly, the other personnel in the immediate area have jobs to do and should not be distracted by what is coming out of a speaker box. If a diver is having a problem and sounds very stressed, the responding backup diver may pick up on the overheard stress. Stress breeds stress. The backup diver needs to respond with calm confidence.

If a speaker box is used and other surface support personnel are talking near the box, the tender may have difficulty hearing the diver, unless the tender is wearing a headset that is attached to the speaker box. This solution creates a problem; for the tender to move around the hole, someone has to carry the speaker box and follow the tender. Speaker boxes are more expensive and much more cumbersome than tender headsets. Boxes make it more difficult for tenders to move around the ice hole to keep their divers' tether lines in direct line access even without an attached headset. If the ice is weak during public safety diving operations and tenders are working from platforms, such as inflatable boats, speaker boxes can get in the way.

Since public safety ice dive operations use solo divers, and recreational divers put a maximum of two divers in a hole at a time, there is no advantage in being able to plug multiple divers into a speaker box. Headset systems do allow two divers to be connected to one tender.

Hardwired systems generally provide clear, strong sound and are usually less expensive than wireless systems. Since the divers are tethered anyway, teams might as well work with the system that provides the clearest communication. Hardwired systems are easier to use and set up than wireless systems. In our experience, hardwire systems give far less trouble than wireless systems.

A wireless system uses sound waves to transmit signals through the water. Although some types transmit and receive signals well, wireless systems can be prone to background noise, such as that caused by boat engines, bubbles emitted by the diver, and air passing through the diver's regulator. Acoustical considerations also come into play, since ships' hulls, rocks, and concrete piers can reflect and distort signals.

Both wireless and hardwired systems require either a full face mask or an oral cup to enable the diver to speak. Before using either type of setup, be sure that you are trained in its use and emergency procedures.

A four-wire system is recommended because it is a full duplex communication system. This means that both the diver and tender can talk at the same time. This is better than a VOX system, which stands for Voice Operational Transmission. As John Hott of OTS™ explains, "X is an abbreviation for trans/mission-ceiver-mitter, *i.e.*, xceiver." Your voice, or any sound you make, keys the microphone in a VOX system. That is good. What is not good is that VOX is only half duplex, which means that only one person can talk or make sounds at a time. So, for example, if a diver is very stressed and is vocalizing continuously, he can prevent his tender from talking to him. Push-to-talk (PTT) is another system that we do not recommend. PTT is half duplex and it requires divers to push a button on the microphone at their mouth to talk. The four-wire system allows divers to talk while keeping hands free. For example, when divers are spidering, a method of crawling on the underside of the ice roof for rapid movement to a target area, push-to-talk systems slow down or fully stop divers whenever they are talking, since they need two hands to pull their way straight across the ice ceiling. Public safety divers using four-wire systems do not lose the ability to search continuously with two hands. Another advantage is that tenders can hear all sounds made by a diver, even if the diver would not normally push a button. This can help a tender to hear a diver's breathing, which could warn of possible trouble.

When using a communication system, it is easy to tell a diver to "go left" or "go right", but it is far wiser to say, "go three" or "go four", since doing so will reinforce the use of basic line signals. If the technology fails at any point, the diver and tender can still resort to using line signals. Their proficiency will depend on their experience and how recently they have practiced. Divers and tenders who use normal language instead of verbalizing line-pull signals often will abort a dive if their electronic system fails. If, on the other hand, verbal line-pull signals are used, when a communication system goes down, it is simple to switch to physical line-pull signals. And what if the dive cannot be aborted because the diver is entangled on the bottom or entrapped between the ice roof and a shallow bottom? Without strong line-signal system proficiency, a small problem can quickly escalate into a large one.

When divers and tenders are first learning to perform tethered diving, they should use physical line-pull signals; then, when they reach proficiency, they can use electronic communication systems. Although it is better if divers have tethered diving experience prior to making their first ice dives, this is not always the case. If not, an electronic communication system can be used to monitor diver breathing while physical line-pull signals are used for diver-tender communication. We have observed that public safety divers who have used electronic communication systems for most or all of their tethered diving tend to be sloppier about keeping their lines taut because they are used to the tender frequently telling them, "Tighten up your line." Starting out with physical line-pull signals, and then every six or seven dives reverting back to line-pull signals, will reduce sloppiness and will prepare divers to safely manage communication system failure.

During search, mapping, or other task-oriented dives, try to talk as little as possible. Sometimes tenders like to start up conversations with divers while their divers are working. This is distracting for divers who are trying to concentrate on a search in blackwater, for example. If a diver is moving quickly from one place to another to reach a victim's point of entry, talking may increase the chances that the diver becomes out of breath. During recreational dives with a buddy system, divers do not need to concentrate simultaneously on maintaining visual communication with their underwater buddy and listening to a talkative topside tender. Likewise, tenders need to maintain their strict attention on tether lines and diver breathing, which may be compromised if part of their attention is focused on discussing plans for the evening. Lastly, divers who are used to lots of talking are more likely to be stressed should the communication system fail and all goes silent.

The exception to this recommendation is when a diver is nervous. Then gentle or even humorous conversation can make a big difference in improving the diver's mental comfort level. If a diver initiates more than average talking, this may be a sign of diver nervousness. If tenders initiate the conversations, this sign could be missed.

Tips for using communication systems

- Speak slowly and clearly. Tenders especially need to speak slowly so that divers do not unconsciously hold their breath to avoid missing words.
- Verbalize line signals. For example, tenders say "4" and "3" instead of "left" and "right," "1" instead of "stop/face the line/take up slack" and "4 plus 4" instead of "time to surface."
- Divers and tenders return all signals as they would when using line-pull signals. For example, a diver says "6 plus 6" to indicate "search object found," and the tender returns the signal by saying "6 plus 6."
- Avoid unnecessary talking.
- Have tenders get into the habit of moving away, or covering, their microphones when talking to anyone other than the diver.
- If tender headsets have two earpieces, the tender should wear one over one ear to hear the diver and lift the other earpiece off and above the ear to hear shore personnel ask questions or give a command.
- Put hardwire communication line in a backpack for the tender to wear. This provides tenders with complete, unhindered mobility, which is particularly important for moving around the hole in ice diving.

Line-Pull Signals

A 3/8in (10mm) multifilament line is the most effective line for sending and receiving signals. In good conditions, a signal can be sent and received 150ft (46m). A taut line is required for signals to transfer across the line. In search diving, namely tender-directed diving, it is the diver's job to keep the line taut; in recreational diver-directed diving, the tenders have to take up the slack when divers come in.

Proper signal sending is a skill that takes practice to attain proficiency. All too often signals are: too soft, too rough, or too quick. Too soft, or light, a signal does not transfer effectively through the line, while too strong, or hard, a signal pulls on the diver, physically moving them in the water column. This movement not only physically disturbs the diver, but also ruins any search or swim pattern. Too strong a signal is a major contributor to search patterns falling apart; one minute the tender has the diver at 80ft, and the next, after signals are given, the diver is at 78ft Due to the lack of line angle, a diver who is close to the tender will be pulled off the bottom if the tender is strong. Hard pulls are also bad for spidering divers or divers in mid-water because they cannot stabilize themselves on the bottom. Tenders need to realize that a hard pull is not necessarily easier to feel than a proper pull.

Signal sending is sometimes referred to as fishing for the diver. Give a signal with a wrist flick, not with an arm pull. Snap the wrist up as if you were setting a fishhook in a fish's mouth. The line should be taut enough for the tender to feel almost every movement of the diver. Water current and other conditions have a direct effect on how sensitive this feeling is. The tender's hand should not have to move more than 4in to 6in to transfer an adequate signal in a non-current environment. If the diver is maintaining a properly taut line, the return signal need not be much more forceful than that of the tender. With light currents, the tender's hand movement may need to be a little more aggressive; a 6in to 8in movement should be adequate. Line length can also affect signal transfer. The line may have to be flicked a bit harder for a diver out 100ft (30m) than for a diver out 20ft (6m), but since most ice divers will not be 100ft (30m) from the hole, this is less important.

Transfer your signals with a slight pause between them to facilitate better understanding on both sides. When deciding how long to make the pause, consider how long it will take the signal to travel the distance of the line. This is one of the things learned in land drill practice sessions. Long pulls combined with fast repetition are only going to confuse both sides.

It would be helpful for recreational divers to learn line-pull signals. Although public safety divers use them more often, recreational divers may need to be directed in blackwater should any of the following occur: they accidentally silt-out the bottom; they must assist another diver in a silt-out; they must perform a search for missing gear or for a disconnected diver; or they must be directed by the tender to re-establish direct line access (a line with no bends or obstructions). Both recreational and public safety ice divers can safely and effectively use the same signal system.

The signal system used in this book provides the diver with the orientation necessary to maintain an effective blackwater search? For divers who have not experienced a black silt-out, it is important to be mentally prepared. When you believe the water is as black as it can get and suddenly the unthinkable happens—it becomes even blacker—you know that something has happened.

The key to blackwater and low-visibility diving is to stay oriented; if the brain cannot maintain a logical sequence of events and locations, it becomes disoriented. Think of the mind's eye as the connector and coordinator of information perceived by our five senses. It is the portion of the brain continually integrating our external perceptions and internal data into logical order and orientation.

In the low-visibility, weightless medium of water, there is a loss or partial loss of our sight, sound, smell, and touch that increases the likelihood of disorientation. When divers do not know where they are, they become confused, uncomfortable, stressed, and perhaps even panicked. Disorientation rarely leads to a job well done or a fun dive.

In blackwater diving, the mind will attempt to enhance the senses that are still active to become centered, focused, and oriented. Blackwater divers need always to know where they are in relationship to their tender and the hole. To do this, the brain seeks a standard orientation of left, right, up, down, forward and backward. It seeks this logic constantly. If the information cannot be obtained, disorientation will most likely occur.

To prevent disorientation in low-visibility, tethered diving, line signals must provide the diver these standard orientations. One of the most common problems we see in public safety diving worldwide is tenders who disrupt a diver's orientation by issuing a "change direction" signal. Without "stop and face the line", "go left", and "go right" signals, divers will lose the mind's eyesight, which tells them where they are. When this happens, the mind cannot help but focus its energy on regaining orientation, thereby reducing search effectiveness and mental comfort.

Many divers use a "change direction" signal instead of a left and right signal. Because this is so commonplace, it is an important issue to address. The "change direction" signal system lacks instant orientation for the mind's eye and will in a short time foster confusion. That is why we sometimes see divers who use the "change direction" signal surface and state, "I just wanted to see where I was" or "How am I doing?" Perhaps not even consciously knowing why, they surface to reorient themselves and regain their bearing, because their mind's eye is confused. Without the "stop/face the line/take up slack/are you ok" signal before each "go right/left" signal, tether lines often become slack and require a "take up the slack" signal. This happens because the physical act of turning often creates slack. If the line is slack and the diver cannot feel the "take up the slack" signal, the tender will have to take up the slack and will destroy the search pattern. This process starts as a mistake to fix a mistake and results in more mistakes.

These are important insights into why public safety divers too often miss what they are searching for. A good search requires a confident, comfortable, oriented diver with a taut line who dedicates the entire functioning of the mind's eye to the search, not to a conscious or unconscious "where am I, help, I feel disoriented," frame of mind. When divers maintain a taut line, the angle of the tether line provides a simple logical orientation to the tender and to home.

Every industry encountering low visibility has some type of left and right signal system, except many public safety dive teams and recreational divers. Think of airplane pilots, underwater rover pilots, commercial divers, military divers, and confined-space rescuers. Imagine what would happen in the air if pilots above the cloud cover were only told to "change direction". How well would rovers find anything without left and right lever directions? Picture what would happen to confined-space rescuers if they lost their mind's eyesight of the entry point. The same consequences hold true for visibility-impaired divers.

As was discussed in chapter six, many divers are permitted, or are actually taught, to hold the line with one hand during the search pattern. These divers are often taught to put a hand loop in the line to maintain their hold. This is done for two reasons, both of which provide evidence of the ineffectiveness of a "change direction" signal system. The first reason is that without left/right orientation signals, many divers feel a conscious or unconscious need to hold onto the line, for orientation and comfort. In an attempt to gain an orientation to the tender, divers feel the need to hold onto the line because the signal system does not refer to the line. By holding the line, these divers are giving themselves a right and left orientation. A left and right signal system does both these things and allows divers to have both hands free in addition to other advantages. "Right" is to the right of the line, therefore, I know that the tender and home are to my left. Conversely, when I turn left, I know that home is to my right.

The second reason why hand-loop holds are used is because divers must have them to effectively give and receive signals when lines are not sufficiently taut. Because "change direction" systems do not use the "stop and face the line" signal, slack is created when turns are made and the signal system does not automatically take it up. The act of giving "1" prior to every other signal not only provides diver orientation, but prevents slack in the line.

Without left and right signals, divers cannot be stopped for standby, then restarted again in the same direction, rather they have to be told "change direction" and then "change direction"—talk about confusing and inefficient! Such divers have less sense of where objects lie within the search pattern, and as a result their comments to profilers later are of limited value in creating accurate profile maps.

A diver in a proper harness diving a taut line will clearly feel signals through a maximum of 150ft (46m) of line without holding or touching the line. There is no reason to hold the line other than helping the mind's eye gain orientation, which the "1", "3", "4" signals provide. Besides losing 50% of their arm-searching capability, line-holding divers create slack in their lines and can lose hand function quicker than divers searching with two hands. In cold water diving, divers should avoid holding anything, because hand circulation will be restricted.

Divers who hold onto their lines swim perpendicular to them. If they swam away from the lines at a 45° angle, their arm would be straight out behind them, and their arm would be pulled out of its socket. Divers who are oriented perpendicular to their lines cannot use their bodies to keep the line taut; rather, they can only maintain tension by pulling the line with their arm.

A perpendicular orientation takes a great deal of muscle work, particularly for longer line lengths, but it also requires divers to be over-weighted to keep their bodies in place as they pull hard on the line. If they were not over-weighted, pulling on the line would only serve to pull their bodies toward the line. After working with literally thousands of divers who were first taught to use hand-loops in the line, we can report that they are typically over-weighted by six or more pounds, which is a significant safety hazard. In addition, a perpendicular position keeps the line away from the diver's body so it cannot be felt as easily, hence the diver feels a need to physically hold it with a hand.

Another problem with hand-hold loops is that anytime the divers bend their arms they are changing their position in the pattern. Most divers bend their arms when transferring the line from one hand to the other during a direction change, which means that a pattern's accuracy can be eroded by 6in to 12in (15cm to 30cm) on every sweep!

In summary, visibility-impaired divers need an orientation as to where they are to remain mentally comfortable and to perform accurate and effective searches. All of a diver's orientation comes from the tether line that leads directly from the diver to the tender and home.

To mentally know where the line is and how it is angled, divers should use a signal system that continuously orients their minds to where their bodies are in relationship to their tether line and that helps keep the line taut. Hence "4" (left), "3" (right), and "1" (stop and face the line) signals are vitally important. The "change direction", "swim", "change direction", "swim", "change direction," "swim..." system has no intrinsic mechanism for providing orientation, mental comfort, or a taut line.

Opting to use a universal "change direction" signal is a mistake that simply causes more mistakes. In addition to creating diver orientation and pattern accuracy, the "right", "left", and "stop/face line" signals enable divers to accurately spider from the hole under the ice out to a target location, such as the victim's point-of-entry (VPOE) or a previously mapped object on the bottom. Without left and right signals, the spidering technique cannot be done. As a diver travels out and begins to veer slightly to the right, how can a correction "go left" be given with a signal system that only has a "change direction" signal? It can't. Also, vertical (in-out) box searches cannot be conducted. Vertical box searches are conducted when there are obstructions such as high weeds or trees that prevent side-to-side sweeps (the obstructions catch on the tether line). In a vertical box search, a diver is sent straight out from the tender along the bottom to a designated point. The diver is then asked to ascend to the surface/ice, move over two feet to the "right" (or "left"), and drop back down. The tender likewise moves two feet to the "right". The diver then proceeds to search back in a straight line toward the tender as the tender gently takes up the slack. When the diver reaches a designated point from the tender, the tender tells the diver to ascend, move over two feet to the right, descend, and the process is repeated. This very effective and useful search pattern cannot be conducted without "3" (right) and "4" (left) signals.

Another consideration is tender orientation. Tenders need to call out their diver distances and directions so that their profiler can check the information and then document it. If tenders use a "change direction" system they themselves do not have an orientation, which means they have to mentally translate each "change direction" into a left or right before calling it aloud.

There is an occurrence we have observed on several occasions that proves the better efficiency of the left/right system. The scenario involves teams who have only received training with the "change direction" system, and who state that they do not want to even consider the "left/right" system because it is too complicated. These teams have electronic communication systems. Guess what verbal commands they give to their divers? "Go to your left...go right." When we observe this and ask them "so you do use a left/right signal system?", they reply "no, left/right signals are too complicated, we like to keep things simple." So of course our next question is, "but then why didn't you just tell your diver to 'change direction'?" The most common answer is, "because I wanted him to go left." The simple fact is that there is no more effort to verbally say, or receive, the word left than there is to give or receive a line signal of a "4". If it makes sense to do it with electronic communication systems, it makes equal sense to do it with line signals.

Basic line-pull signal rules include:

- All signals given are returned verbatim.
- All signals given and received by tenders are called aloud by tenders to profilers.
- All tender signals begin with a "1" pull ("stop/face the line/take up any slack/are you okay?"), which is returned by the diver before the tender gives the next signal.
- All signals are based on the diver's right and left with the diver facing the tender.

Signals are based on the diver's left and right because divers are the ones in the disorienting environment. By giving a "stop/face the line" (1) signal before every other signal, tenders always know which way their divers are facing, so a diver's right is a tender's left, and a diver's left is a tender's right. After a few dives this becomes reflexive to almost all tenders.

A few tenders may need to use a memory device or two if they commonly mix up their own left and right. For one, ask tenders to extend their left hands out in front of them with the palm facing out. Count from the pinky: one, two, three, four, left has four letters. Then look at the direction that the thumb is pointing—that is the direction of the diver's left. Another trick is to write a "4" on the back of the left glove with an arrow pointing to the diver's left, or the tender's thumb, then write "3" on the right glove with an arrow pointing toward the diver's right, again in the direction of the tender's thumb.

If by some chance the signal is misunderstood and the diver heads in the wrong direction, the tender should send a one-pull signal. The diver stops, faces the line, returns the "1", and receives another set of three pulls. The diver should return the signal and head in the correct direction. The key factor here is that, when a signal is misunderstood or a direction needs to be changed, the "1" is given to restore the mind's orientation to where the diver is in relationship to the directions.

Tender-to-diver signals

Any time a line signal is given, it must be acknowledged with a return of the same signal to ensure that no miscommunication takes place. When giving multiple signals, wait for acknowledgement of the first signal before sending the next one.

One pull = "Stop, face the line, take up the slack, and prepare for a new signal/ are you okay?" By stopping and facing the line, divers prepare to receive and respond to a command. Also, their mind's eye creates a mental picture of where the shore and the tender are in relationship to their location. They are in the direction that the tether line goes, they are facing home. This mental picture fosters instant orientation and provides an opportunity for divers to move back and take up any slack in the tether created by the turn.

Once divers have stopped, faced the line, and pulled it taut, divers signal that they are okay and ready by returning the one pull. They are then ready for the next signal.

In general, a single pull at any time from the tender means, "Are you okay?" and a single pull returned from the diver means, "I am okay". All line signals from the tender should be preceded by this one-pull signal.

Once divers have become proficient at tethered diving and keeping a taut line, then "1" gradually loses its "take up the slack" definition because divers automatically, reflexively take up any slack that may be created during a left or right turn, and because they do not allow slack to be created while they are swimming a sweep.

Three pulls = "Go to the diver's right." Divers receive the signal while still facing their lines and their tenders. They stick out their right arms and turn toward that arm until they are at a 45° angle to perpendicular. They should feel the line run down from their chest across to their left side. They then proceed to move to the right.

Four pulls = "Go to the diver's left." Same as for three pulls except that the turn is to the left.

Two plus two pulls = "Search the immediate area." To give this signal, tenders give two pulls in a reasonably quick succession, pause, and give two more. This signal is used when a diver is placed on a known object such as a tree, rockpile, or vehicle that needs to be thoroughly searched.

Three plus three pulls = "Stand by." This signal is used to tell the diver to stop and wait for another command; there may be a change in the operation. Perhaps new witness information has been received that changes the search area. The diver should stop on the bottom and wait for another command. This signal is also used when the target has been located and shore personnel need the diver to stand by while they clear out the family and media or make other preparations. When conducting multiple searches simultaneously, if one diver issues a help signal, all of the other divers should immediately be put on standby with a three plus three signal. It is also used to stop a heavily breathing diver and give the diver a chance to get breathing back under control.

Four plus four pulls = "Come up/surface." The diver is commanded to surface slowly and directly, as long as there are no obstructions or dangers at the surface. Tenders should time the diver's ascent to make sure that it was not faster than two seconds per foot.

Diver-to-tender signals

One pull = "I'm okay." With this signal, the diver is simply replying "I have stopped, faced my line, taken up any slack, and am ready for the next signal".

The one-pull signal can also be used as a "please repeat" by the diver. If divers respond with one pull after any signal instead of returning the given signal, the divers are simply stating, "I am okay, but give me that signal again."

Also, the diver can at any time send a signal of one pull notifying shore that all is okay. The tender simply returns the signal to say, "received". Remember, both diver and tender return all signals.

Two pulls = "Tender, make notation." The diver would like a specific area noted for debriefing.

Six plus six pulls = "Object found." This signal indicates that the diver has found the object of the search and is waiting for the signal to surface. Surfacing could be immediate or could be delayed slightly while command removes family and press from the direct scene. Tenders must remember to alert command discreetly. If the diver does not plan to surface with the found object, perhaps because it is part of a crime scene, the object should be marked with a buoy before the diver leaves it. The lengthiness of this signal helps give divers a chance to normalize their breathing rates before making an ascent.

Diver-to-tender contingency signals

Having covered routine signals, we next need to address diver contingency signals. It is very important to have three different signals to call for help instead of just one. Consider that police use various radio codes to signal a situation, that fire uses multi-alarm systems, and that EMS has different code responses. These systems are all designed to give responders an idea of the situation for which they are being called. What would happen if police were told only to respond at a location, if firefighters were told only that there was a fire, or if EMS were told just to show up? Responses would have to be all-out every time to be truly prepared for any situation. When is someone more likely to be injured, when responding to "officer down" or "help with equipment staging is needed"? There is a positive correlation between the severity of the request for help and the risk for the responding personnel.

Signal systems that have only one signal for help cause the same problem in public safety diving. In the majority of backup diver requests, public safety divers do not need more than minimal assistance. With a one-signal help system, a call for help must automatically be assumed to be life threatening. Such a response significantly increases everyone's stress level. Ironically, increased stress lessens the likelihood that the backup diver will reach the primary diver. Stress causes mistakes and allows ear trouble, equipment problems, and panic to occur and be compounded.

Signal systems with only one help signal also encourage divers to wait until a problem is serious before calling for assistance, and cause topside confusion when a diver has stopped moving to handle a problem. Because a diver cannot alert shore to prepare but not deploy a backup diver yet, because the diver is handling the problem for now, the one-signal system leaves shore personnel ignorant of a diver's situation. A multi-signal system alerts surface personnel if they need to prepare immediately for a possible serious problem or not (*e.g.*, alert EMS personnel, ready the transport device).

So, what is a well-proven contingency signal system that increases, rather than decreases, safety and success when assistance is needed?

2+2+2 pulls = "I'm okay, but have a problem: alert the backup diver." Since the backup diver is fully dressed, they should simply get into the water, snap the contingency line onto the primary diver's tether, and mentally prepare to deploy. However, do not deploy the backup diver until the primary diver requests it or if the primary diver stops responding to signals. At the same time, the 90%-ready diver should complete gearing up, put the mask in place, and move into a fully ready, backup status.

3+3+3 pulls = "I'm okay, but need help from the backup diver." The primary diver is not dying, but cannot get out of this situation alone. Most likely, the diver cannot reach an entanglement to cut it or perhaps has a leg cramp that needs stretching. The backup diver should be deployed and can move slowly down the line knowing this is not a life-or-death situation, but simply a problem that the primary diver needs help with. As the backup deploys, the 90%-ready diver prepares to take the backup diver's place, in case the backup diver cannot for some reason perform the task.

4+4+4+... pulls = "I need immediate help!" For this signal, divers should continuously give four pulls to indicate that they are still alive. The backup diver should deploy as quickly as possible, and the 90%-ready diver should prepare fully. The reason for continuous series of "4" is twofold. First, as long as the primary diver is giving a correct series, topside personnel know that the diver is alive and still functional. Second, it gives the diver something to concentrate on to keep them calm, like when expectant fathers are asked to go boil water. A "4" is a long enough signal to help with focusing without being too long, which can cause confusion.

When this signal is received, you, as the backup diver must be prepared. Have you really trained for this? Are you ready for possibly the one emergency in a lifetime? Practice self-rescue, diver rescue, and underwater cutting so that you will have a chance to save your fellow team member's life.

When multiple searches are conducted simultaneously and one diver gives a help signal, primary tenders always put the other divers on standby with a three plus three "stand by" signal. If the diver in trouble signals for immediate help, have other divers surface and return to shore. This procedure can prevent a situation in which two divers simultaneously need help and prepares the divers for possibly assisting the diver in need.

To avoid miscommunication, divers and tenders should review line signals periodically, such as monthly, especially if an underwater communication system is used. Without the regular practice of using line signals, divers on communication systems can forget what line signals mean. Tenders may also wish to make a "cheat sheet," a laminated, wallet-sized card with all of the tender-to-diver and diver-to-tender signals briefly listed. The card can be kept clipped to a PFD for quick reference. (See appendix A for line signal cards you may copy.)

If the diver fails to return a signal: three strikes, and you're out

Sometimes divers will fail to return a tender's signal. This could happen for a variety of reasons. It could be that divers are so intent on searching that they fail to notice a one-pull stop signal, or it could be that they are busy maneuvering around a tree or other tricky object and have not yet had the chance to respond. In any case, follow the three-strikes-and-you're-out rule: whenever a diver fails three times to respond to a signal, send in the backup diver. That way, if the primary diver has been somehow incapacitated, help is on the way.

Diver-to-diver contingency signals

Primary divers and backups must have a way of communicating underwater. Otherwise, the backup may not immediately understand what the problem is. Having a rehearsed plan helps lower the anxiety of the moment, and better allows the divers to solve the problem, get out, and go home.

After signaling for the backup diver, the primary should stop and put one hand on his/her harness carabiner. The primary diver's only job at the moment is to concentrate on relaxing and slowing the breathing rate with long, slow exhalations. The primary diver will be able to feel the backup coming down the tether and should be confident that help is on the way. On arrival, the backup knows where to find the primary diver's hand, since it is on the carabiner. The two divers should clasp hands so that the primary can indicate to the backup exactly what is wrong.

The divers use one or more of the following hand signals to communicate their needs.

Tapping the backup diver's hand to the primary diver's second stage. This signal means "I am already on my pony bottle and I need more air." Air is always the number one priority for any underwater emergency. If the diver's main tank has been exhausted, this signal should be given immediately. Once this signal is given, the backup diver should place the primary diver's hand on the first stage of his (the backup's) pony bottle. The primary will pull the bottle and put the mouthpiece in his mouth. The question is often asked, "Why not put the diver's hand on the pony mouthpiece?" The reason is that second stages often get disconnected from the backup's BCD, so the primary might be unable to find it. The first stage, however, is always in the same place, and the hose of the first stage will always lead to the second stage.

The backup diver places the primary's hand back on the carabiner and gives three squeezes. This signal means, "I am leaving, but I am coming right back." This signal should immediately follow the handing over of the pony bottle. Air is still the immediate concern, so the backup needs to get another pony, as well as a full tank for the primary. After returning, the backup should clip the contingency tank to the tether line so that it cannot be lost, but on the way down he should handle its weight.

Large circular motion. This signal is used to show the backup diver where an entanglement is. The primary diver takes the backup's hand, traces a large circle with it, and then places that hand on the problem area. Locating it for the backup this way helps the backup avoid getting snared on the problem.

Tapping the backup's hand on the primary's chest. This signal is used in the event of injury. The primary taps the backup's hand on their chest and then guides that hand to the affected area. The backup diver must move slowly and gently to prevent further injury.

Surface Signals

Tender and diver

While on the surface, the diver and tender can communicate using hand signals, although line signals may also be used. Tenders should remember that at night, divers will have trouble seeing hand signals; in fact, tenders may wish to light themselves from the front. Tenders should not shine lights on divers, though, because to do so ruins their night vision and blinds them. While in the water, divers should not communicate verbally unless they are using communications gear, because they will have to remove their regulators to do so.

If the tender wants the diver to move, the tender should point in the appropriate direction. Before divers swim out, and after they return, tenders should signal, "What is your tank pressure?" This signal consists of tapping the open palm of one hand with two fingers of the other hand. The diver should report the pressure with one hand, with each finger representing 100psi.

Divers should give an "okay" signal (one hand touching the top of the head) upon surfacing, unless they are unable to do so because they are holding onto a recovered victim or other object. If a diver surfaces and fails to give the "okay" signal, the back-up diver should be sent out to assist.

Having all personnel on a site be familiar with all signals will facilitate teamwork and a smoother operation.

Tender and shore

Signal	Hand	Whistle
STOP	Arm straight up with a fist	One blow on the whistle
TAKE LINE	Arm straight up, and make a circle	Two blows on the whistle
GIVE LINE	Arm out to the side, up and down	Three blows on the whistle
DIVER HAS OBJECT	Arm straight up and down, similar to a pumping action	Three sets of two blows on the whistle

Table 12-1 Tender Hand or Whistle Signals

Line signal cards

Purchasers of this book have full permission to make copies of this page. We suggest that you cut out the cards, fold them over, and laminate them so that they can be read front and back. The laminated cards can be clipped to tenders' PFDs to ensure that line signals are sent and interpreted correctly. Full-color, laminated cards are also available from Lifeguard Systems.

Line-Pull Signal Overview

The following lists provide standard line-pull signals for dives all year round.

Tender-to-diver signals

1	Okay? Stop, face line, take up slack.
3	Go to diver's right.
4	Go to diver's left.
2 + 2	Search immediate area.
3 + 3	Stand by.
4 + 4	Ascend slowly.

Diver-to-tender signals

1	Diver is okay.
2	Tender, make note.
2 + 2 + 2	Tangled or some other minor problem but okay; alert backup diver.
3 + 3 + 3	Okay, but need help from backup diver.
4 + 4 + 4...	(Continuous signal) Need immediate help.
6 + 6	Found object.

Specialized line signals

For recreational ice diving that is diver-directed:

Tender-to-Diver

2 Resume your activity.

For public safety and recreational ice diving when communication systems are not available:

Tender-to-Diver

2 Give signal on every inhalation until a stop "1" is given.

Diver-to-tender (afterthe signal above)

1 I just inhaled. (This signal is not returned by the tender.)

Summary Questions

1. What are the most proven and tested tender-diver signal systems?
2. What is the best option for tender-diver communication?
3. List at least six advantages of a tender headset compared to a speaker box.
4. ______ (VOX, push-to-talk, four-wire system) systems are preferred.
5. When using a communication system, tenders should say "_____" (4, left) and "______" (3, right).
6. An effective way to store and carry hardwire communication line is in a ________.
7. Give line signals with a ____ flick.
8. The mind's eye continuously seeks an ______ of where we are, requiring an understanding of ____, _____, up, down, backward, and forward.
9. Explain why a "change direction" signal shortcuts a "stop/face line and left/right" system.
10. Describe the diver's position when performing tender-directed diving.
11. Give the three basic line-pull signal rules, aside from "do not pull too hard".
12. A "1" pull from the tender means what?
13. Give line signals for the following tender commands: left, right, search the immediate area, standby, surface.
14. What do the following diver-to-tender signals mean: 1, 2, 6+6, 2+2+2, 3+3+3, 4+4+4?
15. Explain why a three-contingency signal system is safer than a single-signal system.
16. The primary diver taps the backup diver's hand on the primary diver's mouthpiece. What does this mean?
17. The backup diver needs to leave the primary diver to get more air. What is the signal?
18. How does one diver tell another diver that he/she is entangled?
19. The primary diver taps the backup diver's hand on the primary diver's chest. What does this mean?

Chapter 13
Public Safety Diving Operations

Public safety diving teams do not have the luxury of planning where they will dive and nor do they typically have options for when. The first step is to pre-plan what types of operations the team wants to be able to perform. The next step is to obtain the necessary training to conduct these operations. The third step is to write department Standard Operating Procedures and Guidelines (SOP/SOG); it is important to state the "no-go" situations in the team's SOP/SOG. Educate the public on the team's capabilities and limitations and give the reasons why certain operations have to be no-go. If some of these no-go operations could be safely conducted with additional training and equipment, and the public would like public safety diving teams to respond to these kinds of situations, the public needs to provide the department with the necessary financial support.

The first step in pre-planning is to accomplish the awareness-level duties described in chapter 2 to establish the incident command system. These include:

- Perform a scene size-up.
- Assess risks and benefits (rescue or recovery?).
- Secure the scene.
- Set up scene staging areas.
- Assess available resources.
- Dispatch for needed resources.

- Secure and interview witnesses.
- Draw a scene profile map.
- Mark the shoreline in front of the believed VPOE(s).
- Set up an accountability system for personnel in the warm and hot zones.

The incident commander determines whether the operation will be a go or no-go. If it is the latter, make sure to have an effective and experienced media liaison officer ready to move into action. Always remember that the most important question to ask in this decision-making process is, "if it goes bad, are we sure we can get everyone out?"

Photo 13-1 Especially in a rescue mode, do not let a primary diver descend until the backup and 90%-ready divers are ready in place. This is one of the most common standards to be broken in rescue modes.

Incident command needs to develop an appropriate plan for what to do after a "go". If the operation is a go, the incident commander, along with a technician-level advisor and safety officer, develop a plan of action. This committee has to also decide whether the operation will be in the rescue or recovery mode. Unfortunately, many public safety diving teams have the mindset that a rescue mode means that safety standards can be stretched or even broken because the potential benefits are high. The opposite is true. Safety standards need to be exceptionally, stringently followed during rescue modes because that is when personnel are most likely to become injured as they move with more adrenaline and more of their minds are on the victim rather than on themselves or each other.

The main difference between rescue and recovery modes is timing. If the time is 0200, an icy rain is falling, and a recovery mode was designated, the team should decide to perform the operation in the morning, when conditions are safer and less difficult. An exception to this might be a possible crime scene, in which waiting would mean scene contamination or lost evidence. Officers, team members, the media, and families need to understand that the submerged victim is dead, and even with a very rapid response time of less than ten minutes, there is very little chance that the victim will be resuscitated. There is even less possibility that the victim will stay alive with significant quality of life. Yes, there are long-term drowning victims walking around today leading normal lives, but they are very rare. Whenever an operation can be performed in a rescue mode safely, that option should be chosen. All involved, however, need to understand the one mission that must be accomplished every time—when the job is done, divers and tenders need to go home. If that mission is compromised, all of us fail. We cannot stress that often enough.

Some departments state in their SOPs that they will not conduct an operation if the ice is less than four inches thick. These teams are by default strict recovery teams, because the ice is often that thin if a person, especially a child, has fallen through it. These teams document the VPOE on a profile map and wait until the ice is thicker or until it disappears in the spring to perform the body recovery. This type of SOP decision usually results from a diver on the team being injured or killed during an ice diving operation, or the team being sued for not successfully rescuing a drowning victim. If the team SOP is strictly recovery-based, then operating in the rescue mode is not one of their duties or responsibilities. They cannot be held liable for not rescuing an ice drowning victim. But these teams need to keep in mind that if one of their own falls in, they will be in rescue mode. And what will happen if by coincidence a person falls through weaker ice near where the team is training? Will they refuse to respond? They need to have a pre-planned answer to that question, because it is not rare for a team to experience a real call while they are drilling or training.[1]

And what happens when the body or the search object is part of a crime scene? Fingerprints on a weapon will not last much longer than ten days; wounds on a homicide victim may no longer be present when a body is recovered two months later. These factors must be considered in the risk/benefit analysis. Teams can be trained to operate safely on thin ice, but what other hazards may exist? Is the water moving faster than a half knot? If it is and the ice is thin, then "no-go" would be a wise decision. Is the ice entirely separated from the shoreline? If it is and there is minimal wind, the operation can be conducted safely. If, however, there is a strong wind, the risks rise as the ice cover moves.

An important step in the plan of action is to designate who will determine where the hot-spot search area is, or where the victim is most likely to be found. If there are multiple victims, there may be multiple search areas. If that is the case, triage will be necessary if there is not enough equipment or personnel to search each hot spot simultaneously. For example, a person who is on the bottom because of suicide should be of lower priority than the people who are on the bottom because they attempted to save the suicide victim. The rationale for this is that even if suicide victims are resuscitated, their lack of will to live may decrease the chance that they will ultimately survive.[2] Which victim has been submerged the longest? Which victim may have the most trauma? Triage may also involve accessibility. For example, if one victim is submerged 200 feet closer to shore than another victim, it may make more sense to recover the closer victim first.

Where Is the Body Most Likely Going To Be?

Photo 13-2 If the VPOE is small and within 150ft (46m) of shore, a victim recovery can be conducted safely in less than fifteen minutes from the time of arrival. With practice, it can take fewer than twelve minutes with no shortcutting of standards.

Ice gives us a once-in-a-lifetime-guaranteed point-of-entry, so take advantage of it whenever possible. The victim-point-of-entry (VPOE) is the location where the victim punctured or fell through the ice. There may be more than one VPOE, since some victims get themselves out and fall through again. Three multiple immersions seem to commonly occur before the victim can no longer get out. If there are no witnesses, it is difficult to determine which hole is the most likely victim-point-of-descent (VPOD). The VPOD is sometimes the last-seen point given by witnesses. However, sometimes there are no witnesses and therefore no last-seen point. In that case, a designated team member must compile information to figure out the most likely VPOD among multiple holes or in an area of open water.

Ascertaining the victims' destination is an important issue in this assessment. Were they going home by way of a short cut, or were they simply playing on the ice? If they were going home or somewhere special, there is a reasonable chance that they got out and kept going in that direction. If they were playing on the ice, there is a reasonable chance that they went back in the "safe" direction that they came from. If the ice was safe enough to get them there, they often believe it will be their best way out. Were they skating? Where are their belongings? Many people have a tendency to return to their belongings, even if that area is not the shortest distance to shore, and consequently are drowned. Look closely at the broken ice where they climbed out; this may show the direction they intended to go.

Night operations require careful searching to make sure there are no missed holes that may be evidence of additional victims. Do not assume that at night, victims headed toward the closest shore because they may not have been able to see the shore if there was no moonlight or artificial lighting. Rather, the only visible light may have been from a house on the farthest shore, toward which the victim headed.

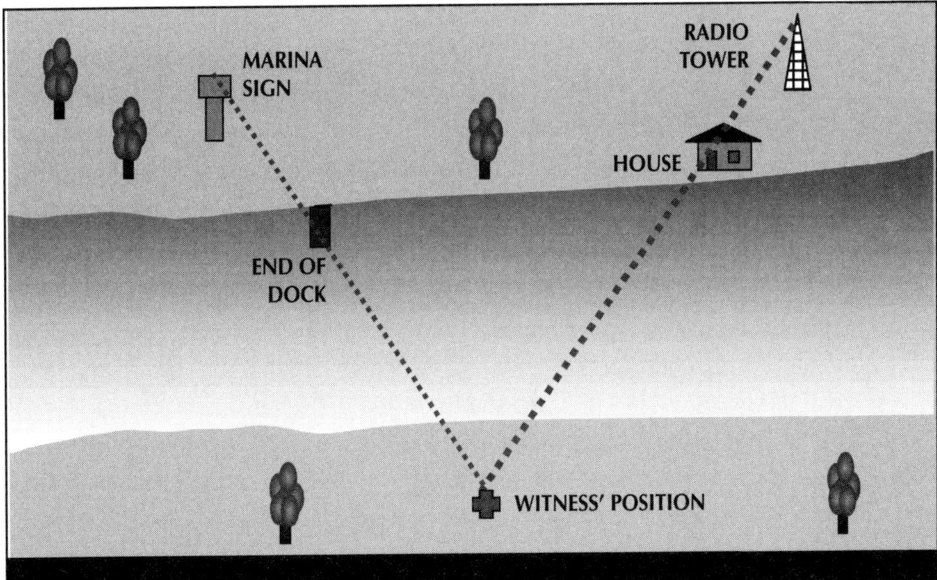

Figure 13-1 Ranges are found by lining up two sets of objects on the far shore. In this case, by lining up the end of a dock with a marina sign and a red-roofed house with a radio tower, a witness was able to return to the exact position from which he saw a boy drown.

If the VPOE is small, the VPOD is obvious. Sometimes though, the victim, would-be rescuers, or even public safety responders destroy the ice to such a degree that the VPOE is barely recognizable, with the result that the VPOD may not be evident. This can hamper a rapid recovery. Train awareness personnel to immediately draw two or more ranges (the lining up of two objects one behind the other) on a profile map to document the location of holes in the ice. If ranges are effectively drawn, anyone should be able to use the map to figure out exactly where the recorder stood when drawing the map and exactly where the hole locations are. If the holes no longer exist because the ice has been destroyed, the locations can still be determined from the map.

If the VPOE is unknown due to ice destruction, witnesses (if they exist) should be effectively interviewed and a survey should be made of the scene. Searchers need to look for any clues that might indicate where the victim was coming from and going. See *Public Safety Diving and Homicide by Drowning* for more information on witness interviewing.[3,4] The key process is to have the witness "show me" a physical reenactment of what happened prior to, during, and immediately after the drowning event. Physical demonstrations are typically much more accurate and valuable than verbal statements. Observe what the witness demonstrates and see if it makes logical sense. If not, you may want to listen to a different witness, ask for another reenactment and perhaps question the witness with the illogical scenario.

A witness's written statement can also be an invaluable tool for many reasons. One reason is that most homicidal drowning incidents are investigated in hindsight, which means that valuable scene and initial witness evidence is lost. Having written witness statements on file can be invaluable to investigators, particularly if they have statement analysis training. Taking photographs of the scene and of people around the scene can also be helpful. Someone in the crowd of bystanders could have been involved in the drowning, so those photographs could be very useful for the law enforcement department investigating the incident.

Children can make very useful witnesses and should not be dismissed. If children were involved, reassure them that telling the truth is what a hero would do. Some children are afraid because they may have taken their friend or little sister on the ice and now that child is on the bottom. They may feel that they are to blame or that they will get into trouble for telling the truth. If a child has fallen through, be sure to find out from parents if other playmates could have been with the child. These children could also be on the bottom or perhaps hiding under a bed after witnessing the incident. The former children need to be recovered and the latter need counseling.

Once the most likely VPOD is established, the next step is to determine whether there is any water current. This can be done in several ways. A neutrally weighted ball attached to 50ft of fishing line secured at one end to a smooth-running reel can be dropped into the water. How long it takes the line to be fully pulled off the reel is timed. Current is reported in knots.[5] There are approximately 6,000ft in a nautical mile. A knot then is equal to 6,000ft divided by sixty minutes, or 100ft/m. If the ball takes one minute to travel 50ft, the water is moving at .5knot, which is fast for ice diving. This technique will also identify the direction of the current. Do not use a positively buoyant ball because it is more likely to become trapped or slowed at the ice roof.

In a still water environment, the body will almost always be within a circle whose radius is equal to the depth and whose center is the point of descent (VPOD). That is, if the depth is forty feet, the hot spot to search would be a circle with a radius of forty feet from the

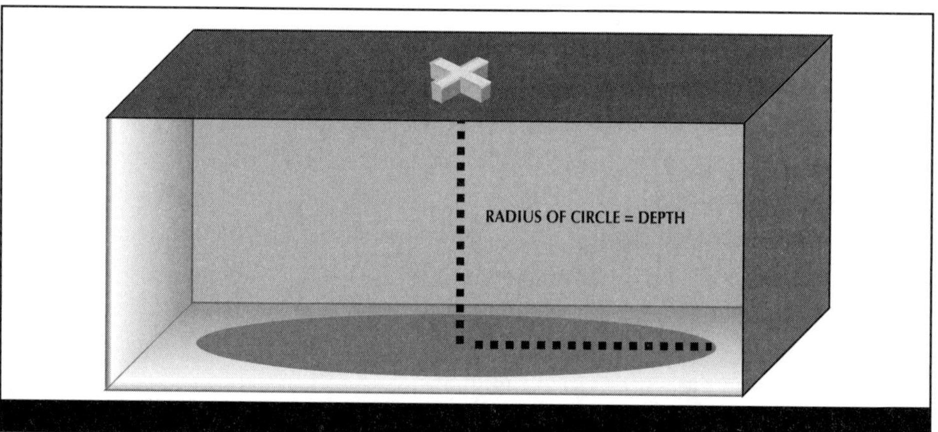

Figure 13-2 In still water, the rule-of-thumb is the hot zone is a circle with the radius equal to the depth.

VPOD. Often, the victim will be in a circle with a radius of half the depth, which would be a twenty foot radius in the example, but it is better to assume the larger hot spot when planning a search. Keep in mind that typically victims let go of the ice roof and drift at the water's surface in their final moments of drowning. Knowledge about current and wind are important. Wind can affect where the victim is pushed while at the surface before drowning. Current affects the victim at the surface and below it as well.

Overhead obstruction environments also may affect the victim's final location on the bottom in relation to the VPOD. A victim who falls through and immediately ends up under the ice roof might frantically search under the ice to find the hole. Not many people know that the hole is the dark spot, and therefore they are more likely to head toward light spots that may take them farther away from the hole. The average adult drowning process takes sixty seconds, and the average child drowning lasts twenty seconds.[6] Consequently, an adult could end up more than twenty or thirty feet from the hole. Even if the water is very shallow, do not doubt the veracity of the assumed VPOD until a large area around the VPOD is searched. If witnesses watched the victim hang onto the edge of the ice roof and then go through the classic drowning processes in the hole before descending, the victim is probably on the bottom within half the depth of the VPOD in a no-current situation. It is doubtful that such a victim would have had the ability or mental capacity to search for the hole. When they submerged, they would most likely have been unconscious.

Public safety divers have reported incidents in which children were found suspended just below the ice roof following a rapid response time.[7] In one case, an infant had been floating in a winter papoose-like outfit at the water's surface for over twenty minutes in an open water situation. The infant died of hypothermia, not drowning.[8] The children in these cases may have been floating because air became trapped in their snowsuits or because babies usually have a low muscle-to-fat ratio. A circle search under the ice may be warranted for a child victim if the response time is very short. Perform an under-the-ice-roof search prior to the bottom search. If the under-ice search is performed after the bottom search, the bottom may need to be searched again, because the child might have descended while the diver was coming up to search under the ice roof.

If there is a current, it is very unlikely that the victim will be below the VPOD. As a body drops through the water column, it is pushed downstream until it lands on the bottom. Unless the water is very fast (more than 6knot or 7knot) and very shallow (less than 5ft), the body will lie on the bottom until enough decomposition gases are created to lift it up. The body will then typically drag downstream along the bottom until enough gases are created to raise it to the surface. If the water is deeper than 75ft or is very cold, the body may never rise. This means that recovery teams who wait until spring to recover a body may find it right where it landed.

If there is a current, a calculation needs to be made to estimate where the current might have pushed the body while it was descending. A body in fresh water typically drops at a rate of 1.5ft/s to 2ft/s, and in salt water generally has a 1ft/s to 1.5ft/s drop

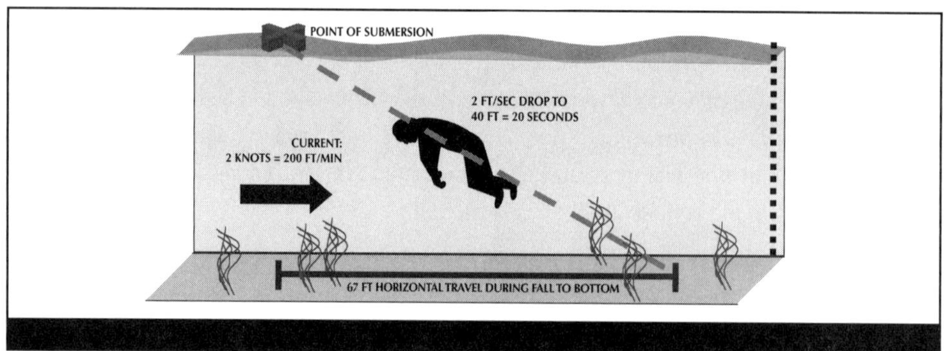

Figure 13-3 When performing the where-is-the-body calculations, always draw the picture of the body's journey, rather than simply putting numbers on paper.

rate. For example, a body drops to a 40ft depth in approximately twenty seconds in fresh water (1/3 of a minute). If the body dropped to its 40ft depth in a 2knot current, it would have traveled 200ft/m × 1/3min for a distance of about 67ft before it reached the bottom.

Calculate the time it takes to reach the bottom in minutes. Multiply that time by the speed of the current and the answer will be the distance traveled from the VPOD to the bottom location. This rule-of-thumb has worked well for teams all over the world for many types of victims wearing various types of clothing. Be sure to draw the scene when performing the where-is-the-body calculations, rather than just writing down the numbers. Drawing the journey of the body from the surface to the bottom can sometimes trigger a realization that something in the witness accounts does not make sense, or that an environmental factor is being ignored. This drawing does not have to be done by an artist; stick figures work well.

Photo 13-3 It may be easier and quicker to spider the primary diver from a hole closer to shore to the VPOE, than it is to transport and stage the necessary divers and tenders at the VPOE.

Once the hot-spot search area is determined, the next step in the plan of action is to decide where the divers will enter the water to reach the hot spot. Should they use the VPOD hole or a hole closer to shore? A closer hole might be located at a point where personnel have unintentionally broken through the ice.

When rescue personnel break through the ice, command must decide whether to continue moving tenders and divers over the weak ice to the VPOD or send in divers from the new hole. In many cases, it takes less time and effort to send the primary diver in the first hole and spider him under the ice to the VPOE than to transport and stage the tenders and contingency divers at the VPOE. If the ice is known to be thin, the plan of action may be to intentionally create a hole to put divers under the ice at a location closer to shore than the VPOD.

Photo 13-4 The first tender goes out to establish the ice board pulley system. Notice how the bow-line is laid over the board, under the tender.

The next step in the plan of action to determine how to get personnel to the area under the ice where the divers will be deployed and how to stage. Will tenders and divers be tethered to the ice or back to shore? Is the ice too thin to support ice screws, and, if so, is the team capable of the human piton procedure? A decision will also have to be made regarding the necessity of using an ice platform, such as an inflatable boat or ramp.

Photo 13-5 Before entering the water, the tender lowers his feet and pushes himself about a foot toward the rear of the board to prevent the bow of the board from submerging when it hits open water. Note that this tender is not using a bow-line; both lines are attached to the stern cable. One line will later have to be switched to the bow.

Sending the First Tender Out

For most public safety diving operations, the ice is weak; otherwise, the team would likely not have been called. This is important to note, because the vast majority of teams primarily (or exclusively) train and drill on strong ice. With proper training and equipment, a team should be able to: arrive on a weak ice scene, dress, develop the plan of action, safely get all five minimum necessary personnel to the diver point-of-entry, recover a child on the bottom 150ft from shore, and gently transport the child to shore—within 10 to 15 minutes.

Photo 13-6 The bow is up as he approaches the other side of the hole. Just before the board hits the ice, he will reposition his weight forward and use the ice awls to pull himself and the board up onto the ice. Notice the bow-bag containing the ice screw and pulley system.

A very safe and effective way to move personnel and equipment is with an ice rescue board. To use an ice board effectively, it needs to be secured to the ice near the diver-point-of-entry with a pulley-type system so that shore personnel can move the board back and forth. To establish this system, a tender is sent out with the board to the hole. The tender needs a full surface ice rescue suit or drysuit, fins, and possibly ankle weights. When using a drysuit made from something other than full-thickness neoprene, use a Class 3 or better PFD under the harness. If the drysuit should flood it may not have enough buoyancy to hold the wearer up.

Photo 13-7 The diver point-of-entry will be forward of the tender, who places a mariner's hitch into the ice to set up the pulley system and tether-attachment point.

The tender should lie on the sled with the bow attachment portion of the line coming over the top of the sled. The pulley should lie just under the top of the tender's chest. At a controlled pace, the tender picks, crawls, climbs, swims, digs, etc. to get the board to the diver point-of-entry.

Once the tender reaches the diver point-of-entry, he/she must decide whether the ice can support an ice screw to which the pulley can be attached. If the ice can support the screw, the tender installs the screw and attaches the pulley with the ice board tether line. This completes the pulley system back to the shore. The tender should stay at the hole as shore personnel use the ice board to transport the primary diver, backup tender, backup diver, and 90%-ready diver, in that order.

If the ice cannot support an ice screw, the next option is to use the human piton procedure.

Human piton

The human piton procedure is specific to rescue operations, but it also works for recovery operations when no platform is available and the search objects need to be recovered without delay. As in all facets of rescue, practice makes perfect. Having everything work as planned and trained for is rare. Sometimes, measures that may appear extreme are needed to get the job done and done safely. Although the human piton procedure may seem a bit radical, when performed properly, it is extremely quick, safe, and simple. However, to be performed properly, it requires redundant training at all levels of operational participation.

The human piton is a tender who uses himself to replace the ice screws used for the pulley system. The human piton wears a well-insulated immersion suit, with heavy enough ankle weights to allow a vertical position in the water. When he arrives at or near the hole, the tender gets off the board, enters the hole, and turns to face the shore crew. The tender's hands go up against the ice and his legs bend up and forward against the ice as well. When the tender gives the appropriate signal, the shore crew slowly takes up tension as the tender readies for the load. Surprisingly, there is very little load on the tender. When coordinated properly, shore operations take up the line so slowly that the human piton barely feels the load coming on. All line pulling is done at a rate of 1.5ft/s to 2ft/s, especially if there is an emergency. Slower is surer when it comes to reducing the risk of injury.

The human piton's job is to remain in place as a piton. As the team becomes more skilled in the technique, the human piton can also function as the backup diver's tender, improving the speed and reducing the number of personnel needed for a safe operation.

The human piton is tethered directly to shore, and, if needed, can be pulled back to shore any time the tender or the incident commander wishes. Tether the human piton the same way ice rescue personnel tether someone who will be used as a human sled for the victims.[9]

For operations far from shore, command should use binoculars to monitor the scene and the human piton. The tender should use whistle signals, a small hand-held flag, or a brightly colored handkerchief to assist in signal communication.

Once the human piton is in place and ready, shore operations pull the sled back to shore and deploy another tender to the site.

Sending Out the Divers

Next comes the primary diver, fully dressed, with mask and fins in place. If the diver is wearing a full face mask without a surface-breathing vent, get the necessary personnel deployed rapidly as the diver will be using air from his tank. If air is a concern, the diver can wear the full face mask on top of his head during transport; it can be rapidly pulled down and tightened into place. Make sure that regulators and power inflator do not drag on the ice or snow.

The diver should carry a deployment tether-line bag just under his chest on the forward portion of the sled. He should be attached to the sled by his harness with a foot of webbing and a quick-release snap shackle. The diver should remain recover-

Photo 13-8 The human piton signals shore that he is ready to receive the primary tender.

Photo 13-8a Using the human piton, shore personnel pull the primary tender out to the diver point-of-entry.

Photo 13-8b The primary diver slides off the board. While his tender assists him into the water, the piton signals shore to pull the board back.

Photo 13-8c The backup diver arrives last. The tenders disconnect him from the board and snap him into his tether line.

able to shore until he has secured his tether line into the carabiner at the main ice screw and the tender has the deployment bag. The diver then disconnects from the ice board, enters the water, and waits. The sled returns to shore to transport the backup diver, and then the backup tender. The primary diver should not descend until the backup diver and tender are in place and the 90%-ready diver is dressed, checked out, and on the sled. At no time should a tender be accountable for more than one diver, except in an extreme diver emergency.

There are times when extremely cold temperatures, ice, and snow may cause the pulley lines to freeze to the ice. The shore crew should move or jiggle the lines every few minutes to keep the lines from freezing to the ice.

One advantage of using the human piton system, besides speed, is that staging in the water rather than on the ice better preserves the ice. If a tender lies on the ice and the ice

Photo 13-8d The diver's lines are secured to the human piton's pulley. The primary tender has his own line back to shore.

Photo 13-8e The 90%-ready diver is stationed on the board and is ready to deploy on command.

Photo 13-8f The primary tender directs the primary diver to the descent point.

Photo 13-8g The tender is held in place by shore personnel as he directs his diver, who keeps the dive tether line taut. If the tender was allowed to move, the search pattern would be inaccurate.

Photo 13-8h The human piton serves as the backup tender.

Photo 13-8i Because all personnel have been in the water, both diver and possibly tender are switched to perform a second dive. Here, shore personnel pull a diver home. He is secured to the board and his tether line is placed between his legs.

Photo 13–9 The two communication (comm) tenders (back) control the tether lines of the in-water tenders, have the diver's hardwire comm lines secured to their harnesses, and wear the comm headsets.

breaks, the accuracy of the search pattern is compromised. Anyone who has experience with tethered diving knows that the tender's location is as important as the diver's location, because the latter is completely dependent on the former. If the diver is 40ft from the tender and the tender is 100ft from shore, the diver's actual position is 140ft from shore. If the tender moves 4ft backward, the diver is still 40ft from the tender, but the diver's actual location is 136ft from shore. Almost two full sweeps of the search area will be missed.

There are three ways for a tender to maintain a designated location on the ice. If the ice is strong enough, tenders can stand, kneel, or lie on it, just as they would on shore. If the ice is too weak to support the tender, or too weak for the hole to stay the same size, two options are available. The first option is to tend from a platform that is secured in place with anchors set on the shore, ice, or bottom. The second option is the human piton system. In the human piton system, a tender is tethered to shore with a line secured to the back of the tender's harness. The tender hangs in an ice hole with his back against the ice as he faces away from the shore. Shore personnel hold the tender's back securely against the ice by keeping his line taut. This keeps the tender in one location while the tender's diver performs a search. The shoreline tenders document the in-water tenders' distance from shore with the marked tether line.

When electronic communication systems are used, the human piton system is modified. Since the tender part of the communication system needs to stay dry, in-water tenders cannot wear headsets or listen to surface station speaker boxes. Therefore, a second set of tenders is necessary. In this modified system, each tender has a communication tender staged on the shore or on stronger ice who controls the in-water tender's tether lines and who wears the communication headset and holds the hardwire line. If a wireless system is used, the second set of tenders have the communication surface station. The in-water tenders monitor the diver and determine what pattern direction commands are needed. The communication tenders listen and talk to the diver. Because the ice is weak, the communication tenders normally will have to lay prone if they are staged on the ice.

Photo 13–10 The in-water tender tells the comm tender that the diver needs to go "3" to the right.

When it is time to stop and turn the diver, the in-water tender gives a hand signal to the communication tender, who then relays it to the diver.

Switching in-water tenders or human pitons

With a little practice, the following steps for changing in-water tenders (or human pitons) can be accomplished quickly.

1. As the tenders meet, the in-water tender should unlock his carabiner. The relief tender then reaches out and holds onto the in-water tender's harness.
2. The relief tender attaches the rescue strap from the sled to the in-water tender's harness.
3. The in-water tender removes his carabiner, attaches it to the relief tender's harness, and locks it.
4. The relief tender disconnects from the snap shackle, slides off the sled, and enters the water.
5. The tender being relieved pulls the sled into the water, climbs on, signals shore, and goes home.

With proper instruction and a couple of practice sessions, this maneuver can be completed in less than sixty seconds. Entire teams can safely be changed using this technique.

Recovering the victim/search object

In real-world rescue and recovery, personnel often find themselves falling through the ice twenty, thirty, sometimes fifty feet away from the VPOD. Often, a lack of training or planning causes teams to believe that they must get to the VPOD to dive. Diving the VPOD might be the ideal, but in an imperfect world, it's not always possible. If another hole location is used, a diver using the spidering technique can cover seventy to eighty feet between holes in a matter of thirty or forty seconds and search everything under the ice roof in between. He can then descend and search the best hot zone very quickly and effectively.

In actuality, falling through thirty or forty feet from the VPOD may be an advantage. Anyone who has ever dived tethered knows it is more difficult to keep a pattern with a vertical line between the diver and tender than it is to dive a line at a greater angle. A vertical line over the diver's head makes it difficult for the diver to distinguish left, right, forward, or back. A diver can spider to the VPOD from the first hole, raise a hand to show he is there, drop down, and perform an arc sweep from one end of the VPOD to the other with a nice line angle. As the tender drops the diver down, the tender gives enough line to put the diver at the furthest edge of the planned search area. The diver then performs tender-directed sweeps inward.

Surface rescue and dive team personnel can devastate the VPOD simply trying to get there. If the plan is to dive from the VPOD, then when the first surface rescuer or diving tender reaches the VPOD, mark their tether lines for the distance they are out and draw it on the profile map. If the VPOD is destroyed or enlarged while deploying other personnel, the original location will still be known.

In most non-ice search dives, out-to-in patterns are used. With out-to-in patterns, divers start at the farthest end of their search pattern and are brought continuously closer to home at the start of each new sweep. This is a preferable method from the diver's viewpoint. As divers use up air, and perhaps become chilled, overheated, frustrated, fatigued, or less focused, they always know that each sweep brings them closer to their tender, their backup diver, and home.

The out-to-in pattern is often used in ice searches when divers deploy from a hole that is not a VPOE or VPOD hole. But if divers are deployed directly from a VPOD hole, then it makes more sense to begin in the hottest target area and then work out from there. Hence, dives directly from the VPOD hole usually use in-to-out patterns.

The tender lies on the ice with hands low to the water as the diver sweeps three sections of DAD in a half-moon pattern. Once the half moon is completed, the tender moves to the opposite side of the hole and performs another half moon. Another option that takes a little more skill is to run the diver in an expanding circle pattern. More skill is required because tenders either have to move around the hole or drop their tending arms low in the water to manage the loss of direct line access while the diver is underneath or behind them.

Photo 13–11 If no transport device is available, the ice piton has to pull himself out to the diver point-of-entry and be prepared to fall through and then get out of the hole on the way.

Once the body is recovered and the diver has brought it to the ice hole, the backup tender should pull the board into position. With the diver's assistance, the backup tender pulls the body on the board. Next, the backup tender mounts the stern of the board and signals shore personnel to take the load and pull the board to shore. It is the backup tender's job to maintain balance and control and not to lose the body if the ice breaks under the board. The victim can be secured to the board with a strapping system. Be sure to have a practiced working plan on patient handling procedures to cover the time from when the patient reaches shore to when the patient is secured in the ambulance.

Photo 13–12 Whether or not an ice board is used, the human piton is the last person off the ice.

If the team is using the backup tender as a human piton, the primary tender will transport the body while the divers care for each other at the surface. If his help is needed at the hole, the backup or primary tender can return to the hole on the sled once the body is on the shore. With proper training, divers can signal shore to send them the board and can use the board to extract themselves from the hole. The human piton is always the last to go unless there is an emergency. As a rule of thumb, in operations 300ft (91m) or less from shore, it should take less than eight to ten minutes to extract and recover all diving personnel.

The human piton system can be used without an ice board or other transport device if the ice can take the weight of one person at a time crossing it. However, without a board, personnel are more likely to be exhausted and cold stressed by the operation's end. Furthermore, working without a board requires more training than if an ice board is used.

Using an ice platform

Ice platforms are used when the ice is weak. The platforms help prevent additional ice from breaking near the diver point-of-entry, keep tenders out of the water or on the ice, and provide a place for contingency divers to sit off of the ice. They are most often used in weak-ice recovery operations.

Inflatable boats without hard transoms or V-hulls and inflatable ramps make great ice platforms. Inflatable boats with a hard transom are often difficult to pull over the ice or through snow. The transom drags, which can create a severe workload for personnel. If transported over long distances, the bottom of the vessel may be destroyed. A Department in New York dragged a hard-transom vessel over a mile across a frozen lake to perform a surface rescue and destroyed the boat in the process. Destroying equipment to save a life or maintain a safe platform should not be a concern, but if the bottom of a vessel is partially or totally destroyed, the boat can no longer function for the operation. Small transom-fitted vessels can be protected by duct-taping an ice board to the bottom. When the vessel reaches its desired location, simply cut the duct tape and the board is back in action.

Aluminum boats do not make good ice platforms because their metal construction absorbs the heat of the personnel in them. If the boat has a U- or V-hull, it will not sit squarely on the ice and will rock from side to side. It may not move in a straight line when being transported to the hole. In addition, aluminum boats are more likely to damage weak ice than inflatable boats. Lastly, it is more difficult to move divers into and out of aluminum boats.

Ice platforms are moved to the dive site with the same techniques used to move ice boards. The human piton can handle a small, empty, and unequipped (except for anchors) vessel. Reasonable ice can usually be found ten to fifteen feet from the hole. If the boat does not fall through, the ice is usable enough to place a double or triple set of ice screws in two, three, or even four directions. This eliminates the need for bottom anchors. If the ice breaks away, anchor the boat with two to three anchors. If there is no ice for the boat to sit on, a three-anchor system will be needed with a depth-to-scope ratio of at least 1:5.[10]

Once the boat is anchored, use an ice board to transport personnel and equipment between the shore and the boat.

To set up a platform, follow the procedures described earlier in this chapter for setting up the ice board pulley system. Then, place the platform on the sled to transport it to the hole.

Recovering a body

If a victim is recovered in a recovery mode, the best way to bring the body up is in a bag. A bag preserves possible evidence and will keep the body out of sight from family and the media. Efficiently bagging and surfacing a body takes practice, first on land with blacked-out masks, perhaps next in a pool, and then in a non-overhead open water environment.

Assemble and fold/roll the bag so that it can easily be carried and used by the diver. If communication systems are used, the diver should describe the position of the body prior to moving it. Tenders can either write the description down or tape record it. In non-overhead environments, lift bags can be used to lift the bagged body. In an ice environment, the bagged

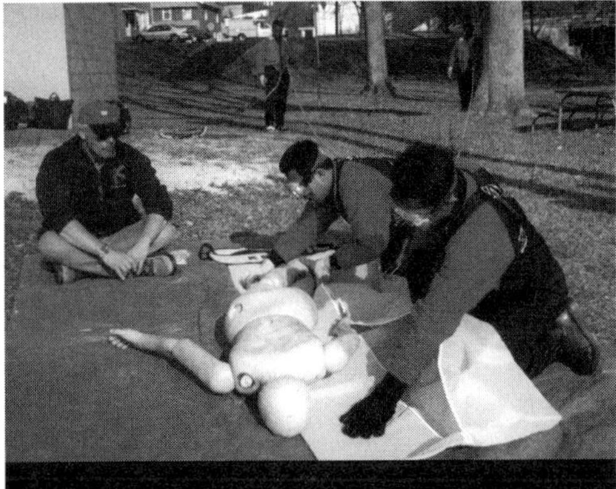

Figure 13–13 If rigor needs to be broken to get the body into the bag and no comm systems are used, divers need to do their best to describe the body's posture as soon as they are back to shore.

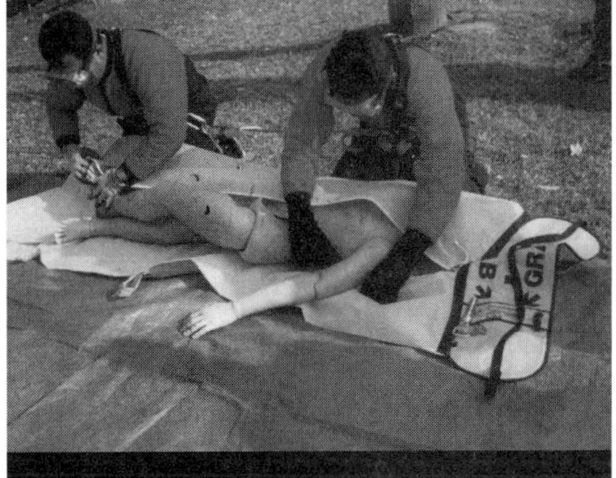

Figure 13–13a Roll the body carefully and gently into the bag. Notice the lift bag and air cartridge at the end of this MARSARS™ water-recovery body bag.

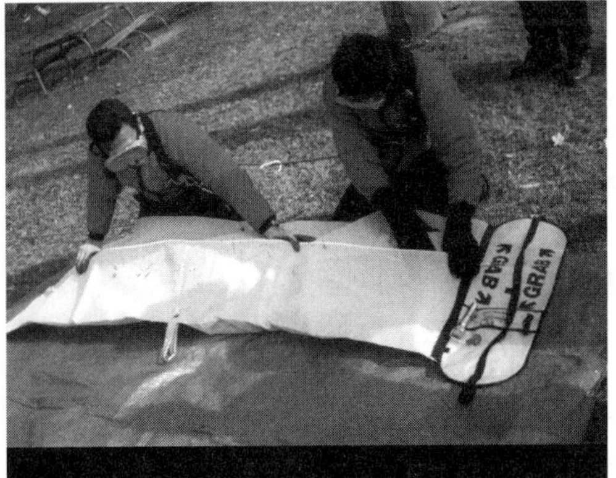

Photo 13-13b Make sure there is nothing in the way of the zipper such as the victim's scarf or hair as the bag is zipped closed.

Photo 13-13c Practice surfacing the body in warmer months prior to training with the bag under the ice.

body should be lifted to the hole, as damage to the body could occur if lift bags cause it to hit the ice roof. An alternative is to load a small-sized cartridge to a small lift bag that will have just enough lift to make the body neutral or only slightly negative.

Water Level Changes

Recreational divers can avoid suspended-ice sites by pre-season planning and information gathering. Public safety divers and surface rescuers may have to work on suspended ice if it becomes an actual rescue site. To limit the hazards at known suspended-ice locations that are manageably small, lay plastic snow-fencing across the water before the first freeze. When the water freezes, pull the fencing taut. The fencing forms a web in the ice that strengthens the ice and could serve as a catch net for someone falling. Display highly visible danger signs to identify areas that may have suspended ice. In addition, install lifeguard stands that are tall enough to allow rescuers full view across the entire body of water. This reduces the time it takes to find an immersion victim's location.

Unless the benefit is very high, public safety diving teams should wait until the ice disappears to perform the search. Advanced training specific to suspended-ice-roof environments is required. If the operation cannot be conducted from shore and there is no visual or auditory sign that the victim is alive, the operation should be a no-go for most teams.

Divers should enter the water from a hole close to shore so that tenders do not have to be on the ice. Once underwater, divers need to stay underwater to protect their heads should the ice roof come down. This means that operations need to be within 100 feet of the shore. If this is not possible, waiting until spring may be a good idea.

When a Snowmobile, Vehicle, or Airplane Goes Through the Ice

Any time there is a fuel-containing vehicle in the water, the dive becomes a hazmat situation and a hazmat team should be requested. Divers need, at minimum, drysuits tested to withstand petroleum product contaminants and full face masks. Communication systems also become imperative because of the increased risk of entanglement and entrapment, the loss of direct line access between divers and tenders, and the need for two divers to coordinate underwater.

If the vehicle is on the bottom and a diver will attempt to get passengers out, two types of training certification are required: underwater vehicle extrication and advanced ice diving operations. The former involves teaching tenders and tethered divers how to safely find, work around, and remove victims from submerged vehicles. Students need to learn about all the hazards of such operations and how to mitigate them. Some of the hazards include:

- Fuels and biohazards from injured-body fluids
- Jagged metal
- Vehicle shift or movement
- Air-filled objects that can become projectiles
- Entanglement and entrapment
- Loss of direct line access between a diver and tender

Because of the last hazard, working around submerged vehicles requires two divers down at one time. The primary diver will locate the car. After the car is located, a second diver will descend down the primary diver's line with a contingency strap to the first diver. The second diver will become the primary diver's line tender if the primary diver loses direct line access. Loss of direct access can occur if the diver enters the vehicle to search for victims in a rescue mode. There is a controversy about whether divers should be taught to enter the vehicle to search for victims or only reach in as far as their arms will extend. The latter procedure is certainly safer, but in reality, even if divers are taught, and repeatedly told, not to enter a vehicle, they often do so anyway during a rescue mode. If these divers are not well trained to enter and work in submerged vehicles, their safety is significantly compromised.

If there is fuel in the vehicle point-of-entry hole, it may be advisable for divers to enter the water from a hole farther away. This may require cutting a new hole. Before a diver enters or even reaches inside a vehicle, the vehicle door should be secured with strong bungee cords to prevent it from closing accidentally. Before entering a vehicle, divers need to assess the outside of the vehicle to see if there is any possible vehicle instability, opened doors, or blown out windows. There are many procedures for such operations not covered in this book. The purpose of this section is to make teams aware that they should not perform submerged-vehicle operations without proper training, certification, SOP/SOG, and equipment, and that until they have this support, their SOP should call vehicle operations a no-go.

When snowmobilers compete to see who can jump over the largest area of open water, when they unknowingly drive onto snow and ice-covered water, or when they operate under the influence of alcohol, they seriously risk breaking through the ice. In snowmobile accidents on ice, there may be two VPOD holes, one for the snowmobile and one for the victims. Upon arriving at the scene, take a moment to attempt to determine which hole is which, because the victim hole should be searched first. One clue is a fuel slick in one of the holes. If the direction of movement of the snowmobile can be determined, it is more likely that the snowmobile is in the first hole and the victims in the next holes. Some snowmobile suits have enough flotation to keep the wearer afloat. Divers need to a search under the ice roof before searching the bottom for this reason.

When a vehicle punctures the ice and descends, the heavier front descends first. The rule of thumb is: if the water depth is greater than the length of the vehicle, it will land on the bottom upside-down. If the water is shallower, the vehicle will land right-side up. If the vehicle is still on the ice, if possible, gently open all the doors to create more surface area in case the ice breaks.

When airplanes hit ice and water, the nose stops milliseconds before the tail, which can cause the tail to J-hook. Water is 800 times denser than air and is not compressible, therefore, hitting water is very similar to hitting land. The damage to an aircraft can be so devastating that it may be recognizable only to a crash expert. If a plane hits the water at a high velocity, the chances of rescue are nil, while the risks to divers are extremely high. In fact, the risks are far greater than those for submerged vehicles. Possible problems for aircraft recovery operations, as opposed to other vehicles include:

- More dangerous fuel
- Extensive wiring that can easily entangle divers
- More severe biohazards, because passengers can sustain extreme damage
- Electrical systems that could still be live
- Higher degree and greater amount of sharp, twisted metal
- Ejection systems on some military planes that can injure divers
- Greater possibility that cargo could contain unknown hazards

When large items go through the ice, the issue of how to get them out is often raised. In some states, vehicle owners are fined for each day the vehicle remains submerged. In New Hampshire, for example, snowmobile owners can be fined $500 per day and car/truck owners can be fined $1,000 per day.[11] Public safety diving teams are often asked, even begged, to get these vehicles out of the water and back on land. A department should determine a policy for this type of operation and adhere to it. Who will provide the funds for the training, equipment, and overtime necessary to do such operations? Lifting and extraction operations can be very exhausting and time consuming. For example, if a car punctures the ice 1,000ft from shore, the skills, equipment, and personnel hours required can be tremendous.

Photo 13–14 The Subsalve VRS 2000™ pillow bag system is an excellent system for lifting vehicles, especially with the remote filling attachment.

When getting a vehicle back to shore, the goal is to use the least amount of cable possible. Taut cable can break over long distances and cause tremendous injury. There are several safer options. Everyone involved should have lifting training and certification with proper ice environment lifting training, as there are many hazards to mitigate. Improper handling of a tow cable chain can result in a diver being pulled rapidly down to the bottom. The use of open, rather than pillow, lift bags can result in a diver fatality if the bags accidentally empty and divers are in the path of the lifted item. The team should know how to calculate the size of the lift bags to avoid the danger of using bags that are too big. Personnel must also understand how to work with suction from mud, and know where to attach the lift bags safely and efficiently. Whenever possible, remote filling is the preferred technique, although it involves additional equipment and training. Lifting involves far more than simply attaching bags and adding air.

If a snowmobile is less than 100ft from shore, pillow lift bags can be attached to provide just enough buoyancy to lift the snowmobile easily and bring it under the ice to shore. When the bottom becomes sufficiently shallow for the snowmobile to be just below the ice roof, cut a hole and bring it up with a ramp and tow cable. If the snowmobile is more than 100ft from shore, a 1ft wide trench can be cut back to shore. Apply lift bags and pull the snowmobile to shore with a tow cable. A distance of 200ft places so much stress on a tow cable that it is not worth the risk; additionally, cutting such a long trench is a great deal of work. An alternative is to cut holes every 75ft to 100ft, depending on water depth. This method requires the most diver lifting training. Divers secure lift to the snowmobile, secure a towline to the vehicle, and transport the tow-line to the next hole by spidering. Surface support personnel then pull the snowmobile to their hole, and a tender changes the diver's tether to one that is tethered to the new hole. This process should be repeated until the shore is reached. Be sure to watch the diver's downtime and to move the necessary backup divers and tenders to each new hole promptly.

If a snowmobile is a great distance from shore and the ice is relatively strong, a confined-space tripod technique can be used to lift the snowmobile onto stronger ice and transport it back to shore on the surface. In actuality, there is rarely a reason why the recovery operation cannot wait until spring.

When it is necessary to lift items, particularly large items, and transport them back to shore, relevant questions are: "Who needs this done?" and "Who should be paying for it?" Is it the responsibility of the local public safety diving teams, or should military or commercial divers do the job? Public safety diving teams in the U.S. should be aware that some lifting operations can place them under OSHA regulation. Department officials should speak with local OSHA representatives about this issue before the need to perform one of these operations arises. If items are to be secured for a criminal investigation, it is imperative that personnel are trained in underwater investigation techniques and that they maintain the evidence chain of custody with all necessary underwater documentation.

Summary Questions

1. Clearly state the "no-go" situations in the team's _____.
2. Standards can be broken during rescue modes. True or False?
3. What is the primary difference between rescue and recovery modes?
4. When possible, after witnesses perform the "show me" re-enactment, have them make a _____ statement before they are interviewed.
5. What is the rule of thumb for locating a body in still water?
6. How might a drowning victim end up much farther from the hole than expected in still water?
7. A victim descends to sixty feet in a 2knot current. Where should the victim most likely be from the VPOD?
8. Once the hot spot search area is chosen, the next decision is where will the diver ____ the ____.
9. Explain why it may be preferable to send the diver out from a hole closer to shore rather than from the VPOD.
10. Use a(n) _____ _____ pulley system to move personnel and equipment between the hole and the shore.
11. When is a human piton used?
12. Describe a human piton's equipment, location, and function.
13. If the ice is too weak to operate on directly and the team does not want to use the human piton system, the other option is to use a _______.
14. Can electronic communication systems be used with the human piton procedure?
15. List advantages of using platforms on the ice and describe an effective platform.
16. Before attempting to use a body bag, you should do what?

Notes

1 Ulster County (NY) Sheriff's office URT, for example, has experienced three rescue calls during training sessions, one of which occurred less than 200 feet from where the team was working and while the media and sheriff were present. If they had not responded, especially to the last one, in a decisive mode, the consequences would have been severe.

2 Martin Nemiroff, M.D., personal communication.

3 Walt Hendrick, Andrea Zaferes and Craig Nelson, *Public Safety Diving* (Tulsa: PennWell, 2000)

4 Walt Hendrick, Andrea Zaferes and Craig Nelson, *Homicide by Drowning* (1998).

5 U.S. Coast Guard, navies around the world, commercial divers, and other maritime agencies.

6 Frank Pia, *The Reasons People Drown LSA Productions* (1987)

7 Magne Overrein and Roy Larsen. Oslo Fire Department, personal communication.

8 Corporal Robert Tether, *Encyclopedia of Underwater Investigations* (Flagstaff: Best Publishing, 1994).

9 Walter Hendrick and Andrea Zaferes, *Surface Ice Rescue* (Tulsa: PennWell, 1999).

10 Walt Hendrick and Andrea Zaferes, *Public Safety Diving* (Tulsa: PennWell, 2000).

11 Lt. Todd Hansen. Hudson N.H. F.D., personal communication.

Chapter 14
Contingency Planning for Underwater Emergencies

All divers know they need an emergency plan in case things go wrong underwater. A realistic, well-practiced, proven plan is best. Ice diving is certainly no different from any other type of diving in this regard, except that planning needs to be even more professional, tested, and practiced.

All too often, an emergency plan is prepared by tabletop discussion, and there is no testing or practicing to make sure the plan works. The plan has been discussed, but when something goes wrong, everyone discovers that the plan was not reality-based.

The first step is to consider what could possibly go wrong. The following list provides some of the possible problems. Keep in mind that these problems could take place in combination. Injuries or deaths are rarely due to a single problem; rather, they are caused by a series of mistakes and problems, with three being a common number.

- Entanglement or entrapment
- Equipment free-flow
- Drysuit or BCD self-inflation
- Accidental loss of weightbelt
- Over-exertion with resulting respiratory distress
- Unexpected blackout conditions
- Leg cramps

- Injury (*e.g.*, fish hook puncture, head banging into ice, and falling on or through the ice surface)
- Ice breaking with accidental tender immersion
- Mask flooding or dislodgement
- Fin loss
- Drysuit flooding
- Cold stress or hypothermia
- Contamination exposure
- Cardiopulmonary problems (*e.g.*, heart attack, stroke, asthma attack, or laryngospasm)
- Ear and sinus problems (*e.g.*, inability to equalize, alternobaric vertigo, round window rupture, or tympanic membrane break)
- Lung over-expansion injuries, water aspiration, or drowning
- Mental stress, anxiety, or panic
- Out-of-air emergency
- Diver disconnect

Divers and tenders should have the knowledge and tested skills to manage each of these problems. Sadly, many divers are taught that the contingency plan is for the backup diver to descend to the primary diver, figure out what the problem is, and then manage the problem. That is not a proven, tested plan—it lacks specifics. It leaves both the primary and backup divers without practiced behavior, which greatly increases the chances that these divers will become stressed and possibly panicked when the original problem is not resolved quickly, when new problems are created, or when they realize there is no plan.

To fully appreciate the importance of practiced, proven contingency plans, let us take an imaginary tethered dive. You are searching for the body of a seven-year-old boy in frigid blackwater at a 45ft depth in a hole 130ft from shore. You see even less than if you were in a closed closet with your hands over your eyes. In fact, if you shine a flashlight in your face, it does not penetrate the utter blackness. Your mind's eye envisions everything that your fingers and body touch, leaving only a few items unidentifiable. Something bumps into you and you give a little shout into your full face mask. You wish the team had a budget for a communication system.

As you continue on, your legs begin to feel restricted. Now you feel something pulling on your tank valve. You're entangled. Okay, stop, get your breathing under control, you are okay. Give a "2+2+2" signal telling your tender that "I'm okay, but am tangled: alert the backup diver." Your tender returns the signal while telling the backup tender to ready the backup diver for possible descent and to make a note on the profile slate that the diver's current location contains an entanglement problem.

You attempt to find the entanglement. No, you can't reach it, because it is behind you and around your legs. No problem, you give "3+3+3" signal, signifying *"I'm okay, but need help from the backup diver."* Your tender returns the signal and tells the backup tender to deploy his diver. The backup diver calmly descends down your line with a contingency quick-release strap, providing full freedom for both his hands. The backup tender notes the time and diver location on the profile slate. The 90%-ready diver moves to a backup-diver status by fully suiting up and getting into position at the hole.

While you are waiting, your tender asks for a quick breathing-rate check, which makes you feel good that you are being monitored. You give a pull for each inhalation for thirty seconds, which slows your breathing. You also feel secure knowing that there is a contingency main cylinder waiting for you if you need more air. In less than sixty seconds from the time you requested assistance, your backup arrives. He firmly clasps your hand, which is resting on the carabiner securing your tether line to your harness. He gives a reassuring squeeze as you make a circular motion with your hands clasping his hands to tell him you are entangled. You then place his hand on your tank valve where you feel one of the entanglements. He removes the shears from his harness and begins cutting away the fishing line as you place your hand back on the carabiner, giving your tender one pull, stating that you are okay. When the backup diver is finished with the tank valve area, he again clasps your hand. You give another circular entanglement signal and place his hand on your knees. The procedure is repeated. The next time he clasps your hand, you move a few feet to check that you are free. You squeeze his hand three times, telling him that you are okay, and then gently push him away to let him know that you no longer need him. He disconnects his contingency strap from your line and heads back toward shore with his tender taking up the slack. Your tender asks if you are okay (one pull), you respond (one pull), and you are sent to your left or right to continue on your search pattern sweep.

This sounds pretty simple and logical, right? But, unbelievably, the vast majority of dive teams do not have practiced, pre-planned blackwater contingency procedures. Let us imagine what would happen to you in the same situation if your team did not have a practiced contingency plan. Since the majority of teams have only one "help" signal, we will follow that procedure in the following scenario.

You realize that you are entangled and stop to solve the problem. The tender does not understand why you stopped and gives you an *"are you okay?"* signal. You return the signal even though you're not really sure if you are okay. You continue to try to disentangle yourself, when the tender, still confused, gives you another *"are you okay?"* signal. You decide you can't do this by yourself and become frustrated that you have to give the *"I need immediate help!"* signal. You give the signal. "Oh God," your tender says, "Something is wrong. He gave the emergency signal. Send in the backup diver. Hurry!" The backup tender rushes to help the diver get in the water. The diver grabs onto your tether line and starts swimming down headfirst. He forgot to fully burp his drysuit so he is having trouble getting down. He starts to kick harder, which increases his breathing and buoyancy. You are now getting jerked forward and deeper into the entanglement. You hope there aren't any hooks in this mass of fishing line that is tying you up.

The backup tender pulls the backup diver up and yells, "Get the air out of your drysuit!" He does so, then re-grabs your line and swims toward you. In his haste, he accidentally lets go of the line when he performs a Valsalva ear equalization maneuver. The line floats and he is unable to see anything, so he has to resurface to relocate your line. To further add to the stress, there is no 90%-ready diver available; therefore, everything is riding on the now stressed and heavily breathing backup diver.

Finally, after three and a half minutes he reaches you, and by this time both your breathing rates have tripled. Suddenly you feel something all over your head, chest, and arms. He has no idea what your problem is so he is feeling around your body to "identify the problem." At least you hope it is your backup diver feeling you up. His elbow hits your mask and almost knocks it off. You reach up to try to grab his hands and he gets spooked and backs away. Both your air consumption rates rise even further. You try to grab his hands to put them on your tank valve, but he keeps moving around. If only.

If only you had a practiced contingency plan!

Before we present a well-proven contingency system, let's first debunk a few systems that we have seen taught that are particularly ineffective or dangerous. As illustrated by our last scenario, the most common system of going down, identifying the problem, and attempting to manage it without a specific, well-practiced communication system is ineffective. In addition to wasting time, a backup diver without a specific plan could easily: accidentally knock off the primary diver's mask or regulator, become entangled himself, accidentally drive a fish hook deeper into the primary diver's leg, or greatly stress the primary diver by blindly groping around him.

Another response to a diver problem is to pull on the diver's tether line to get the diver back to the hole when she is not responding to line signals or communication system inquiries. Never pull on divers to get them out. Ask yourself what may be the cause of the diver's lack of response. It is usually because the tether line is fouled. In that situation, pulling on the line can only foul it more, and in the worst-case scenario, could result in an accidental disconnect. Other reasons why a diver may be unable to ascend or respond is because the diver is entrapped in a vehicle, is between the ice and bottom, or is severely entangled in something. Attempting to pull such a diver out by pulling on the tether line could cause serious injury, dislodge the diver's mask or regulator, or greatly increase the diver's stress level. Instead, the tender should gently take up any existing slack and deploy the backup diver.

One of the most dangerous contingency plans we have ever heard of is to have the back-up diver inflate the primary diver's BCD to lift the diver off the bottom and then cut away the entanglement. As the primary diver feels the fishhook or Spiderwire™ cut into his leg as he is raised off the bottom, he may start to wonder what his head is going to feel like when he hits the ice roof at 200ft/m once the entanglement is cut loose. Please do not try this; take our word for it that it is a dangerous and ineffective plan.

First we will discuss help signals, and then self-rescue techniques, which will be followed by procedures for diver-to-diver assists.

How To Ask for Help—Why Three Signals Instead of One

As stated in chapter 12, having three signals for help lessens stress and provides greater opportunities for a smooth, safe, efficient, and effective assist. Since the majority of recreational and public safety divers today still use a one-signal system, it is important to stress the need for multi-signal help systems.

Advantages of a multi-signal system

Multi-signal systems provide the following safety advantages:

- The tender will know whether this is a simple entanglement problem or a life-and-death emergency that requires alerting EMS and shutting down other dives going on simultaneously.

- Topside personnel will not treat 2+2+2 or 3+3+3 problems like life-and-death emergencies, thereby preventing stress levels from increasing significantly, as they would with a one-signal system.
- High stress levels reduce the odds that the backup diver will make it to the primary diver in time to help. Stress causes mistakes, ear problems, panic, equipment problems, etc. That is the reason why a 90%-ready diver should be available to quickly replace a failing backup diver.
- Divers will not wait until the entanglement is a more serious problem before giving a "help" signal because there are options other than "I need immediate help!" available.
- The 2+2+2 signal prevents topside confusion when a diver stops moving to handle a problem.
- All military and commercial diving operations use different codes for different types of problems. Likewise, all public safety operations other than diving use multiple codes. The one-signal system is an example of shortcutting, not simplifying.

Divers using communication systems should still call out the line signals to prepare both themselves and the tenders for a possible communication system failure. The primary tender calls out the signal to the backup diver and the dive coordinator/supervisor at the hole. The officer in charge on shore is notified and kept up to date by this officer.

The following signals were adapted from the US Navy contingency signals by author Hendrick in the late 1960s.

Primary diver to tender

2+2+2 "I'm okay, but am tangled: alert the backup diver."

3+3+3 "I'm okay, but need help from the backup diver."

4+4+4+... "I need immediate help!"

***Note**: Continuous sets of 4 pulls tells the shore that the diver is still alive and able to concentrate and gives the diver something to focus on while waiting for the backup diver.*

Self-Rescue: STOP, THINK, and ACT

The words are simple and familiar to all divers, but what do they mean? How do we put them to our best use? Let us take a look at some of the other ways we can put the STOP, THINK, and ACT plan into ACTion.

Breathing

Before you can begin to fix specific problems, you must first address the number-one problem—breathing. Breathing will at some point be an issue in almost every underwater situation. If you cannot control your breathing, all else is almost irrelevant. The first concept is to STOP, which means more than simply stop moving. The term STOP means that if you can, stop doing whatever it is that is causing the problem. That could be as easy as catching your breath. But for a diver who is overexerted or panicked, the words "catch your breath" do not mean very much. We need to understand how to STOP and catch our breath so that we can then THINK and ACT.

Hypoventilation (shallow, rapid, labored breathing) results in a buildup of carbon dioxide. Increased arterial carbon dioxide levels create an urgency to breathe. Hypoventilation can therefore become catastrophic as the cycle of increased carbon dioxide levels, increased urge to breathe, more rapid and shallow breathing, increased carbon dioxide levels...spirals into panic. Divers tend to use improper, shallow ventilation techniques while under stress, perhaps not realizing that carbon dioxide is a major contributing factor not only to breathing problems, but to many stress-related problems, including panic. When breathing is out of control and the blood contains high carbon dioxide levels, the survival window shrinks because you have less time to respond to a situation before the urge to breathe causes you to REACT, instead of ACT, in an emergency.

To start, STOP, THINK and attempt to make your ventilations a little fuller and longer. Concentrate on long, gentle exhalations to reduce the level of carbon dioxide in the blood. Three-second exhalations are a good goal. Exhale just a little more volume than the normal tidal exhalation and do so in a longer than normal amount of time. Do not attempt to blow out large volumes of air, because the reaction will be a large inhalation that may not help get breathing back to a comfortable, normal state.

Photo 14-1 Author Zaferes works with a student to teach vertical hovering in shallow water to prepare him for managing problems in mid-shallow water. Hover vertical and neutral and see what happens when you tense up your arm muscles.

If you discover that slowing down your breathing is difficult, find something to visually focus on, such as a watch or gauge. Again, concentrate on long, slow, gentle exhalations. If there is no visibility, focus on a task such as slowly counting your fingers. Only by reducing carbon dioxide levels and bringing breathing volumes and rates back into normal ranges can you begin to catch your breath. When breathing is under control, you can start to get other problems under control, including your heart rate and your ability to THINK.

Basic buoyancy

As you gain control of your breathing, the next important issue is your location in the water column. If you are near or on the bottom, gaining physical control will be relatively easy, as most problems are easier to fix when you are stationary. If you find yourself suspended in the water column and have difficulty stabilizing yourself, the course of ACTion will be a little different. Do not over-inflate or deflate the BCD, as rapid changes in buoyancy can only increase the problem. You need to make small, fine-tuned adjustments in buoyancy, that can later be altered easily as diver position in the water column changes. During every dive, every adjustment to the BCD should be performed slowly and with control; perfect practice makes perfect skills. In any case, you need to direct your attention to stabilizing yourself in the water

Photo 14-1a Also practice horizontal motionless hovering at a variety of depths. Then in colder months, practice these buoyancy skills in the same gear you will wear under the ice.

column. Continued descent or ascent must be by choice, not by accident.

Ascending may not be the best option under ice; going to the bottom may be preferable when a problem must be managed. This is especially true if the backup diver will be called. Too few divers can competently hover in shallow water, and even fewer can hover while fixing a problem with another diver who may or may not be stabilized at one depth. With regard to tether lines, it is more difficult to keep them taut when not on the bottom, and slack lines can only lead to worsening conditions. And lastly, working to hover at one depth in shallow water will also make it more difficult to get a breathing problem under control because it is task loading for most divers and pressure changes are greatest in shallow water.

Photo 14-1b Practice hovering with no mask and with flooded mask to prepare for accidental mask flooding or dislodgement and for switching to a pony regulator from a full face mask. Notice gauges secured under BCD.

If the diver is not entangled, the best plan of action usually is to head toward the hole and let tenders assist at the surface. As tenders quickly realize that you are heading home at an earlier than scheduled time, they will gently assist you in.

If a problem is air related, such as that you had to switch to your pony bottle, or one or both of your regulators is free-flowing, going back to the ice hole should definitely be the plan of action—unless of course you are also entangled.

What is the problem?

To "STOP" is initially to stabilize breathing, buoyancy, and body movements. Next, establish what is going wrong, maintain your composure, and avoid rushing. THINK, are you still connected to your tender with a tether line? Do you have air, and if so, how much? Is your pony regulator in place in your neck holder? Are you entangled, and if so, in your tether line or something else? If you are not entangled, can you ascend to the ice hole in a controlled manner? Is there a gear problem? If so, do you know what it is, and can you fix or control it during your ascent? Where is your buddy if you have one? Can you get his or her attention? Think, "What is the problem and what can I do to fix or control it?" What did you learn in your basic training that can help you here and now?

If your buddy is the one having difficulties, often a simple firm hand on the arm can begin to make a difference. Look your buddy directly in the eyes if visibility will allow it. Use your eyes and the gentle, yet firm, grip to tell your buddy "I am here; STOP, breathe, THINK." ACT to bring the situation under as much control as possible, and then decide what it will take to fix it. If you are a backup diver, a firm handclasp with the primary diver will be the first step to helping the primary diver stop and regain control.

When you first learned to dive, you should have learned and performed many of the skills that will help you stay calm, think and act in real situations.

Leg cramps

Leg cramps are not uncommon in diving. Leg cramps do not discriminate as to who can or will be affected by them. Lack of exercise, sitting in a poor position before the dive, dehydration, fatigue, cold-water temperature, restrictive equipment such as drysuit legs without enough air, diet, and carbon dioxide are all factors contributing to leg cramps. If you have a history of leg cramps while diving, perform exercises regularly that will prepare your leg muscles for the use of fins, such as lunges and calf stretches. Make sure you are well hydrated, and eat a plum or two in the morning.

If you have a tendency to cramp often, have your potassium levels checked, change your kicking style, make sure you have air in the legs of your drysuit, seek training for a stronger or more efficient kick, or change the type of fin you are wearing. Contingency divers should frequently move their legs, fidget, and stretch while the primary diver is diving.

No matter where you are in the water column, leg cramps are not only painful, but can become disabling. The first imperative is to control your breathing, remembering that high levels of carbon dioxide are only going to increase the problem. Long slow breaths will help control the carbon dioxide levels and therefore the lactic acid levels. STOP, move slowly, and stabilize your position in the water column. THINK, slow movements will assist in both controlling your breathing and helping to stabilize your position.

ACT, reach for the tip of your fin and pull it gently toward you for about sixty seconds. If the cramp is in the back of the leg, create a slight amount of tension in the upper thigh muscle to help release the back thigh muscles. If possible, use the other hand to massage the cramp. While doing this, be sure to breathe good, full respirations. Next, let the fin fall forward and fully relax the leg. Hold this relaxing position for ten to fifteen seconds. Repeat this alternating stretching and relaxing process slowly until the cramp begins to show relief.

Entanglements and entrapment

Entanglements are anything that hold you in place or restrict your ability to move freely in the water column. There are two different types of entanglement: direct and indirect. Direct entanglements snag divers' bodies and the equipment they are wearing. Direct entanglements can include: fishing line that snags untaped fin-strap buckles, weeds that catch a dangling gauge, a branch caught between a tank valve and the diver's body, or a gill net encompassing much of a BCD tank assembly. Indirect entanglements occur when the tether line becomes fouled, such as when the line gets stuck between two ice chunks at the surface, the line is bent around a rock or trapped under a vehicle because the diver went around instead of over it, or perhaps the line is tangled in a submerged fir tree.

Photo 14-2 A backup diver (left) brings up a primary diver (right) after freeing him from entanglement.

To help prevent direct entanglements, divers should move slowly and should have their gear set up properly with:

- fin straps taped
- pony regulator second stages and cutting tools secured firmly in the golden triangle area
- pony regulator hoses worn under their arms
- pony bottles secured against BCD backpacks
- regulator hoses trimmed downward alongside main cylinders as much as possible
- gauge and drysuit hoses secured under arms through BCD armholes
- no stacked weights on the belt
- nothing worn on the legs
- no snorkels worn

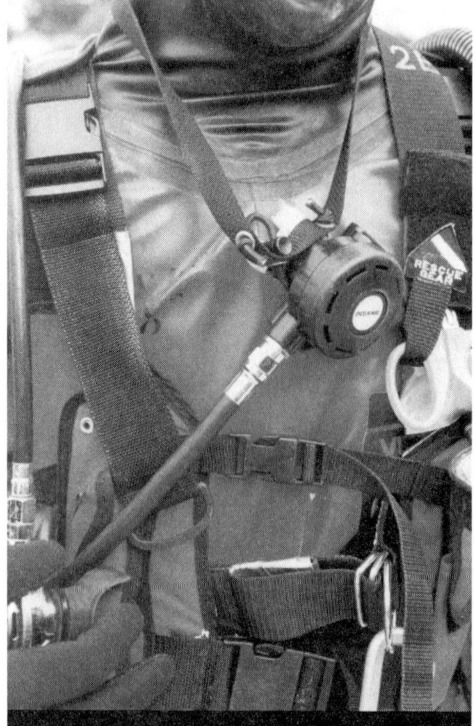

Photo 14-3 Proper trim to decrease entanglement. Gauge is secured through BCD armhole under arm. Pony hose, with exhaust tee removed, is worn under the arm and the mouthpiece is secured in the neck strap. Shears on harness in holster. Knife, second pair of shears, and standard mask secured in BCD pockets.

- tools and other items worn in BCD pockets rather than hanging off the divers. Tenders should examine fully dressed divers to see if there are any potential snag places and what can be done to eliminate them.

The risk of indirect entanglements can be reduced by keeping the tether line taut, which is done by the diver in tender-directed diving and by the tender in diver-directed diving. Tenders should continually work the line to make sure it is not caught in chunks of ice against the underside of the ice roof. Tenders may need to lie on the ice and put their tending hand low into the water to release the line from an ice snag. If the line does become slack, the tender should take up the excess line. In tender-directed diving, the diver should then give one pull and give enough line to put the diver back at the right distance.

Entrapments are rare, but when they do occur, it is usually because the diver is wedged between a shallow bottom and the ice roof. A diver can also become entrapped under part of a snowmobile or vehicle after using improper lifting techniques. Preplanning to ensure that water is deep enough and proper lifting training will help prevent these occurrences. Author Hendrick experienced a blackwater entrapment in a submerged van when the shag carpet separated from the van's ceiling and fell down, covering him completely. Be prepared for the unexpected by using the STOP and THINK techniques.

Photo 14-4 Tenders need to do whatever it takes to make sure there is no slack in a diver's line. The recreational diver's tender on the boat did not take up the diver's line quickly enough as she ascended so a 90%-ready diver's tender immediately managed the problem to prevent an entanglement.

When you relax, STOP, and THINK about the problem; usually you can fix it (ACT). Conversely, when you pull and tug and try to rush, problems have a tendency to escalate. If the use of a knife or shears becomes necessary, slow methodical movements are required. The slower you move, the longer your air will last; the better you can think, the better you can act and the easier the job becomes. "Rush makes slow and slow makes fast" is a good saying to repeat to yourself regularly.

When you first feel entangled or entrapped, STOP, and give the "2+2+2" signal to tell topside that you have a problem, you are okay, you are going to begin managing it by yourself, but alert the backup diver. The tender should gently take up the line tension if it is not already taut, alert the backup tender to ready the backup diver and make a notation on the profile map. THINK about how you are restricted. Is it a particular part of your body that is restricted? If so, search it with your hands to figure out what is snaring you. If you cannot reach the entanglement, send the "3+3+3" signal to say that you are okay, but you need some assistance. If you can use your shears to cut the restricting object, do so. If you use a knife, leave it on the bottom for later retrieval with an evidence container. Do not resheath knives with cold hands or in low visibility to prevent injury to self or equipment.

If you cannot move your body, are you entrapped? If this is the case, try backing out the same way you went in. On her first blackwater search dive, author Zaferes swam into a broken shopping cart. Not able to go forward, up, left, or right she backed out, but not in the exact same way she went in. Her regulator hoses at the first stage caught on some part of the broken cart, preventing retreat. After she stopped, moved a little forward, and then went back the same way she came in, she was free. If backing out is not a possibility, send the "3+3+3" signal.

Is forward motion prevented because the tether line is caught? If so, attempt to retreat a few feet, tighten the line, and then dance the line over whatever it may be snagged on. The tender will feel this line motion and can give a one-pull signal to check that you are still okay. If the line cannot be freed, ask for assistance. If the tender realizes that the line is snagged, his first action is to make sure that the line is taut. Second, the tender moves a few feet in the opposite direction that the diver was headed. This second step is done to regain direct line access, to make sure line-pull communication is still possible, and to perhaps release the snag.

In a non-overhead environment, it is easy for tenders to notice an indirect entanglement because the diver's bubbles are no longer in line with the tether line. Divers who have continued to move forward when their lines are snagged should feel that their body is pulled slightly toward the line to a more perpendicular position, rather than the earlier described position 45° away from the line. Sometimes tenders will give a signal through a taut line and the diver will not reply. If the diver does not reply after two more signaling attempts, there is a good chance that the line is fouled. The tender should move a few feet in the opposite direction of the diver's direction and attempt to signal again. The tender then makes an attempt to dance the line over the snag. If this fails, the tender should send in the backup diver to man-

Photo 14–5 Diver surfaces with his cutting station "artwork". In black water, cut twine, wire, fishing line, and a tie-wrap and link each of these items together with knots and by re-using the tie wrap.

age the problem. The profiler needs to make a notation of where the diver was when the snag occurred. If the snag recurs more than twice, split the pattern and work either side of the snag until it is removed, and then go back to the normal length sweep.

If you find that you are entangled directly and can reach the entanglement, use shears to cut yourself free. Do you have the skills to do this? How often have you practiced with your cutting tools? Have you practiced cutting fishing line or twine underwater with hands in thick gloves? In zero visibility? If your answers are no, then practice in a controlled environment. Take some fishing line, twine, light wire, and plastic tie wraps, go underwater, and cut them. Gain skill and experience in the learning environment. Then practice cutting entanglements off another diver. Perform these skills with thick gloves and a blacked-out mask.

Backup divers should not use knives when managing entanglement problems on a primary diver. There is always a strong probability that the knife could cut equipment or the diver, especially in limited or no visibility or when hands are cold or shaking. Another reason is that the cutting motion of a knife will pull on the entanglement, which can cause injury if there is a fishhook involved. Become proficient with shears, wire cutters, or whatever tools are necessary for the entanglements in your local waters.

Accidental weightbelt loss

In non-overhead diving, the accidental loss of a weightbelt can result in a lung overexpansion injury or, if enough nitrogen is ingested, decompression sickness. An ice roof adds the possibility of head or neck injury and getting stuck under the ice roof.

The heavier the belt, the more likely it is to slip down over narrow hips, and the faster the ascent will be because the wearer will have more air in the BCD/drysuit to compensate for the overweighting. Proper weighting is important.

These guidelines will help prevent weightbelt loss:

- Be neutrally weighted.
- Use metal, rather than plastic, belt buckles.
- Make sure the free webbing comes straight out of the buckle. If it is at an angle, the buckle cannot properly close.
- Have 10in to 11in (25cm to 28cm) of free webbing beyond the buckle when it is worn to give the wearer an opportunity to grab the webbing should the buckle accidentally open.
- If you forget to put the weightbelt on before donning the BCD, do not shortcut by trying to put the belt on afterward. Rather, remove the BCD, don the belt, and then don the BCD.
- If you have experienced the belt sliding off your hips in the past, use a weight harness instead of a belt, but make sure the harness allows the weights to be removed at the end of each dive.
- Some divers wear double buckles, which require ditching practice. We have not found the need to use double buckles, but it is a valid technique.
- Wear a weight harness system if your body shape allows weightbelts to slip down past your hips.

Should the weightbelt come off underwater, do your best to flare out tank-side down, hit the drysuit exhaust valve with your head, pull the BCD inflator hose down to dump it and exhale continuously. The act of dumping the drysuit exhaust valve against the side of your head allows you to cover your head with the upper arm. If you cannot get the exhaust valve against your head, use the arm to protect the head and face anyway.

Once on the surface, allow surface personnel to transport you back on a sled and perform field neurological examination to check for any decompression illness or head injury problems. Seek medical attention if any signs or symptoms are present.

Photo 14-6 A tender checks that the webbing comes straight out of the buckle to ensure that the buckle is properly closed.

Photo 14–7 If the belt is slipping off, do not shortcut by trying to get it back in place with the BCD on. Rather, remove the belt, pull the diver out, remove the BCD, put the belt on properly, and then don the BCD.

Sudden loss of visibility

Diving is most enjoyable when we have the opportunity to witness the underwater world; the sudden loss of that ability can be unnerving, to say the least. Many new divers state that their greatest concern without their instructor is murky or silty water. Low visibility exists in many parts of our diving world but should not be a deterrent to enjoying diving. Because sediment settles in ice-covered water, ice divers typically have better visibility; but they also have greater risks of silt-outs causing zero visibility if the bottom is stirred up. Ice divers, like all divers, should learn how to move without any use of their hands. Sculling hands stir up bottoms.

During warmer months, practice swimming a few inches off the bottom without kicking up any bottom sediment. Practice slow 360° rotations one foot off the bottom; without moving your navel past a chosen spot on the bottom, make a big circle with your fins. Do this while holding your hands together against your abdomen. Practice gently settling down on the bottom on your knees and stomach and then gently lifting off the bottom with minimal sediment agitation. Being over-weighted will greatly increase the occurrences of silt-outs, as will ankle weights that are too heavy. Find an instructor who can demonstrate these skills, and then teach students to successfully perform them.

In warmer weather, take a few minutes to play with silt. Kneel in shallow water on a silty bottom area, move your hand slowly over the silt, and watch what happens. Next, place your hand into the silt slowly and pull it out to see what happens. The silt will come up and then settle. Now do it faster and notice how much more silt is kicked up. See how much movement, or lack of movement, it takes to stir up a silty bottom and how quickly it clears up.

If you suddenly find yourself in a silt-out, STOP, and try to keep in mind that it was most likely your movements that caused the disturbance; additional movements are only going to make things worse. Settle down and relax, breathe, and THINK, "There is nothing here that is going to harm me, it has simply gotten dark, and the bottom has simply become stirred up." If you can sit quietly, with no quick or unnecessary movements, visibility will usually return in a relatively short period of time, generally in sixty to ninety seconds. You can then ACT: ascend or move slowly away from the silted area. If the silt does not clear up in a reasonable time, you can still slowly ascend if you are not entangled. Once you have learned how to deal with silt-outs they become simply a momentary inconvenience.

Photo 14-8 Author plays with swimming a gauge-length away from silt.

Public safety divers often search in low, zero, or blackwater visibility and should be trained accordingly. Practice all basic, search, self-rescue, and diver-to-diver assists/rescue skills with blacked-out masks to be prepared to successfully perform these skills with confidence in blackwater. Use your mind's eye to picture what you are touching, just as you would when searching. If you become spooked, repeatedly THINK, "I have plenty of air in my cylinder and pony bottle. I have three cutting tools that I have practiced with and can always reach; my equipment was fully checked by my tender and the safety officer; I have direct line access to my tender and a fully ready, trained backup diver and a trained 90%-ready diver. I am continuously being monitored by my trained tender and the dive coordinator. I am okay." All blackwater divers have become spooked at least once, most likely more often. It is perfectly normal, so do not let it upset you. Having that ice roof overhead increases the chances of becoming spooked, so be prepared for those feelings. Communication systems help prevent such feelings and help manage them should they occur.

Current

The best ice typically does not form over moving water, so recreational divers rarely find themselves in current under the ice. If they do, they should abort the dive. Currents add new dimensions that can initiate problems or make minor ones into larger ones. Currents can render entanglement management more difficult, increase the chances of free-flowing equipment, and cause mental and physical overexertion. A tender holding the tether line in a current may be pulled into the ice hole if the current is strong enough and other surface support do not take hold of the tender's tether webbing. For these reasons, drop a slightly weighted object into the water to see if there is a current prior to putting in divers. Public safety divers should set a safety standard of half a knot (50ft/m) of current as the maximum allowed to put divers in the water. Extensive training is required to dive under the ice in faster currents.

Out-of-air emergencies

Divers are more likely to have an air-related emergency in frigid water than in warmer water because of regulator free-flow problems and possible higher air consumption rates. Because divers may be 60ft, 80ft, or even 90ft from the hole, they may have to travel a distance to reach air at the surface should a low-air or out-of-air emergency occur. For these reasons, dive times are kept short, typically fifteen minutes with a possible five minute extension. Depths are generally kept above 60ft. All divers wear a true redundant air source, or quick-release pony bottle. Divers preplan their air according to their SAC rates, and tenders regularly check and document diver breaths per minute. To prepare for simultaneous out-of-air and entanglement/entrapment incidents, there are a full main contingency bottle and a full pony bottle ready for the backup diver to take down if necessary.

Should the main cylinder empty for some reason, a diver should switch to the pony regulator and make an ascent back to the hole. As the diver heads toward the hole, the tender should pick up the slack. Because this is an unexpected return to the hole, the tender should aid the diver's return by gently taking the line in, and the backup diver should be put in ready status.

Figure 14-9 Train for reflexive self-rescue procedures on every dive. Implement the procedure of switching to the pony regulator prior to the final ascent and ascend on the pony regulator. Notice that the pony mouthpiece did not have to be removed from the neck tubing.

The question often arises: When should the primary diver request backup diver assistance? The answer is: Only if the backup diver is actually needed. If nothing is wrong with the pony bottle air supply, and if there is no problem swimming to the hole, then it is best for the out-of-air primary diver to go back to the hole without backup diver assistance. The minute a backup diver is put on the primary diver's line and heads out, the primary diver's return may be slightly or significantly hindered. It takes a great deal of practice for a backup diver to head across and down an unanchored line in shallow water without pulling on that line at all. This becomes even more difficult when there is no visibility or if there is stress because it is a real call for help, not just a drill.

The maximum distance the primary diver should be from the hole is 100ft, and that is only if the diver is an experienced ice diver and conditions are good. It should not take more than ninety seconds for the out-of-air primary diver to reach the hole. How long will it take the backup diver to deploy and reach the primary diver? Without sufficient practice, it will likely take that much time or more. Then, what happens when the two divers meet? If you have ever watched two divers attempting to hover together in shallow water with visibility, you know it is not a pretty sight. Now imagine what happens if there is limited or no visibility. Such an occurrence should be avoided unless absolutely necessary. An example of this might be if the diver's pony bottle begins to free-flow when the diver is still far from the hole. If there are two divers down as primary divers, both will have to go back to the hole in a low-air or out-of-air emergency.

Transferring to pony bottles, especially in contaminated water

If public safety divers run out of air in their main tanks and must switch to their pony bottles, what is the best way for them to do so? There are four methods to handle an out-of-air or planned air-exchange procedure that are explained in chapter 5:

- Standard primary regulator switch to pony regulator
- Commercial block system
- A full face mask with ports for two separate regulators
- RSV-1 public safety diving block

Whatever options and equipment your team chooses to deal with out-of-air emergencies, one factor remains critical: every diver must practice handling these problems! Divers should practice being the one who runs out of air, as well as the backup diver who provides more air. Only consistent, frequent rehearsal will enable a diver to perform adequately and safely in an out-of-air situation. Procedures for a simultaneous out-of-air and entanglement occurrence are covered in the diver-to-diver assist section.

Free-Flow Problems

Regulators at the surface

Recreational divers usually keep a bucket of warm water nearby for free-flowing equipment. Turn the air off, place the item in the warm water, and gently slosh it around. Gently turn the air on, gently put it back in the cold water, and give a few long, slow exhalations and gentle inhalations. Do not remove it from the water until the dive is over.

Public safety operations do not usually have warm-water bucket options, unless it is a recovery dive. During surface free-flows, the best option is to turn off the tank. Make long, slow exhalations through the regulator while keeping it out of the water. Four to five exhalations can often clear a second-stage free-flow. Slowly turn the air back on. If the regulator did not thaw out, repeat this process until it does, or better yet, get a new, dry regulator. A heat pack wrapped in a piece of cloth can also be used to thaw out the regulator. Do not warm the regulator too much, though. Keep in mind that a second-stage free-flow will often become a first-stage free-flow if the air is not turned off quickly enough.

With a first-stage free-flow, which sometimes appears as a thin coat of white ice around the first stage, immediately turn off the air. Next, submerge the first stage and make at least ten long, slow exhalations through the second stage, as that will most likely be frozen as well. Have the tender wrap his gloved hand around the first stage and gently rub it, creating a little friction. If caught soon enough, this sometimes works. A heat pack can also be applied, but without warm water, it is usually better to simply change the regulator and not waste time. Use the transport device and a dry container to bring in the replacement regulator.

Submerged regulator free-flows

Though surface free-flows are more common, sub-surface free-flows can also occur. To prepare, ice divers must practice breathing off a free-flowing regulator in warm water. To simulate a free-flow, depress the purge button fully. Keep in mind however that there is a chance that depressing the button for a minute or more will cause an actual free-flow; therefore, be sure to have a pony bottle as a contingency air source. Place one hand flat and horizontal between the upper lip and the nose to divert the air and prevent mask dislodgement. With the other hand, pull the second stage slightly out of your mouth so that you are not choked by the rush of freezing air and water vapor. Breathe gently as you return to the hole. If you feel like your mouth is freezing and will soon be nonfunctional, switch to the pony regulator. High-pressure air mixed with water produces frozen water crystals. Breathing off a free-flowing regulator under the ice can be more than disconcerting for someone who has not properly practiced the skill.

Some divers have been given an opportunity to breathe off a free-flowing regulator in their entry-level class, but they most likely only did so while kneeling on the bottom. Ice divers should practice this skill in warmer months in mid-water while making a swim back to a designated "ice hole".

During the ascent, divers most likely will need to release air from their BCDs and drysuits. If you have to use a hand to make these adjustments, use the hand that was protecting the mask. Tilt the face downward so that the bubbles will have less effect on the mask. If the main cylinder empties prior to reaching the surface, switch to the pony regulator.

Full Face masks

Full face masks are masks with a built-in second-stage regulator that requires the same basic care as any other standard regulator. Demand masks are less likely to free-flow, but they need to be kept warm and out of the water until ready to go. Once wet or submerged, they should be kept in that condition. Positive pressure masks are more likely to have free-flow problems.

The exhalation port can freeze on some units after becoming wet. This can make it difficult or nearly impossible to exhale through the mask. If tenders and divers are unobservant during gear checks, a frozen exhalation can be missed, because the diver's exhalation can escape from the mask seal instead of the frozen exhaust port. It is important to listen carefully to the diver's exhalation to ensure that it is expelled through the exhaust port.

If the port is frozen, have divers make several warm, long, slow yet forceful exhalations through the mask to warm the ports. Warm exhalations are made by making an "ahhh" whispering noise. Try it now. Blow through pursed lips on your hand. Then open your lips and exhale with the "ahhh" whisper. The latter should feel warmer. If the exhalations do not work, try placing the mask in water and repeat the exhalations. If warm water is available, use it.

Once underwater, the best action to take for a full face mask free-flow is to breathe off the mask until the main cylinder is empty, then switch to the pony. If a block is used and the free-flow is from the main regulator first stage, the diver may avert a second-stage free-flow if he makes the switch quickly enough. Then, the diver can make a comfortable egress to the hole. If the second stage is free flowing, switching to the pony will not help, because the same free-flow will occur since the same second-stage regulator is used. Breathe on the mask until the main cylinder runs dry and then switch to the pony.

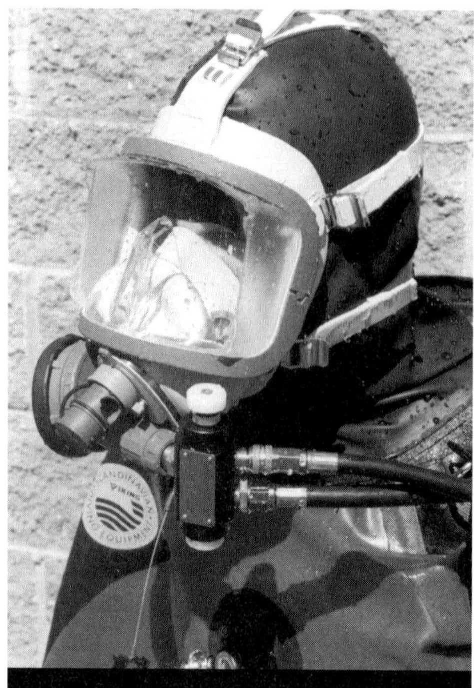

Photo 14–10 For a second-stage free-flow, slowly open the SBV to allow excess air to vent. When the main cylinder empties, pull off the block cable and push up the slide to switch to the pony.

Another option is to wear a second-stage regulator on the pony in addition to the connection to the block. Then if a full face mask, second-stage free-flow occurs, and the diver is unable to breathe from the mask because the free-flow is too strong, the diver can remove the mask and switch to the separate pony second-stage mouthpiece. If the water is contaminated, this is not a good option. If the switch is made, don a standard facemask and make sure the block on the mask is switched back to the main cylinder, or the pony will be emptied much more quickly. If the diver wears an SBV, gently pull it out to let the excess air vent.

Full face mask divers should keep their eyes closed when breathing off a free-flowing regulator, because air and water drops will be blown into their eyes. Blacked-out mask drills are thus imperative. Also, tenders should be aware that they might be unable to receive verbal communication from divers. Since the communication microphone is located in the oro-nasal mask the sound of free-flowing air may overwhelm voice communication. Therefore, line signals become essential.

Keep full face masks indoors until needed and always hand dry and hang them in a warm, dry area after use. During cold weather, store the masks in a warm, dry felt or wool bag. Keep the mask in the bag until the diver is ready to go.

Power inflators: BCD and drysuit

Power inflators are treated basically the same way as regulators. Keep them off the ground and out of snow and ice. During extremely cold conditions, test them in a dry environment. Once they are wet, keep them wet. BCD oral inflation reduces the occurrence of freeze-ups on the surface. Do not make oral inflations underwater, because that requires removing the regulator from your mouth. That is not a good idea.

If you use the BCD power inflator to make buoyancy adjustments, use short, gentle bursts. Obviously, in an emergency at the surface, use it as necessary. Use power inflators sparingly underwater—simple to do when neutrally weighted. Elevator or BCD-buoyed ascents are major no-no's in ice diving, as in all diving. Therefore, divers need to be neutrally weighted, or if public safety diving, be a maximum of 2lb negative.

Tenders can hold divers up at the surface so that they do not have to use their BCD power inflator to fully inflate their BCDs. The tender should go down on one knee, lay the diver's tether line over the tender's upward knee, then pull down on the line with the opposite hand.

Photo 14–11 Author Hendrick takes control of student's tether line over his knee to talk with the student who is showing early signs of stress. Notice that Hendrick holds the diver up and keeps the regulator submerged to help prevent a regulator and BCD inflator free-flow.

Should a drysuit or BCD free-flow occur, the best defense is to immediately disconnect. For a BCD free-flow, disconnect and dump enough air to become neutral, and do not forget to keep breathing. Flaring the body may be necessary if a rapid ascent has already initiated. A drysuit free-flow is a serious problem. Hopefully, the drysuit did not already have too much air before the free-flow occurred. Your line-of-defense action is to very rapidly disconnect, dump, and flare out in a supine position. If your arm exhaust valve has a domed top, practice dumping the air by hitting it against the side of your head. Do not forget to breathe normally and continuously.

Attempting to reach up into a hood to release air from a neck seal while wearing thick gloves is not a realistic option. Even if it can be done, it will take a long time. If the diver is wearing a full face mask, it's even harder to reach the neck seal. If a wetsuit hood is worn and the seal is neoprene, there is more probability that the diver will be able to grab lower down on the neck and pull it away, but it would take much practice to do this quickly.

If a free-flow results in a diver inversion, do what you were taught in your drysuit course: tuck and roll to become right-side up in order to dump and flare to slow the ascent. If you are in shallow water and there is a good deal of air in your suit legs, it is likely that you will not have time to become right-side up. Therefore, be prepared to exhale gently and continuously and to hit the ice roof with your feet. Once your feet have made contact with the ice roof, you can stay upside-down and let the tenders pull you home. Under-ice hockey or soccer games provide the necessary skills for this kind of egress. Remember to breathe gently, because upside-down regulator second stages may breathe wet.

Disconnected diver

In more than thirty years of teaching tethered diving, including ice diving that encompasses more than 10,000 dives, we have had only one case of an accidental disconnect. This incident occurred in warm water diving and happened because the carabiner was not locked and the diver accidentally opened it underwater. It was rumored that the diver disconnected purposely as a goof. After that occurrence, we made it a rule that all carabiners be locked, even in warm-weather diving. Since all carabiners in ice diving are taped, the odds of tenders forgetting to lock the carabiner and the safety officer missing it in the second equipment check are almost nil.

The other possible cause of a diver-disconnect is that the line hits a sharp object that cuts the line. We have never heard of such an occurrence but that does not mean it cannot happen. The following discussion on diver-disconnect is academic. However, we still need to plan and train for diver-disconnect.

Diver-disconnect deals with an ice diver's worst nightmare: finding themselves untethered and disconnected from the surface. At this point, the diver needs to make a very specific series of decisions and take specific actions. It is difficult for the rescue crew divers and tenders to think along with the diver unless the diver has an exact, practiced pre-plan or a plan that everyone has agreed on and is 99% sure will happen. The diver, if not tangled on the bottom, has only one job—go directly to the surface and hang. The diver does not swim around looking for the hole, because the chances are that the diver will progress farther from the hole and out of the DAD area that the tender knows the diver was last in.

Going directly to the surface greatly reduces the area that the backup diver has to search. In shallow water, it also decreases the diver's air consumption and increases his visibility. Being on the surface just under the ice roof might also cause the diver to be snagged by the backup diver's line as the backup diver spiders in a circle directly under the ice.

Some ice divers shovel snow in bars off the ice to make spokes of a wheel, all leading toward the hole. The problem with this approach is that the divers, if they can see the bars, may progress away from the hole. A better plan is to simply make a direct ascent to the surface and wait. The tender can use the profile map to locate the last known distance out and DAD quadrant. If the tether line is kept taut, both diver and tender know immediately when a disconnect occurs; therefore, the last known diver location would be fairly accurate. For example, the diver may have been 37.5ft out to the left side of quadrant 2, in line with the blue house on the opposite shore. If the diver moves around, that accuracy is lost.

The disconnected diver can use an ice screw to anchor into the ice if he is carrying one. An ice pick can also be used as an anchor and can chip a hole through the ice relatively quickly if the ice is thin. If the main cylinder empties and the diver transfers to the pony, the weightbelt should be dropped to ensure that the diver remains at the surface even if consciousness is lost. The diver must not go back to the bottom for any reason. Remember, the surface crew and backup divers want to rescue the diver as quickly and as safely as possible. If forced to search both the bottom and the surface, the attempted rescue may take too long.

In this situation, some training programs present the use of backup-diver tether lines 250ft to 300ft in length. This is not recommended! The idea is that the backup diver will swim directly out 300ft and then swim a 360° circle to find the missing diver. As the diver swims, the line must be kept taut. This is a difficult skill, and for some divers it is nearly impossible. To swim a circle of that circumference, including the 300ft out and back, the diver must swim nearly half a mile. It would be difficult for any backup diver not to start out at least a little stressed when searching for a disconnected teammate. In addition, these divers are swimming under the ice roof, not on the bottom, which is more difficult, especially because most of them are overweighted.

During summer trials in a variety of conditions, divers repeatedly failed at this task. Leg cramps, low air pressures, fatigue, long duration, respiratory overload, and other problems occurred, including regulator free-flows from hard breathing. Often the divers were not able to maintain a taut line long enough to swim one-third of the circle. The drag on the line would make the line move more slowly than the diver, forcing the diver to continually back feed to the center point. The slower the line moved, the greater the drag, and the greater the loop in the line. Failure was recurrent, even when testing divers who said they had successfully performed this skill during ice training programs. Even more frightening is the fact that many instructors who teach this plan suggest 1/2 inch diameter tether line or heavy Kernmantle water rescue line for tether line!

After many dozens of trials, we found that almost any diver with proper training, but less than six ice dives, was able to make a successful rescue operation at distances up to 110ft to125ft most of the time. It takes an additional ten dives to make them capable of successfully working out to 150ft under the ice. These trials established 150ft as the maximum backup-diver line length for safe diving operations.

In warm months, we discovered that it can take a backup diver three or more minutes to comfortably reach a still tethered, trapped, or tangled diver on the bottom 200ft to 250ft away. The arriving backup diver often would not be capable of doing more than trying to catch his breath for the first ten or so seconds. These results demonstrate a lack of safe diving practices. These distances are not acceptable in non-overhead dives and are certainly not safe for overhead or confined-space diving.

As a diver, if you discover you are disconnected, perform the following procedures:

In non-moving water:

1. STOP, THINK, and then ACT.
2. Do an air check.
3. If you know that something is wrong, so does the surface crew, and the backup plan should already be in motion.
4. If you have an extreme amount of excess line, do not become entangled in it. Move slowly. Do not follow the line, as you cannot be sure which direction the line came from. It might go back to the hole or it may have moved. Although keeping the remainder of the tether line attached could cause an entanglement, the backup diver could find the floating line and perform a more rapid rescue. Therefore, keep the tether attached until you are snapped into the backup diver's contingency strap. Be careful of any loose line.
5. Make a direct, continuous safe ascent to the ice roof or surface. Make yourself slightly positive, so that you can hang in the upper water column directly under the ice. Make yourself as large a surface area as possible so that the safety diver can increase his possible visual contact as well as his tether-line catch capability. Hang vertically to increase the probability of the backup diver's line catching on your body as it goes by you. Make sure that a hand or your head remains in contact with the ice to prevent the backup diver's tether line from slipping between you and the ice.
6. While waiting to be found, breathe normally, with slightly longer, gentler exhalations as described earlier. Do not attempt to conserve air, as that typically results in higher carbon dioxide levels and greater air consumption. All you can do to minimize air consumption is breathe properly and move as little as possible.
7. Rotate slowly while staying in place to look around for visual contact with the backup diver or his tether line.
8. Remind yourself that without any free-flowing regulators you should have plenty of time. You have at least 1,000psi in your main tank and a full pony bottle. This provides you twenty to thirty minutes for the safety diver to make contact.
9. Wait, do not move, do not go anywhere, keep telling yourself that you have a plan and stick with it. You will be found!
10. If you must transfer to the pony bottle, remain calm; again, do not go anywhere.

11. If you have to switch to the pony bottle, drop your weightbelt. The team is looking for you at the surface of the ice roof, not on the bottom; hence, you need to stay in one place at the surface even if you become unconscious.
12. If the backup diver's tether line does snag you, grab hold of it and give the "6+6" object found signal to tell both the backup diver and tender that you are on the line. The backup diver will then head toward you as the tender gently pulls the line in. If you still have your harness carabiner, snap the line into it.

To recap, in the extremely unlikely circumstance of being disconnected, make an immediate, controlled ascent to the surface and wait for help to arrive. Do *not* swim around in an attempt to locate the hole!

Diving under the ice in moving water is extremely advanced technical diving and requires advanced specialty training. Divers in moving water should carry an ice pick. As a diver in moving water, if you discover that you are not attached to the tender, perform the following procedure:

1. Perform steps 1 through 4 of the preceding procedure.
2. Do not leave the bottom, and if possible do not move. Hold onto something if possible to give you physical attachment to the bottom. You can jam the ice pick into the bottom and then bring your body as high off the bottom as possible without losing contact with the ice pick. An increased surface area is easier to find.
3. Follow steps 6 through 10 of the preceding procedure.
4. If you must transfer to the pony bottle due to a lack of air in the main system, remain calm; again, do not go anywhere.
5. When you know you are on the last few breaths of your pony or alternate air source, *do not drop your weightbelt.* Let all the air out of your BC and become as heavy as possible. The backup diver will be looking for you on the bottom.

Backup diver procedures

In the case of a lost or disconnected diver, the search area must be as small as possible.

In a non-moving water environment, where entanglement is not part of the problem, the surface (under the ice roof) is the shortest distance and smallest area to search. Having the diver come to the ice roof reduces the need for depth and slower bottom-type searches. The diver spiders out and then proceeds with a spider search.

Photo 14–12 An ice pole is placed in a weep hole in DAD quadrant-1 to simulate a disconnected diver. The backup tender gets low to the ice, with hand in the water if possible, as the backup diver spiders a little beyond the primary diver's last known location to begin a search.

The question of where to locate the hot spot to begin the search should be provided by the back-up tender, who has accurately documented the "last-known" area. In rare clear-ice conditions, a topside search by surface personnel may visually pick up the lost diver hanging just under the ice roof. Do not cut the ice (unless no other choice is available) to retrieve the diver from the water. Send the backup in for the rescue.

When the backup diver has located the primary diver, the backup diver snaps the contingency strap into the primary diver's harness tether point, gives the backup tender a "6+6" and home they go, gently and slowly. If visibility allows it, the divers should remain in close eye-to-eye contact. The backup diver may have to assist the primary diver with maintaining neutral buoyancy. The two divers can swim side by side or, with practice, the backup diver can rotate in a supine position directly under the primary diver maintaining eye-to-eye contact. If the diver's weightbelt is gone, the diver needs to be pulled back to the hole by the tender. This must be done very slowly to prevent pieces of uneven ice from hitting the diver in the head.

When the Backup Diver Has To Respond

Everyone should be aware that deploying the backup diver potentially creates problems. These can include:

- Congestion is created in a small space—the ice hole.
- Without special hands-on training, all divers involved can be injured.

- More divers on the bottom cause more silt-outs, more chances of entanglement, and if not well trained, more confusion and panic.
- Backup and primary tether lines may become entangled.
- The operation is further complicated if there are two primary divers down.

Ice dives should never take place without a backup diver, but having a backup diver does not alter the fact that the primary diver must be as self-sufficient as possible. When the backup diver is deployed, alert the shore operations and move the 90%-ready diver into backup-diver status.

Take a moment to reread the scenario at the beginning of this chapter. Now it is time to learn well-proven procedures for a backup diver to assist. To start the assist operation, the backup diver needs to have a contingency strap, as described in chapter 5. A contingency strap, secured into the backup diver's harness carabiner, is snapped into the primary diver's tether line when the backup is called. This strap provides several important advantages.

Backup divers should not hold onto the primary diver's tether line, because it is likely that the tether line will be yanked and pulled as the backup descends to reach the primary diver. We have had many teams tell us that their divers make a circle with their fingers and thereby never pull on the tether line, but when we throw an unexpected "I need immediate help" call, more often than not, the circled fingers become a strong grip on the line. Remember, stress can increase buoyancy for a variety of reasons. Increased backup diver buoyancy or over-weighting will cause the primary diver's line to be pulled. If the backup diver must use a hand to maintain contact with the primary diver's tether line, the backup diver becomes a one-handed diver. When one of our own is in trouble, we want the backup to have both hands free to equalize and adjust buoyancy during the descent, communicate and reassure the primary diver, and handle the presenting problems. Without a contingency strap, if the backup, for whatever reason, loses hold of the primary's tether line in no or low-visibility water, the backup must resurface and swim back

Photo 14-13 The backup diver (right) brings up a disconnected primary diver (left). The backup diver's contingency strap connects the two divers.

Photo 14–14 A backup diver can rapidly disconnect from the primary diver's line by using the quick-release in the middle of the contingency strap.

to the tenders to regain a hold on the line. This delay could greatly affect the outcome of the assistance/rescue.

If the backup diver must use a hand to maintain contact with the primary diver's tether line, the backup diver becomes a one-handed diver. When one of our own is in trouble, we want the backup to have both hands free to equalize and adjust buoyancy during the descent, communicate and reassure the primary diver, and handle the presenting problems. Without a contingency strap, if the backup, for whatever reason, loses hold of the primary's tether line in no or low-visibility water, the backup must resurface and swim back to the tenders to regain a hold on the line. This delay could greatly affect the outcome of the assistance/rescue.

Photo 14–15 The 90%-ready diver moves to backup position.

The contingency strap can also be used to tether a disconnected primary diver to the backup diver. For example, the primary diver may have an indirect entanglement. Both the primary diver and tender did not pay enough attention to the line, it became very slack, and it ended up becoming wrapped around a submerged tree as the diver changed direction. In this case, it may be easier for the backup diver to disconnect the primary diver from the tether line than to attempt to untangle it.

In thousands of dives, we have never seen this happen, but it could happen to teams who do not know how to dive a taut line, or it could happen to diver-directed recreational ice divers and tenders who have received poor instruction. When the backup diver comes to the point at which the tether line is fouled, he simply disconnects the contingency strap from the tether line, goes to the other side of the entanglement, and reconnects into the tether line. When the primary diver is reached, the backup first makes sure that the primary is not directly entangled in any way. Then the backup diver connects the contingency strap directly into the primary's harness carabiner and cuts or disconnects the primary tether line. The two divers then make a direct ascent to the surface together. Without the strap, the primary is no longer tethered after the tether line is cut, and no backup can guarantee not to let go of the primary diver during the cutting, ascent, and surface swim.

A quick-release clip in the middle of the strap enables the backup to quickly get away from the primary diver, if for some reason that becomes necessary. The clip also allows the backup to disconnect and move around the primary diver to deal with harder-to-reach entanglements. When the job is completed, the backup can reconnect to the other half of the contingency strap.

When a backup diver is called, the backup tender notes the primary diver's location on the profile map and passes off the profile slate or simply puts it down if there is no other person available. If a full face mask is worn, have the diver quickly splash some water onto his face before donning the mask while sitting on the edge of the hole. The backup tender assists the backup diver into the water. He snaps into the primary diver's line with his contingency strap and lays two fingers in a peace sign on the line to maintain knowledge of where it is so that he does not pull too far away from it with the strap, which would pull on the primary diver. He makes a controlled, feet-first descent with his body angled forward to follow the direction of the line. He puts his other hand out in front of his head to protect it and to prevent running into the primary diver.

As the backup diver descends, the primary and backup tenders stand 6 feet apart to help prevent the lines from intertwining. Meanwhile, the primary diver has his hand on his carabiner while waiting for the backup diver to arrive. As the backup diver reaches the primary diver, he follows the line down to the primary's hand and grasps it in a strong thumb-to-thumb grip. The backup knows that the primary diver's hand will be on his carabiner because that is part of the practiced plan. There is no downtime spent searching for his hand. The primary diver tells the backup diver what is wrong by using hand signals, which are particularly important in blackwa-

Photo 14-16 The contingency diver places two fingers gently on the primary diver's tether line during descent to prevent moving in a direction that would pull in the line, and therefore pull on the diver in need. All other fingers are tucked away to prevent the contingency diver from accidentally grabbing on the line and pulling on it.

Photo 14–17 Tenders separate themselves when the backup diver enters the water to decrease the risk of lines intertwining. Here both divers have surfaced.

Photo 14–18 When the contingency diver makes contact with the primary diver's hand that was waiting on the primary diver's harness carabiner, the two divers engage in a strong, confident hand grip that allows effective communication between them.

ter or if the bottom is stirred up. When the primary diver gives the signal and the backup diver goes to work on the problem, the primary puts his hand back on his carabiner so that the backup can always readily find it.

Underwater signals between backup and primary diver

Large circular motion = "I am entangled here." The primary diver indicates where the entanglement is by putting the backup diver's hands on it. The entanglement location information helps prevent backup-diver entanglement. Without such information, the backup can become tangled in the same entanglement as the primary, can accidentally knock off the primary diver's mask or regulator, and will waste time trying to identify the problem and its location.

Taps backup's hand on primary's chest = "I am injured here." This indicates the injury location. The backup diver now knows to proceed very gently and cautiously to prevent injuring himself and further injuring the primary diver.

Taps backup diver's hand to primary's second stage = "I am already on my pony bottle, I need more air." The backup puts the primary diver's hand on his pony regulator first stage. While the primary diver removes the pony bottle from its holder, the backup diver releases the pony second stage from its neck-strap holder. The primary diver switches to the backup diver's pony bottle, thereby saving the remaining air in his pony bottle. The reason the backup diver puts the primary diver's hand on the pony first stage instead of the pony

Photo 14–19 and 14–19a The backup diver has passed his pony bottle off to the out-of-air entangled diver. He surfaces. The backup tender puts a new pony bottle in his BCD pony pocket and then passes him the contingency 80ft^3 cylinder to bring to the primary diver.

mouthpiece is because the latter could accidentally have disconnected from the strap during the backup diver's descent. The pony regulator first stage will always be in the same place.

Backup places the primary's hand back on his carabiner and gives it three squeezes = "I am leaving, but I am coming right back." The backup goes back to his tender, who puts another pony bottle on him and hands him a full 80ft^3 contingency cylinder with a regulator. He returns to the primary diver and hands him the 80ft^3 contingency cylinder. The primary diver then ditches the pony bottle he received earlier from the backup diver. The two divers then clasp hands so that the primary can tell the back-up what and where the problem is. With 80ft^3 of air, there is plenty of time to handle the entanglement, entrapment and/or injury problem that is preventing the diver from surfacing. Once the problem is cleared, they can ascend together, with the backup helping to carry the contingency cylinder.

Squeezes hand three times and gently pushes the backup diver away = "I am fine now, thanks, you can leave me." The backup diver then disconnects the contingency strap and heads to the hole. The primary diver gives the tender one pull to say *"I'm okay, let's continue with what we were doing."* The primary tender returns the signal, and the search is continued in the same direction the primary diver left off.

Photo 14-20 Train to manage an injured/unconscious diver situation. Head blocks can be secured onto the ice board with velcro to immobilize a head or neck injury. Remove an injured diver's gear in the hole and use the board to remove the diver from the hole.

Photo 14-21 Tenders need to be prepared for accidental immersion. See how the ice pole is laid horizontally to help prevent facial immersion.

Brings hand in an upward motion three times = "Time to ascend back to the hole." Generally, it is the primary diver who gives these hand signals to the backup diver, but sometimes the backup diver gives them. For example, in an indirect entanglement situation, the backup diver would come down to the primary diver and give the big circular motion signal to indicate that the primary diver's line is entangled. Either diver can give the time to ascend signal.

There are specific techniques for each of the above procedures that should be taught and practiced in a good training program.

Unconscious diver

If an unconscious diver is found underwater, check for entanglements, clear the entanglements, and then bring the diver to the surface. In non-overhead environments, ditch the weightbelt of the unconscious diver to bring the diver to the surface, and if the diver becomes too buoyant, simply let go and continue with a safe controlled ascent, meeting up with the injured diver at the surface. Under the ice, though, if possible, use the BCD rather than drop the weightbelt. If the needy diver has no air left to inflate the BCD, the primary diver can orally inflate the BCD, unless he is wearing a full face mask. If the backup diver is wearing a full face mask, he could use his own BCD inflator hose on the needy diver's BCD, but this takes training and practice to prevent a rapid ascent by the backup diver. Drop the weightbelt underwater only as a last resort. This should rarely be necessary because the injured diver's tender can assist by gently and slowly pulling the diver in, while the backup tender assists the backup diver's egress. Avoid bringing the 90%- ready diver down as well, because that will only create confusion and intertwined tether lines.

Once in the ice hole, ditch the diver's weightbelt and gear, secure the head and airway, ventilate twice, turn the diver to face the ice roof, and use the harness to pull the diver from the water. Place the diver on the transport device and initiate CPR. If the ice is strong enough, continue with CPR during the egress to shore.

Surface support accidental immersion

Surface support personnel must have contingency plans for accidental immersion. Prevention is the key.

1. Test the ice with poles when first going out on the ice.
2. Move slowly and stay low to the ice if it is weak, and use transport devices if necessary.
3. Use an ice pole to help prevent full submergence.

Photo 14-22 All tenders and divers should learn self-rescue procedures. Ice awls make getting out of a hole much easier. Carry them in wrist cases.

The moment you feel as though you are going to fall through, put one arm or the pole out, tip your face upward, and cover your airway with the other hand. Once immersed, face your line tenders, who will take tension on your front tether point. Keep your face low, keep kicking, and use your arms to wriggle out. The ice pole can help. Lay it across the ice and put your hands on it to distribute weight. Ice awls make getting out of a hole very easy.

Photo 14-23 Keep face low to the ice and arms bent, while kicking feet and pulling forward.

It cannot be said enough that training by an experienced, properly certified instructor is crucial. You cannot plan and train for everything that can go wrong, but if you do train and have a series of well-practiced plans, usu-

Photo 14–24 Once out of the hole, slowly and gently roll away from it.

ally, the rest can be worked out logically, without severe task loading. Without well-practiced contingency plans, you are task loading from the first minor problem onward!

The information presented in this chapter and in the entire book is a tool that can be used during training and as a way to refresh knowledge. Practice contingency procedures regularly, from the diver's request for assistance to a CPR transport to shore. *Never* allow yourself to be lulled into the "nothing is going to go wrong" attitude. Be on the alert and be prepared at all times.

Photo 14–25 The safety officer (on ice board) checks the contingency equipment; a primary tender never removes his eyes from his diver (left). The 90%-ready diver and tender check their communication system (right). Prevent problems and be ready for them if they occur.

Summary Questions

1. Too often emergency plans are tabletop discussions that have never been tested or practiced realistically. True or False?
2. The first step for a diver when experiencing a problem is to ______ and get ______ under control.
3. What should you do if you are hypoventilating?
4. Hypoventilation results in _____ (increased, decreased) carbon dioxide levels.
5. What is involved in the "STOP" self-rescue step?
6. Define direct and indirect entanglements.
7. Keeping the line ____ will help prevent indirect entanglements.
8. If the diver does not reply after sending a signal twice, what should be done?
9. If you suddenly find yourself in a silt-out, what should you do?
10. Should the primary diver call the backup diver if the primary diver's main cylinder is empty and the diver is breathing on the pony bottle?
11. List the advantages of using a block.
12. Why did Hendrick design a block for public safety divers to be directly mounted onto the full face mask?
13. A regulator free-flows at the surface. What can be done?
14. Divers should practice breathing off free-flowing regulators. True or False?
15. What should be done if the drysuit inflator valve self-inflates while the diver is submerged?
16. What should a disconnected diver do in still water?
17. What should the backup diver do to find a primary diver?
18. The maximum backup diver line length is ____ ft
19. Backup divers should make sure that they do not ___ on the primary diver's line.
20. Backup divers descend down the primary diver's line by making a _____ sign on the line.
21. To bring up an unconscious primary diver, backup divers should first make the attempt without ditching the primary diver's weightbelt. True or False?

Afterword

From Author Walt Hendrick

I have been teaching ice diving operations for twenty-three years. After several hundred personal ice dives, I still consider each ice dive as potentially dangerous as any cave or deep penetration wreck dive. I have learned much from the hundreds of students I have taught over the years.

In some ice diving courses, divers make one or two dives, sometimes only lasting for a couple of minutes, and then become certified to dive under the ice. You may find that hard to believe, but it happens. We get such divers in our courses all the time, already certified for ice with less than three dives and a total duration under ice of less than twenty minutes. It's even more common to see certified ice divers who have performed three ice dives, but never trained to safely set up a site. I have seen tenders who were never taught how to cut a hole, how to be a tender, or even what a backup diver should be capable of. Often, divers have never practiced breathing underwater without a mask and have not been trained to remain calm while breathing off a free-flowing regulator in frigid water. Many do not know that a pony bottle should be mandatory in ice diving operations. Some have never demonstrated that they could plan an ice dive, nor have they ever practiced contingency procedures. Sure, these divers descended, breathed under the ice roof for fifteen minutes or so, came up, and survived. Does that make them capable of performing an ice dive without the instructor there setting up the site and running everything? Absolutely not! So why are these divers certified as ice divers?

We believe that there should be levels of certification for ice diving, much as was developed for standard diving twenty years ago, with basic and advanced designations. This recommendation is discussed later in the book and includes the concepts of ice divemaster and ice rescue certification as well.

I consider myself and my staff to be experts in the field of overhead obstructed diving as it pertains to ice. If I were going to do a deep penetration wreck dive tomorrow, I would want to do that dive with Gary Gentile or Steve Bielenda. If my staff and I were going cave diving tomorrow, I would want to do that with Steve Gerrard or Wes Giles. The reason it is called extreme tek diving is that things can and will go wrong. Ice diving can go bad no matter who you are or how well planned the dive is, so let us at least do our best to prepare for the worst.

Does ice diving belong in the advanced realm? Yes. Can ice diving be dangerous? Yes. Can we as an industry do better? Yes.

We are often an industry of mediocrity that has no intention of changing. With a multitude of new diving and public safety diving agencies and their one-day wonders, each with a lesser standard of care than the one before, our industry must do something to protect innocent students.

Heaven protect our industry from those who plague us with ignorance and greed. Let us all work together to make ice diving safer, to prevent needless deaths, and to save the drowning victims for whom we are often diving.

Safe Diving Always,

Walt "Butch" Hendrick

Appendix A

Tender Surface Hand or Whistle Signals

Signal	Hand	Whistle
STOP	Arm straight up with a fist	One blow on the whistle
TAKE LINE	Arm straight up, and make a circle	Two blows on the whistle
GIVE LINE	Arm out to the side, up and down	Three blows on the whistle
DIVER HAS OBJECT	Arm straight up and down, similar to a pumping action	Three sets of two blows on the whistle

Appendix B

Line-Pull Signal Quick Reference

Tender-to-diver signals

1	Are you okay? Stop, face line, and take up slack.
3	Go to diver's right.
4	Go to diver's left.
2 + 2	Search immediate area.
3 + 3	Stand by.
4 + 4	Ascend slowly.

Diver-to-tender signals

1	I'm okay.
2	Tender, make notation.
2 + 2 + 2	I'm okay, but have a problem: alert the backup diver.
3 + 3 + 3	I'm okay, but need help from the backup diver.
4 + 4 + 4+...	(continuous) I need immediate help!
6 + 6	Found object.

Specialized Line-Pull Signals

For recreational ice diving that is diver-directed

Tender-to-diver	2	Resume your activity.

For PSD and recreational ice diving when communication systems are not available

Tender-to-diver	2	Give signal on every inhalation until a stop "1" is given.
Diver-to-tender, after the above signal	1	I just inhaled. The tender does not return these signals.

Appendix C

Surface Air Consumption (SAC) Rate

How to calculate and use surface air consumption rates

You now know that you should always carry a pony bottle. Why? In case you run out of air. That bottle could save your life, or another diver's in the event of an entanglement by providing the time you need to get rid of that entanglement. But, do you actually have any idea just how much time that bottle will provide? For that matter, do you even know how long your own primary tank will last? Sure, we all know that we could make that twenty minute dive to 15ft to make a search and have plenty of air to do it. Do you know whether you could make a dive to 45ft for twenty minutes and still have enough air for a five minute extension on your dive? Would it still allow you to be back on the boat or shore with 1,000psi in your tank as a safety measure? Your gauges only tell you how much air you have used; however, knowing how much air you *will consume*, not just how much you have consumed, is a critical part of diving safety. By including air consumption in your dive planning, you and your dive team can make safer, better-calculated dives, and greatly lower the risks of an out-of-air emergency. So how do you figure your air consumption? This is where you factor in your SAC.

Your surface air consumption rate, or SAC rate, is calculated to tell you how much air in pounds per square inch (psi) or cubic feet (ft^3) you would normally consume at the surface, a depth of zero feet. Your SAC rate can then be used as a basis to give you a good idea of how much air you will consume based on a dive with a certain size tank, and of a particular duration and depth.

It is important for your tenders to know divers' SAC rates, not only to decide whether to allow them a 5 minute extension on dive times, but also to decide whether divers may be experiencing problems during a dive. An unusually high SAC rate may indicate risk factors for lung overexpansion injuries or that a diver is not entirely focused on searching properly. An unusually low SAC rate may indicate that divers are breath holding, which will later result in a hypercapnia-induced headache. Either way, based on the SAC rates, tenders will be able to help divers correct the problem and avoid other potential problems.

The first key to finding your SAC is your logbook. You should record your starting tank pressure, ending tank pressure, depth, bottom time, and the size of the tank used. Doing so will give you all of the information you need to calculate your SAC rate. As an example, a working public safety diver began a dive with 3,000psi of pressure, and ended with 1,100psi. She dove to 25ft for a total of twenty minutes with an 80ft^3 tank.

a. First, determine the amount of tank pressure you used on your dive by subtracting ending tank pressure from starting tank pressure:

$$\text{STARTING } psi - \text{ENDING } psi = psi \text{ USED AT DEPTH} \tag{A.1}$$

From the above example,

$$3000\,psi - 1100\,psi = 1900\,psi \text{ USED AT DEPTH}$$

b. Next, you can determine how many psi you were using each minute during the dive:

$$\frac{psi \text{ USED AT DEPTH}}{\text{TOTAL LENGTH OF DIVE } in\ minutes} = psi\,/\,minute \text{ USED AT DEPTH} \tag{A.2}$$

From the example,

$$\frac{1900psi \text{ USED AT DEPTH}}{20\ minute \text{ TOTAL LENGTH OF DIVE}} = 95psi\,/\,minute \text{ USED AT DEPTH}$$

Although that number tells you how much air you used at depth, it is only valid for that depth.

c. You probably remember that the increase of pressure as depth increases also means an increase in the density of inhaled air. That increase means that a diver uses more air from a tank when breathing at depth than at the surface. Since we are trying to find surface air consumption, we must determine how much air you would use if you had breathed for the same amount of time at the surface. Before that can be found, you must calculate your depth as absolute pressure (ata). Remember that 33ft equals one atmosphere (atm).

(A.3)

$$\frac{\text{DEPTH } \textit{in feet}}{33 \textit{ feet}} + 1 \text{ ATM} = \text{ABSOLUTE PRESSURE } \textit{in atmospheres absolute (ATA)}$$

From the example,

$$\frac{25 \textit{ feet}}{33 \textit{ feet}} + 1 \text{ ATM} = 1.76 \textit{ ATA}$$

The 1atm is added to account for the pressure of the earth's atmosphere.

d. Now, we can find how much air you would use per minute if you breathed for the same amount of time at the surface:

(A.4)

$$\frac{\textit{psi/minute} \text{ USED AT DEPTH}}{\text{ABSOLUTE PRESSURE}} = \textit{psi / minute} \text{ USED AT SURFACE}$$

From the example,

$$\frac{95 \textit{ psi/minute} \text{ USED AT DEPTH}}{1.76 \text{ ATA}} = 54.0 \textit{ psi / minute} \text{ USED AT SURFACE}$$

In the example, the diver used 1,900psi at a depth of 25ft, or 1.76ata, and used 95psi per minute. Had she breathed the same amount of time at the surface as she did at depth, she would have used only 54psi/m, because she would have been breathing air of a lower density.

If you always dive with the same size tank, you can use *psi/minute used at surface* as your SAC rate. However, if you use different size tanks, you will need to know the actual amount of air you use, not just the amount of pressure you use with a certain size tank.

e. Although we do know the amount of tank pressure the diver used, we still do not know how much air she actually used. Calculating psi only tells us the amount of pressure used, not the actual volume of air. What if she was breathing from an $18ft^3$ pony bottle on that dive? Would she still use only 1,900psi? No way! With her air consumption, could she have even made that dive on just an $18ft^3$ bottle? To find out the actual amount of air used, we must convert *psi used at surface* to *cubic feet (ft^3) of air used at surface.* To do that step, we must first find how many cubic feet are used for every psi used, and make a *conversion factor.* The amount of cubic feet used for every psi used is dependent on the size of the tank:

(A.5)

$$\frac{\text{TANK SIZE } \textit{in ft}^3}{\text{TANK'S WORKING PRESSURE } (\textit{psi})} = \text{CONVERSION FACTOR}$$

In the example,

$$\frac{80 \; \textit{ft}^3 \text{ TANK SIZE}}{3000 \; \textit{psi} \text{ WORKING PRESSURE}} = 0.027 \; \textit{ft}^3 / \textit{psi} \text{ CONVERSION FACTOR}$$

f. The next step is to find the number of cubic feet used per minute. Any time you need to convert psi to cubic feet, simply multiply in by the *conversion factor:*

(A.6)

psi / m USED AT SURFACE $\times$ CONVERSION FACTOR = FT^3/M USED AT SURFACE

From the example,

54.0 *psi/m used at surface* $\times$ *0.027 ft^3/psi used at surface = 1.46 ft^3/m used at surface*

You now know how much air you use at the surface both by pressure (easy to work with if you are diving the same size tanks) and by cubic feet, which is necessary if you are using different size tanks. (By multiplying ft^3/m used at surface by the total length of dive, you might notice that the diver in the example would use $29.2ft^3$ of air at that rate at the surface, far more than an $18ft^3$ pony bottle can hold. Breathing 1,100psi from an $80ft^3$ tank allows much more air than breathing 1,100psi from an $18ft^3$ pony bottle.)

In the example, the diver uses $1.46ft^3$ of air every minute at the surface. You can use your SAC rate to determine how much air you will need to do a particular dive. Let's take the example one more step, and find out how long our diver could have breathed from an $18ft^3$ pony bottle at the surface:

That pony bottle would last her a little more than 12 minutes at the surface, but how long would it last if the diver were at her depth of 25ft? The air a diver breathes at 25ft is denser, so divide the time she could breathe from the pony bottle, 12.3 minutes, by the *absolute pressure,* as calculated earlier:

$$\frac{12.3 \textit{ minutes}}{1.76 \textit{ ATA}} = 8.5 \textit{ minutes}$$

Obviously, the sample diver could not have made her 20 minute dive on just a pony bottle! The pony bottle will last only $8^1/2$ minutes at 25ft. You can follow this example to calculate how long various tanks will last at different depths.

You should calculate your own SAC rate, and use those rates to determine how much air you will use on dives. It is also a good idea to use SAC rates to determine how long you could breathe from a pony bottle in the event that you must use one.

Keep in mind that the SAC rate is an individual number: even at the same depth and conditions, air consumption varies widely between individuals due to body size, cardiovascular health, smoking, and other factors. Because of that variation, you should keep track of your own air consumption. It is also extremely important to remember that breathing rates on some dives can be much, much higher than on others. For example, a public safety diver working an actual emergency or recovery will be pumped up and breathing rapidly from exertion. An entangled diver breathing from a pony bottle will be nervous and using more air than normal too.

Using SAC rates is one of the most important aspects of dive planning. They give an excellent idea of how long a dive will last, and will make your diving and your dive team safer and more efficient. However, because you can use more air than you predict, don't rely on them entirely: use SAC rates with caution, and always **watch your gauges!**

Appendix D
Quick Reference Material

Awareness Level Duties

1. Know where to go to assess the scene and stage the operation.
2. Don PFD, gloves, & other necessary PPE.
3. Perform Rapid Scene Assessment
4. Identify life & health hazards.
5. Assess no. and ASAPS status of victims.
6. Establish communication w/ victims
7. Mark a spot on shore in front of each victim if possible.
8. Implement IMS. Establish command post, Incident Commander, Staging Areas, Staging Officers, Safety Officers, and other officers and modules required by the size and complexity of the incident.
9. Direct positioning of rescue vehicles to form windblocks & exits.
10. **STAY SAFE! Follow Procedures**

Aggressive, Self-rescue, Alert, Passive, Sub

11. Call for appropriate agencies, including the dive team, EMS, and law enforcement.
12. Secure the scene, & determine and mark Hot, Warm, and Cold Zones.
13. Secure witnesses for interviewing by law enforcement & check their condition.
14. Draw a profile map of the most likely victim location. Find a safe place to deploy divers.
15. Re-asses ASAPS
16. If higher-trained personnel have not yet arrived, begin interviewin g witnesses

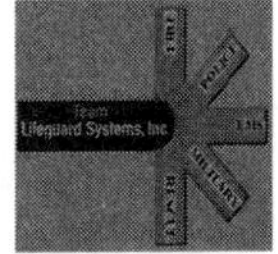

Life guard Systems
P.O. Box 548
Hurley, NY 12443
Tel/Fax:
(845) 331-3383
www.teamlgs.com

Awareness Level Duties

1. Know where to go to assess the scene and stage the operation.
2. Don PFD, gloves, & other necessary PPE.
3. Perform Rapid Scene Assessment
4. Identify life & health hazards.
5. Assess no. and ASAPS status of victims.
6. Establish communication w/ victims
7. Mark a spot on shore in front of each victim if possible.
8. Implement IMS. Establish command post, Incident Commander, Staging Areas, Staging Officers, Safety Officers, and other officers and modules required by the size and complexity of the incident.
9. Direct positioning of rescue vehicles to form windblocks & exits.
10. **STAY SAFE! Follow Procedures**

Aggressive, Self-rescue, Alert, Passive, Sub

11. Call for appropriate agencies, including the dive team, EMS, and law enforcement.
12. Secure the scene, & determine and mark Hot, Warm, and Cold Zones.
13. Secure witnesses for interviewing by law enforcement & check their condition.
14. Draw a profile map of the most likely victim location. Find a safe place to deploy divers.
15. Re-asses ASAPS
16. If higher-trained personnel have not yet arrived, begin interviewin g witnesses

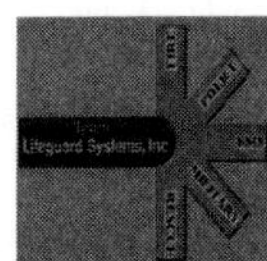

Life guard Systems
P.O. Box 548
Hurley, NY 12443
Tel/Fax:
(845) 331-3383
www.teamlgs.com

Awareness Level Duties

1. Know where to go to assess the scene and stage the operation.
2. Don PFD, gloves, & other necessary PPE.
3. Perform Rapid Scene Assessment
4. Identify life & health hazards.
5. Assess no. and ASAPS status of victims.
6. Establish communication w/ victims
7. Mark a spot on shore in front of each victim if possible.
8. Implement IMS. Establish command post, Incident Commander, Staging Areas, Staging Officers, Safety Officers, and other officers and modules required by the size and complexity of the incident.
9. Direct positioning of rescue vehicles to form windblocks & exits.
10. **STAY SAFE! Follow Procedures**

Aggressive, Self-rescue, Alert, Passive, Sub

11. Call for appropriate agencies, including the dive team, EMS, and law enforcement.
12. Secure the scene, & determine and mark Hot, Warm, and Cold Zones.
13. Secure witnesses for interviewing by law enforcement & check their condition.
14. Draw a profile map of the most likely victim location. Find a safe place to deploy divers.
15. Re-asses ASAPS
16. If higher-trained personnel have not yet arrived, begin interviewing witnesses

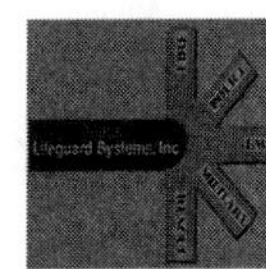

Life guard Systems
P.O. Box 548
Hurley, NY 12443
Tel/Fax:
(845) 331-3383
www.teamlgs.com

Figure A–2 Awareness Cards

The Tender should have and check:

1. USCG-approved, Class V **PFD** closed properly
2. Appropriate footwear and **exposure protection**
3. Proper tending **gloves**
4. An appropriate **time keeper**
5. A water-activated **flasher**
6. **Eye** and sun **protection**
7. Back-tethered harness if working from a steep embankment
8. If acting as the back-up tender, a **profile slate and writing utensil**

Water site must have:

1. **Contingency** reg. on full **cylinder**
2. **Contingency** reg. on full **pony bottle**
3. At least one **rescue throw rope bag** per diver-tender pair
4. At least BLS **first aid** personnel and supplies

Life guard Systems
P.O. Box 548,
Hurley, NY 12443
845-331-3383
www.teamlgs.com

Tender check of diver

1. **Air on!**
2. **Hood** in place.
3. **Mask** ready, **gloves** on
4. **Primary regulator**: exhale first, **breathe** 3 times; check **tie-wrap** and **mouthpiece**
5. **Pony reg** : same as #4; **hose under arm, secured** w/ mud **mouthpiece cover** and **holder**
6. **Drysuit hose** on/functional
7. Diver, **press BC inflator button** 3 times while tender checks **pressure gauge**; **gauges secured**; all hoses trim to body and secured
8. Diver, without looking, find **1st shears, 2nd shears, & 3rd tool**
9. Diver, find **carabiner** at harness tether point; tether line attached
10. Diver, find **weightbelt release** – is it **right-hand**? **10" extra webbing**?
11. **Fins** ready, w/ **ankle wts** if needed
12. Back-ups have **contingency lines**

The Tender should have and check:

1. USCG-approved, Class V **PFD** closed properly
2. Appropriate footwear and **exposure protection**
3. Proper tending **gloves**
4. An appropriate **time keeper**
5. A water-activated **flasher**
6. **Eye** and sun **protection**
7. Back-tethered harness if working from a steep embankment
8. If acting as the back-up tender, a **profile slate and writing utensil**

Water site must have:

1. **Contingency** reg. on full **cylinder**
2. **Contingency** reg. on full **pony bottle**
3. At least one **rescue throw rope bag** per diver-tender pair
4. At least BLS **first aid** personnel and supplies

Life guard Systems
P.O. Box 548,
Hurley, NY 12443
845-331-3383
www.teamlgs.com

Tender check of diver

1. **Air on!**
2. **Hood** in place.
3. **Mask** ready, **gloves** on
4. **Primary regulator**: exhale first, **breathe** 3 times; check **tie-wrap** and **mouthpiece**
5. **Pony reg** : same as #4; **hose under arm, secured** w/ mud **mouthpiece cover** and **holder**
6. **Drysuit hose** on/functional
7. Diver, **press BC inflator button** 3 times while tender checks **pressure gauge**; **gauges secured**; all hoses trim to body and secured
8. Diver, without looking, find **1st shears, 2nd shears, & 3rd tool**
9. Diver, find **carabiner** at harness tether point; tether line attached
10. Diver, find **weightbelt release** – is it **right-hand**? **10" extra webbing**?
11. **Fins** ready, w/ **ankle wts** if needed
12. Back-ups have **contingency lines**

The Tender should have and check:

1. USCG -approved, Class V **PFD** closed properly
2. Appropriate footwear and **exposure protection**
3. Proper tending **gloves**
4. An appropriate **time keeper**
5. A water-activated **flasher**
6. **Eye** and sun **protection**
7. Back-tethered harness if working from a steep embankment
8. If acting as the back-up tender, a **profile slate and writing utensil**

Water site must have:

1. **Contingency** reg. on full **cylinder**
2. **Contingency** reg. on full **pony bottle**
3. At least one **rescue throw rope bag** per diver-tender pair
4. At least BLS **first aid** personnel and supplies

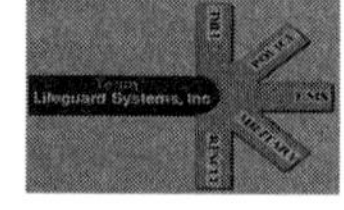

Life guard Systems
P.O. Box 548,
Hurley, NY 12443
845-331-3383
www.teamlgs.com

Tender check of diver

1. **Air on!**
2. **Hood** in place.
3. **Mask** ready, **gloves** on
4. **Primary regulator**: exhale first, **breathe** 3 times; check **tie-wrap** and **mouthpiece**
5. **Pony reg** : same as #4; **hose under arm, secured** w/ mud **mouthpiece cover** and **holder**
6. **Drysuit hose** on/functional
7. Diver, **press BC inflator button** 3 times while tender checks **pressure gauge**; **gauges secured**; all hoses trim to body and secured
8. Diver, without looking, find **1st shears, 2nd shears, & 3rd tool**
9. Diver, find **carabiner** at harness tether point; tether line attached
10. Diver, find **weightbelt release** – is it **right-hand**? **10" extra webbing**?
11. **Fins** ready, w/ **ankle wts** if needed
12. Back-ups have **contingency lines**

Figure A–3 Equipment Check Cards

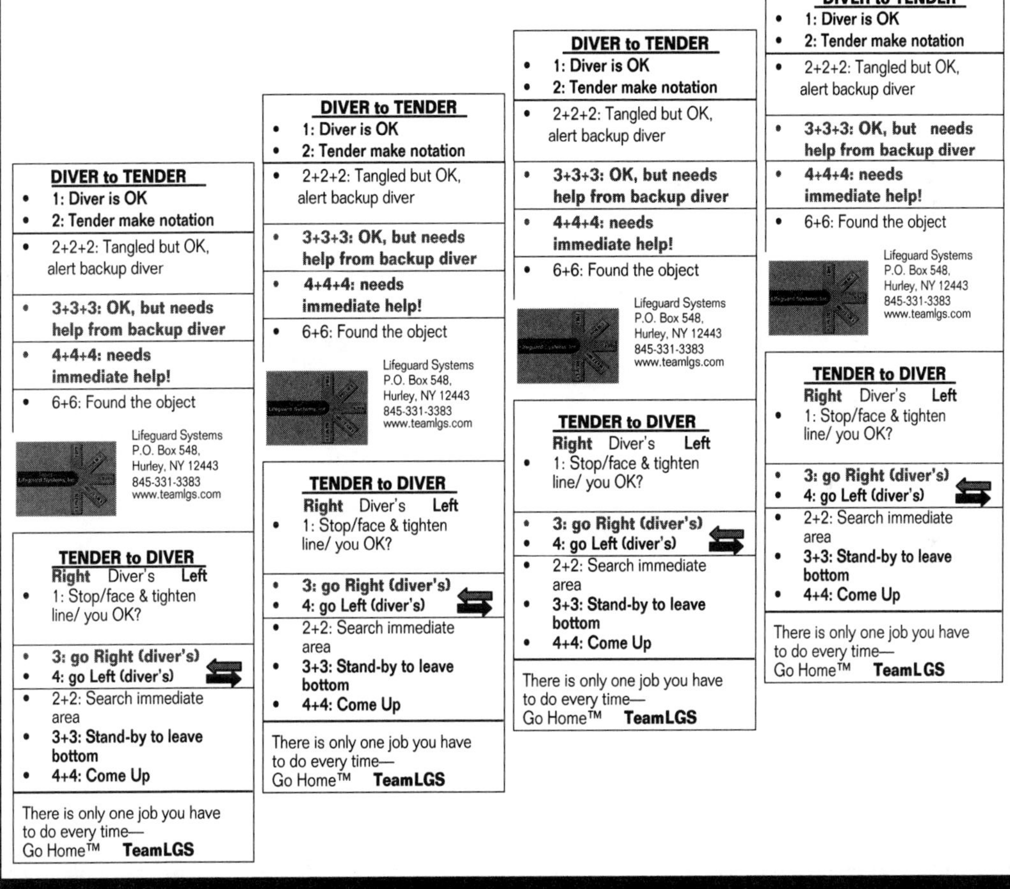

Figure A-4 Signal Cards

Answers to Summary Questions

Chapter 1

1. Recreational divers pick where and when they will dive and only dive on strong ice; public safety divers typically dive solo; the mission of recreational divers is to have fun and of public safety divers is to find search objects/bodies.
2. Weak.
3. Rapid dressing; securing the scene; incident command; risk/benefit analysis; plan of action; executing the plan; transport the victim; incident documentation; equipment maintenance; debriefing.
4. True.
5. Unsafe tethering techniques; lack of effective contingency divers; lack of pony bottles; lack of trained tenders.
6. No direct access to the surface due to overhead obstruction(s) between the point-of-entry and exit.
7. Requiring more than basic skills, preparation, planning, or equipment.
8. A line securely attached to a diver harness that is controlled by a trained tender who is tethered to the shore or ice.

Chapter 2

1. 5 – Primary diver and tender; backup diver and tender; 90%-ready diver.
2. Enough training to do no further harm and perform generic, cold-zone duties.
3. Enough training to directly assist without direct exposure to the hazardous atmosphere – water.
4. Enough training to be directly exposed to the hazardous atmosphere – submersion/immersion in water.
5. Don PPE; know scene-staging location; perform rapid scene assessment; implement the IMS; call for appropriate resources; secure and manage scene safety; draw scene maps; interview witnesses.
6. All awareness-level duties; risk/benefit analysis; identify available resources and ensure adequate response; develop response plan; determine diver deployment location; test the ice; determine whether divers are fully ready; assist divers; serve as tenders and other shore support positions; carry out contingency plan; perform shore-based rescue techniques; throw a coiled and bagged line; maintain appropriate documentation.
7. Report any problems; clean and store equipment; documentation.
8. Tender.
9. Fully.
10. 90%-ready.
11. It is usually quicker to deploy a 90%-ready diver than to fix backup diver equipment problems. Sometimes a backup diver becomes incapacitated due to equalization, or other problems.
12. Scene safety; scene setup; diver blood-pressure check; dress divers; diver equipment checks; diver safety; dive documentation; diver breathing checks; directing divers; post-dive diver care; search quality evaluation.
13. Map of the scene; diver search profiles; diver blood pressure and BPM; water conditions; dive times; diver SAC rates; personnel names; search area total running feet; if the search area can be secured.
14. Backup tender.
15. To maintain constant awareness of the primary diver's condition and searching; be more capable of directing the backup diver to make an assist/rescue; know exactly what areas can be secured or need to be re-searched when they take the primary tender position.

Chapter 3

1. CPR; first aid; hypothermia; cold stress; diving injuries; drowning.
2. Training and certification.

Chapter 4

1. False.
2. Experience with cold water diving; comfortable wearing hoods and warm gloves; breathing underwater without a mask; calculate air consumption; comfortable with basic skills; neutral buoyancy.

Chapter 5

1. Consistent.
2. Free-flowing.
3. Environmental protection; lower the first stage intermediate pressure; blow air through regulators after washing them to decrease internal moisture; dry the dust cap with cloth before replacing it; dry the tank valve orifice before attaching regulator first stage; keep the second stage wet once it is wet; keep regulators off the snow and ice; keep regulators protected from outside cold before use.
4. It does nothing for the wearer if the wearer runs out of air or has a free-flow problem; it attaches the donor to an out-of-air entangled diver; it increases the chances of a first stage free-flow.
5. Quick-release.
6. Contingency; contingency.
7. Exhaust tee; hose wrap.
8. Block.
9. Cutting tool; standard facemask; marker buoy.
10. Hard; soft.
11. Twice.
12. Thermal protection; rapid dressing; ease of storage; protection from hazardous materials.
13. False.
14. Carotid sinus reflex.
15. False.

16. True.
17. Cost; less stretch than neoprene suits; no intrinsic insulation.
18. Easy cleaning; can offer protection from contaminants; can be towel-dried and then put away; are intrinsically negatively-buoyant, depth does not affect buoyancy of the suit itself; easiest in-field repairs; suits less likely to freeze after use; can be worn outside after use; easy to clean.
19. More comfortable; no front flap of material is required; suits generally fit better; easier to don.
20. False.
21. Remove.
22. Depth.
23. To avoid decreasing circulation to the extremities.
24. Making feet neutrally-buoyant; preventing feet from pulling out of drysuit boots when divers invert.
25. False.
26. True.
27. Pony regulator.
28. Two; shears.
29. Golden triangle; legs.
30. True.
31. 10-11.
32. Taped.
33. Contingency strap.
34. Prevents loss of the primary diver's tether line; decreases the chances of the primary tether line being pulled by the backup diver; give the backup diver two hands free; can be connected into the primary diver's tether point to connect the backup and primary diver should the primary tether line need to be disconnected.
35. Drysuits.
36. Ice cleats.
37. Pulling oneself and the ice board across weak ice; pulling oneself out of an ice hole; divers pull themselves under the ice when spidering.
38. Is protection from contaminants needed; will they need to be waterproof for in-water tenders; what kind of fine-motor dexterity is necessary?

39. Ears; nose; mouth/lips; neck.
40. Larger.
41. Wool.
42. Easily carried by one person and stored in a vehicle without assembly needed; easily used across weak ice and openwater and snow; low wind susceptibility; function as a backboard; not made out of metal, soft rounded edges; effective for getting divers out of ice holes; serves as an effective surface ice rescue tool.
43. To test the ice in front of you; decrease the chances of full submergence if the ice breaks; assist someone out of an ice hole.
44. Personal flotation device.

Chapter 6

1. Confined space.
2. Unsafe; unacceptable.
3. Never.
4. Solo.
5. The lines can become intertwined; two tenders, rather than one are required; it requires more tender coordination and training.
6. Harness, locking carabiner, duct tape, tether line, and ice screw or strap to secure the end of the line to a stationary object.
7. The loop can be dropped resulting in a disconnected diver; it leaves the diver only one hand free to search or perform self-rescue; holding a loop restricts hand circulation; it does not allow for an effective, accurate taut line search pattern.
8. A front point on a chest harness.
9. Across the solar plexus; to prevent diaphragm restriction; to afford a horizontal diver position; to allow the ribs, rather than abdominal organs to take the snap of line signals.
10. Adjustable girth and shoulder straps; webbing that will not roll into itself and bind up; come in different sizes; solar plexus and back D-ring attachment points; will not move around when the diver keeps a taut line and changes position.
11. Tie a figure-eight or a rescue-eight and duct tape the bitter end to the line. Attach the loop(s) to a locking carabiner attached to the solar plexus tether point on the harness. Place a strip of duct tape over the lock in the tightening direction and leave a folded-over tail on the duct tape for easy removal while wearing gloves.

12. 3/8in (10mm); It is thick enough to tactically feel, will not entangle as easily as thinner line, and will not create the drag of thicker line; allows effective line signaling for up to 150ft.

13. Line that is less than 300ft can be stored in bags while longer line can be stored and deployed from reels with brakes.

14. Laid line consists of twisting fibers in opposite directions to form a yarn, which are then twisted together to form a strand, which are then twisted together to make the finished rope. Laid line is too stiff and may have too much stretch.

15. The yarn and then the strands are braided rather than twisted, which produces a far more flexible line with less stretch.

16. The fibers run the full-length of the rope, rather than the rope being made of short pieces of fibers spun together into a fuzzy-looking rope. Filament rope is less elastic.

17. Polypropylene; it is the only fiber that floats well; abrasion resistance; resistant to sunlight and hydrocarbons; cost effective; holds knots if braided construction.

18. 100ft (30m) line length; 50ft (15m) with a possible 10ft (3m) extension; 150ft (46m).

19. 5ft (1.5m).

20. Knots weaken lines; and loops and knots created increased drag and diver workload and entanglement hazards; less distance accuracy.

21. Dunk and agitate in a bucket of warm water, and if necessary, use a hose washer; dry and check the line thoroughly before storing.

22. Attach a strap to the end of the line with a carabiner, and then secure the strap with a non-locking carabiner to a stationary object such as an ice screw or tree.

23. Tenders wear a harness. Attach one end of a length of webbing to the harness back D-ring and secure the other end to an ice screw. The webbing length should allow movement around the hole.

24. Avoid sharp edges or rough surfaces; continuous tender observation; tenders keep hands low to ice when divers are directly under the ice roof; tenders may need to lie on the ice; tenders need mobility to keep divers from going underneath them.

Chapter 7

1. Prevent; recognize; pre-hospital care.
2. Gasping.
3. Laryngospasm.
4. Muscle cramps; decreased cerebral blood flow; tetanic convulsions; decreased level of consciousness.
5. Painful.
6. Tensing; weakness.
7. Disorientation; lack of judgment; fear; decreased survival skills; panic.
8. Constriction.
9. Fifty.
10. Irregular.
11. Oxygen Consumption.
12. Reduce; negative.
13. Increases; decreases.
14. Immune system.
15. A body core temperature of 95°F (35°C) or less.
16. Noradrenaline/norepinephrine.
17. Cold-induced hyperventilation can decrease cerebral blood-flow; memory can be impaired; hallucinations can occur; personality can change; mental stress can be increased.
18. Absence.
19. 80.
20. When objects are touching, heat will transfer from the warmer objects to the colder objects until they are all the same temperature.
21. Hands lose heat to cold equipment; divers lose heat to the ice they lay on.
22. Twenty-five.
23. Heat capacity.
24. Wool or polar fleece.

25. Gasp.
26. Insulation; insulation; insulation.
27. Increase; increases.
28. Metal.
29. The transfer of heat by wind or fluid movement.
30. Body movements.
31. Ears, nose, lips, carotid artery neck region.
32. False.
33. The process of a liquid changing to a vapor.
34. Twenty; increases.
35. Poor.
36. Remove.
37. Chamois.
38. Breathe.
39. Moisten; warm.
40. Radiation.
41. Apply cold cervical collar; do not cover the patient's head, hands, or feet during extrication; administer cold, dry oxygen; expose the patient outside; lay patient on a cold, uninsulated backboard; lay the cold oxygen bottle between the patient's legs; administer cold IV fluid.
42. Rescuers are cold; lack of proper training; lack of knowledge about immersion hypothermia physiology; lack of an effective SOP/SOG; lack of inter-agency pre-planning and drills; weak incident command system for water scenes.
43. One; two.
44. 25; 30.
45. Lower.
46. Increases; increase.
47. Peripheral vasoconstriction increases the thickness of the body's protective shell and hence decreases core heat loss.
48. Lower.
49. Shivering; exercise; chemical reactions.
50. Earlier; higher; higher.

51. False.
52. Inhibition of shivering and peripheral vasoconstriction.
53. Increased heat production; a layer of fat in the protective outer shell; peripheral vasoconstriction; behavior.
54. Appropriate exercise; wearing PPE; removing wet clothing; drinking appropriate warm beverages; moving to a warmer environment; using heat packs; eating sufficiently and getting enough sleep prior to cold exposure.
55. Never apply heat packs directly to hypothermic skin.
56. True.
57. We may be more at risk in early morning hours because we normally have a lower core temperature during those hours.
58. Low; high.
59. Large.
60. Effective when wet and allows wearer to perform fine motor tasks without having to be removed.
61. Praises or supports.
62. Lower.
63. True.
64. Increased.
65. Cold; immersion diuresis.
66. Cold causes peripheral vasoconstriction that keeps blood in the core; the approximate .5psi/ft of water pressure counteracts to lower veins from dilating as they normally would when standing on land.
67. Since immersion and exertion can cause increased blood pressure, it is important to make sure a technician does not begin an operation with an already high blood pressure. A post-check helps makes sure that a technician did not have a too large blood volume decrease from immersion diuresis or other processes.
68. When a person's blood volume has decreased from immersion/cold diuresis, the person can experience a significant drop in blood pressure when peripheral vasodilation occurs from external rewarming. The blood vessels increase their volume, but there is not enough blood to fill them and maintain the same blood pressure.
69. Cold and immersion can decrease thirst, and can simultaneously cause circulatory fluid loss.
70. Cramps.

71. Tissues.

72. Hydration; cardiovascular fitness; minimal exertion; pre- and post-dive blood-pressure checks; be mentally, physically, and emotionally ready prior to suiting up; enter the water slowly; kiss the water; appropriate PPE; effective and safe training; keep a well-trained tender with the technician through the check-out process during rehabilitation; thirty minutes of rest after the dive.

73. Human oxygen consumption increases rather than decreases; human heart rates decrease by 50% if at all while marine mammals can drop theirs by 90%; human breath-holding time is shortened rather than lengthened.

74. 0.7°C (1.0°F).

75. Widen; increase.

76. Increased.

77. Increased levels of carbon dioxide.

78. Reduces them.

79. A bony growth in the external ear canal that develops from repeated ear immersion in cold water.

80. True.

81. Mechanical pressure on carotid sinus receptors causes the receptors to continuously lower blood pressure until the pressure is removed.

82. Overly-tight hood or drysuit neck seals.

83. Lightheaded, nausea, weakness.

Chapter 8

1. Shivering and peripheral vasoconstriction.

2. Difficult.

3. Delay.

4. Reluctance to leave a shelter; not removing exposure protection when indoors; withdrawn or short-tempered behavior; memory problems and mistake making; difficulty with fine motor tasks; closed postures or warming movements; high SAC or breathing rates.

5. True.

6. Lack of coordination; inability to hold a standard second stage mouthpiece in place and perform self-rescue procedures.

7. Medical emergency.

8. Skin lesions; repeated unprotected exposure to dry cool environments between 32°F to 60°F (0°C to 16°C).

9. Redness; swelling; hot to the touch; itching; tenderness; burning sensations; protect the area from further exposure and seek medical attention.

10. First stage of frostbite; a redness that later turns white; numbness; soft underlying tissue; burning and tingling during rewarming; blow warm air over it; hold it with a warm hand; protect from cold; seek medical attention.

11. Blood leaves peripheral tissues causing them to become very cold; ice crystals form in cells and in the interstitial place; skin and subcutaneous tissue freezes; skin appears white and waxy, cold and hard to the touch; underlying tissue is soft.

12. Handle the area as if it was a serious fracture; cover it gently with a clean, dry dressing; transport immediately to a medical facility. Do not attempt to rewarm!

13. Frostbite that has progressed to deeper tissues such as muscles, bones, and organs. Tissue will feel hard to the touch. Skin is mottled white, blue, gray, yellow.

14. Rewarming greatly increases the tissue's need for oxygen, but the ability to deliver oxygen is severely compromised, which can result in cell death, infections, and gangrene.

15. A cold allergy.

16. Arterial gas embolism.

17. A water sample.

18. Increases.

19. No urge to urinate when immersed; dry mouth and thirst; dark urine; inability to sweat; muscle cramping, headache; irritability.

20. Can cause dehydration; decreases judgment; increases risk taking; vasodilator; impairs decision making; suppresses the laryngeal reflex; masks cold stress signs/symptoms; hypothalamic dysfunction; decreases awareness of environmental condition.

21. Increase, decrease.

22. Medications may require a minimum temperature to work; freezing and thawing medications may cause strength-loss; they can accumulate and become toxic; different dosages may be needed; establishing an IV line can be difficult.

23. Avoid non-essential procedures.

24. Need to be at least 15.5°C (60°F) to function properly; adhesive pads may not stick to cold skin; benzoin can help to maintain leads in place; conduction of electrical signals may be impaired across cold skin; the QRS amplitude should be maximally amplified if not initially seen.

25. Hypoglycemia.

26. Seek medical testing; load up on carbohydrates before diving.

27. Using the person's own metabolism to raise core temperature; remove wet clothing and dry off; wrap in blanket; move to warmer environment; exercise.

28. Applying heat to peripheral tissues; applying insulated heat packs; heated blanket; warm bath.

29. Applying heat internally; drinking warm fluids; warm, humidified breathing gases; warmed IV fluids; pleural or thoracic lavage; diatherm; extra corporeal circulation.

30. Cold tissues need less oxygen, and rewarming causes a significant increase in tissue oxygen needs. If oxygen is not being delivered to the tissues prior and during rewarming, the tissues may die and become gangrenous.

31. Allows rapid rewarming; not practical in the field; patient may feel warm prior to reaching normothermia and may stop the rewarming procedure and re-enter a cold environment too early; causes significant peripheral vasodilation with resulting afterdrop shock.

32. It is practical for the field; most research has found it to speed up the rewarming process; causes minimal or no afterdrop, does not cause peripheral vasodilation; warms the hypothalamus and lower brain stem to prevent further decrease in respiratory and cardiac functions; if done improperly, can cause respiratory burns.

33. Before.

34. 2; urine; sweat.

35. Handle very gently and horizontally; move to warm and dry environment; dry off; wrap in nonconductive materials from head to toe; administer warmed, saturated oxygen; if IVs are administered, prewarm them; if only mildly chilly, apply insulated heat packs to truncal region and do mild exercise; if fully alert and not nauseous, drink warm, appropriate beverages; monitor vital signs; check pulse for sixty seconds prior to initiating CPR compressions; seek medical attention.

Chapter 9

1. Do you feel mentally and physically capable of doing this dive safely?

2. Excuses typically come in multiples; when the problem is fixed another one comes up; the diver does not seem very relieved that the problem was fixed.

3. Moving slower than normal in gear assembly/dressing; readily accepting when someone suggests that they do not have to dive today; repetitive gear checks; out-of-character behavior; becoming flustered; moving hands with jerky motions during gear checks.
4. Contingency divers.
5. Three mistakes are a red flag that something is wrong, and a fourth mistake, even if it is small, can become a serious task-loading problem. After three mistakes, remove the person and send him or her to rehabilitation.
6. Don PPE and PFDs; receive plan of action; assist setting up staging areas; set up equipment in staging area; turn cylinder air on in BCD-cylinder assemblies; make sure divers are mentally/physically ready and BPs are checked; dress divers; gear check.
7. Diver readiness; is contingency equipment ready in place; are contingency divers ready in place; is backup tender ready to record information.
8. The backup diver tender gives the primary start- and end-times to count the number of diver's breaths in one minute. This is done every five minutes and is recorded by the backup tender.
9. Every ten minutes.
10. Eyes.
11. Pattern quality; rate of search; breathing rate, ability to find the test search objects; the diver's post-dive statement; and whether the diver was cold-stressed or not.
12. Looking away for even three seconds is enough time to miss a change in line angle that showed that the diver is in trouble or is about to be.
13. True.
14. False.
15. Hot, warm, and cold zones are maintained; demarcation line is established; EMS have a clear exit route; shelters/windbreaks are set up; all warm-zone personnel wear PFDs; equipment laid out properly; personnel accountability; ICS and post established; sufficient resources are available; lines from hot zone are secured on shore.
16. Put on clean tarp; powder seals; paraffin zippers; roll top of suit inside-out to inflator valve.
17. Stretch.

18. False.

19. Enough to be warm, comfortable, and to have full extremity movement.

20. A BCD can be rapidly purged; a drysuit can hold more air; air can end up places in a drysuit where it can't be purged easily; inversion in a drysuit prevents air purging; too much air in a drysuit can cause carotid sinus reflex; a drysuit self-inflation is far more serious than a BCD self-inflation and increased inflator valve use increases the risk of self-inflation.

21. Carotid sinus reflex.

22. False.

23. Place it through the BCD armhole and then under the diver's arm. Secure the gauge console to the BCD with a quick-release buckle that matches the BCD chest strap buckles.

24. The diver places a flat hand over his solar plexus and takes a large inhalation as the tender secures the girth strap snugly over the hand.

25. Divers get down on right knee with arms behind them. Tenders place BCD assemblies behind the divers and then on them. Tenders raise tank as divers stand up. Divers pull down shoulder straps. Tenders hold gauge and drysuit hoses up as divers close cummerbunds. Divers secure chest straps so they remain loose with a fully-inflated BCD. Gauge console and drysuit inflator valve are secured. Divers place arm between BCD low-pressure inflator hose and corrugated hose.

26. Check all gear is in proper place; diver knows how to find gear without looking; tenders fully comprehend a diver's gear; diver practices self-rescue procedures.

27. All air on; hood in place; mask/gloves on; exhale and breathe off primary regulator; check mouthpiece and tie-wrap; same check for pony regulator; pony hose under arm and mouthpiece secured on neck strap; diver reach pony regulator; diver disconnect and reconnect drysuit inflator hose; inflator valve at 45° angle downward and is clear of obstructions; diver finds and depresses BCD inflator valve three times as tender checks gauge; gauge hose and drysuit hose secured through BCD and under arm; gauge console secured; what time coming home; diver finds all cutting tools, carabiner, and weightbelt release; fins and ankle weights; contingency strap; mental and physical readiness.

28. Ice board; attach tether line to bow D-ring and lay over top of board; attach tether line to stern underside cable; attach bag to bow containing two ice screws, 5 feet of webbing loops at either end and a carabiner, and a pulley; snap shackle strap attached to top of board at the bow to tether the rider; ice awls.

29. A controlled, seated entry also known as a roof turn.
30. Diver removes mask at surface and breathes off primary, or pony regulator (if wearing a full face mask), with submerged face to acclimate the face and making mouth breathing reflexive.
31. 1.
32. 1.
33. 1000; thirds; 80.
34. Breaths per minute.
35. Diver inverts horizontally, facing the ice roof moves backward by pulling himself forward with ice awls and gently kicking.
36. Victim point-of-entry.
37. Divers choose where to dive and tenders keep the line snug.
38. "1" is at the corner of the triangle pointing away from shore, "2" is at the right side of the triangle going clockwise, "3" is the side facing shore, and "4" is the remaining side.
39. 3; to keep divers out from that area and eroding it with their bubbles.

Chapter 10

1. Reviewing past incidents and potential problem situations.
2. Representatives from each agency who would respond to an ice call.
3. Too few or too many responders? Did they have the right training? Was the operation organized or chaotic? Was the necessary equipment available? Was there any unnecessary risk? Effective debriefings? Review patient handling.
4. Ice diving underwater vehicle extrication operation training and equipment; an appropriate SOP/SOG; hazmat training and equipment; vehicle removal plans; who pays for the operation.
5. Will artificial lighting be needed? Will happy hour be a concern? Will it be difficult to assemble a crew?
6. Entanglement or entrapment; air-filled objects; jagged metal; fuels; bio and cargo hazards; vehicle shifting.
7. Biological or chemical contamination that exceeds personnel PPE and training; suspended ice roofs; moving ice blocks; severe weather conditions.
8. Write sizes on suit and bag; check suit integrity; paraffin zippers; powder both sides of seals.

Chapter 11

1. One that is familiar; shallow depths; few underwater obstructions/entanglements; underwater landmarks; easy parking; 6 or more inches of ice; not far from shore.
2. Suspended.
3. Entrapment.
4. Find the site in the winter; discover water table changes; document underwater hazards.
5. Use the distance-marked diver's tether line; buoy the location and measure distance at the surface.
6. Safe.
7. Salt; geese excrement; springs; bubblers; rain; previous human use; pollution; weather changes; snow.
8. Four inches.
9. Six inches.
10. Flooded.
11. Thunderous; reverberating; cracking.
12. Effectively cycle personnel; have redundant equipment available; enough surface personnel and divers; flexible, quick, effective decision making; rapid dressing abilities; scene organization.
13. Safety holes.
14. Ice screw.
15. It is easiest to cut; provides efficient corners for diver exit/entry; 6 feet.
16. Right angle; 3-4; behind; parallel.
17. Downward; 90.
18. Wrong.
19. Saves time and energy; keeps unnecessary weight off the ice; supports the ice roof that it is placed under.
20. Ice screw.
21. Weep; 3 to 4; 20 to 25.
22. Never step on it; store in bags if less than 300ft; changed and checked for every diver, tie an end off, never wrapped around a hand or body part; cut at maximum diver lengths.

23. $1^1/2$ hours.

24. Responsible.

25. Beaufort.

Chapter 12

1. Military and commercial.

2. An electronic system backed up by a line-signal system.

3. Divers may not want everyone to hear them; speaker boxes may distract other surface personnel; contingency divers should not hear primary diver distress; allow the best ability for the tender to hear the diver when other outside noises are present; headsets allow tender mobility and are less cumbersome; headsets work better on weak ice and take up less room on transport devices/platforms.

4. VOX.

5. 4; 3.

6. Backpack.

7. Wrist.

8. Orientation; left; right.

9. It lacks instant diver/tender orientation; increases risk of a slack line; divers are more likely to hold onto their lines; does not allow divers to be stopped and then continued in the same direction; cannot be used for spidering or vertical grass/obstruction searches.

10. The diver is at a 45° angle from the line pointing away from the tender.

11. All signals are returned verbatim; all tender signals begin with a "1"; all signals are based on the diver's left and right.

12. Stop, face the line, take up slack, prepare for a new signal, are you okay?

13. 4; 3; 2+2; 3+3; 4+4.

14. I'm okay; make notation; I have a problem and I am okay and alert the backup diver; I need assistance, but I am okay; I need immediate assistance and I am not okay.

15. Divers do not wait until their problem is serious enough to need assistance before signaling; reduces tender and contingency diver stress; decreases problems associated with tender and contingency diver stress; tells surface personnel whether they need to immediately prepare for a possible serious problem or not (*e.g.*, alert EMS personnel, ready the transport device).

16. I am on my pony, I need more air.
17. The backup places the primary diver's hand back on the primary diver's carabiner and squeezes it three times.
18. Two or three larger circular hand motions.
19. I am hurt.

Chapter 13

1. SOP/SOG.
2. False.
3. Timing; whether the team will dive now or later.
4. Written.
5. In a circle with a radius equal to the depth.
6. If the victim was conscious while under the ice and attempted to find the hole.
7. 200ft/m × 1/2min (descent time) = 100ft
8. Enter the water.
9. It may take less time and effort; the tether line will be at a good angle for searching.
10. Ice board.
11. When the ice is too thin to use ice screws; when ice needs to be preserved; when time is of essence in a rescue mode.
12. A tender in a drysuit wearing a harness with a pulley system attached to the harness front tether point, who is in the hole facing shore and who serves as an ice screw for the ice board pulley system.
13. Platform.
14. Yes.
15. Good for weak ice; keep personnel off the ice and out of the water if not the primary diver; inflatable boat.
16. Train with it on land with blacked-out masks; fold and roll it in an effective manner.

Chapter 14

1. True.
2. Stop; breathing.
3. Make long, slow, gentle exhalations.

4. Increased.
5. Stabilize breathing; buoyancy; body movements.
6. Snag the diver's body; snag the diver's tether line.
7. Taut.
8. Move a few feet in the opposite direction the diver was heading and signal again. If that does not work, send in the backup diver.
9. Kneel on the bottom and stop moving; relax; wait until the silt settles; slowly lift out of the silt.
10. No, only if the primary diver cannot get to the hole because of entanglement or cramping, or if the pony is free-flowing or is empty.
11. It prevents the diver from having to remove the full face mask to access pony air.
12. Panicked divers often reach up to rip their masks off. If this is done, their hand will hit the block and increase the chance that the diver will access pony air instead of removing the mask air source.
13. Turn the air off; put second stage in a bucket of warm water or make long, slow gentle exhalations through it; warm the first stage with an insulated heat pack or hand friction or tender exhalation.
14. True.
15. Immediately disconnect and dump the exhaust valve and flare out and remember to breathe normally.
16. Immediately make a controlled ascent to the surface, become positively buoyant and hang in a vertical position while making contact with the ice roof; if the main cylinder runs dry, switch to the pony and release weightbelt.
17. Perform an arc or circle spider search, beginning in the last known DAD quadrant.
18. 150.
19. Yank or pull.
20. Peace.
21. True.

Index

A

B

C

D

E

I

K

L

M

N

O

P

T

U

V

W

Z